Check Dam Designs

Check dams are small barriers built across the direction of water flow on shallow rivers and streams for the purpose of water harvesting. This book is a review of basic principles and practice-oriented approaches for check dam designs. It offers a hands-on approach with a strong, practical bias helpful in solving problems that are likely to be encountered in real-life field situations. It has been divided into nine chapters dealing with design details for gabion, masonry, concrete, and rubber check dams, including briefs about topographical surveys, geotechnical, and hydrological investigations.

Features:

- Gives insight into the principles of design and construction of check dams.
- Includes data assumptions and design principles along with design philosophy.
- Discusses design of gabion, masonry, and concrete check dams.
- Contains illustrative examples along with 20 engineering drawings and 140 quick solution tables for design of gabion, masonry, and concrete check dams.
- Explores easy-to-use tables for design of masonry and concrete retaining walls.

This book is aimed at professionals, students, researchers, and practitioners in civil engineering, hydrology, and water conservation.

Check Dam Designs

A Practice-Oriented Approach

Rajendra Chalisgaonkar and Mukesh Mohan

CRC Press
Taylor & Francis Group
Boca Raton London New York

CRC Press is an imprint of the
Taylor & Francis Group, an **informa** business

Designed cover image: www.shutterstock.com

First edition published 2025
by CRC Press
2385 NW Executive Center Drive, Suite 320, Boca Raton FL 33431

and by CRC Press
4 Park Square, Milton Park, Abingdon, Oxon, OX14 4RN

CRC Press is an imprint of Taylor & Francis Group, LLC

ISBN: 978-1-032-81755-2 (hbk)
ISBN: 978-1-032-99713-1 (pbk)
ISBN: 978-1-003-60562-1 (ebk)

DOI: 10.1201/9781003605621

Typeset in Times
by Apex CoVantage, LLC

To
Our Wives, Children and Families

Contents

Figures

Tables

Preface

Check Dams are small barriers built across the direction of water flow on shallow rivers and streams for the purpose of water harvesting. These small dams are built in a range of sizes using a variety of materials, including wooden logs, stones, boulder, gabions, masonry, concrete, etc. While designing check dams, the hydrologic, structural, and hydraulic design of structures need to be carried out. The design vigour depends on the type of permanent check dam structures. However, no complex engineering design is involved in the design of permanent check dams, yet proper design of permanent check dams is very seldom carried out to fulfill their objective.

WHY THIS BOOK?

This book is primarily written for the practicing engineers who have become a bit rusty on the subject of check dam designs. It is a review of basic principles and practice-oriented approach and will also be useful to the students of civil, agriculture, construction technology, architecture, soil, and water conservation, assuming they have already acquired a basic knowledge of statics, geotechnical, and design of simple masonry and concrete structures.

During more than 35–40 years of professional service of the authors in the state government water sector, it was noticed that many engineers often design check dams, presuming a very simple structure. Thus, it was felt that the professionals engaged in building/designing check dams be made more aware and informed about the design principles and processes satisfying serviceability, safety, and economy of structures to keep them abreast with the new technological advances. A hands-on approach with a strong, practical bias has been adopted, which will be helpful in solving problems that are likely to encounter in real-life field situations. This is our attempt to write a straightforward reference book that will appeal to practicing engineers and students.

Although some literature is available in scattered form on the subject, but a single volume for professional practitioners could not be found. Hence, an attempt has been made by the authors to condense and simplify a practice-oriented, easy-to-follow, and thorough approach for the robust design of gabion, masonry, and concrete check dams from many sources, including the experience of the authors. Hopefully, the tables and drawings presented with practical examples will make the book a good resource for civil engineers, agricultural engineers, soil and water conservationists, practicing engineers and NGOs (nongovernmental organisations) working for water conservation, etc.

CONTENTS AND COVERAGE

The book has been divided in to *nine* chapters dealing with the basics of check dam design and design details for gabion, masonry, concrete, and rubber check dams.

Chapter 1 provides a brief description of conventional/traditional check dams and discusses functions, advantages, disadvantages, and failures of check dams.

Chapter 2 discusses in brief about the topographical survey, geotechnical and hydrological investigations, classifications of streams, factors affecting runoff and estimation of design flood by slope area, velocity area, rational and empirical methods for planning, and constructions of check dams.

Chapter 3 describes the components and design philosophy of check dams with design loads and functional, structural, and serviceability requirements. The design procedures are based on basic principles and are commonly used by most engineers involved in the check dam designs. These methods have consistently provided structures that have performed well in service. Fixing of design parameters, like the length and height of the check dam, the design of the weir, energy dissipation, etc., has also been discussed in detail in the chapter.

Various aspects of gabion, masonry, and concrete check dams, including the components of check dams, their advantages, and the comparison/selection of the type of check dams for construction, are discussed in Chapter 4. The classification and components of rubber check dams, their advantages, and their disadvantages have also been discussed-in-brief in the chapter.

Forty-four easy-to-use tables for the design of gabion check dams and abutments/wing walls with heights varying from 1.50 m to 3.0 m for different river widths and shapes have been presented along with examples illustrating the application of tables in Chapter 5.

In Chapter 6, 48 tables for the quick design of masonry check dams with heights varying from 1.50 m to 3.0 m for different river widths and shapes have been presented along with examples illustrating the application of tables. Charts and tables have also been provided for masonry abutments/wing walls for sections having water face vertical or earth face vertical.

Forty-eight tables as design aids for concrete check dams with heights varying from 1.50 m to 3.0 m for different river widths and shapes have been provided along with examples illustrating the application of tables in Chapter 7. Charts and tables have also been provided for concrete abutments/wing walls for sections having water face vertical or earth face vertical.

Chapter 8 describes the site preparation and construction aspects of gabion, masonry and concrete check dams. Comparison of geo-synthetic clay liner and geo-membrane, provision of drainage arrangement and weep holes in wing walls, and maintenance aspects have also been discussed briefly in the chapter.

Ready-to-use engineering drawings have been provided in Chapter 9 for gabion, masonry, and concrete check dams with heights varying from 1.50 m to more than 1.50 m for different shapes. Specifications for the construction of gabion, masonry, and concrete check dams have also been presented with weep hole, joint details, etc.

Though care has been taken to present error-free material, some errors might have crept in inadvertently. We would highly appreciate if these errors and suggestions for improvement are brought to our notice at *chalisgaonkar@gmail.com/*mukeshmohan3@gmail.com to enhance the usefulness of the publication.

Finally, we would like to offer the following personal dedications:

I owe an enormous debt of gratitude to my wife, *Deepa*, who was always supportive to this endeavour, otherwise this would not have become possible and to my children,

Rahul, Rohan, Madhurima, Nandita, Grandchildren Reyansh, my family and friends for their valuable support and encouragement.

November 22, 2024, Rajendra Chalisgaonkar
Thane, India

I express thanks to my wife, *Alka* and to my children, family and friends for their enthusiastic support and encouragement during the preparation of this book.

November 22, 2024, Mukesh Mohan
Dehradun, India

Acknowledgements

We acknowledge many engineers, who constantly challenged for an answer, and the book has evolved from such discussions. Our understanding of the subject was greatly influenced by the publications of several authors. The authors wish to acknowledge and express grateful thanks to the following for according permission to reprint some tables and figures from their standards and codes of practice in the present book:

1. The Bureau of Indian Standards, New Delhi, for according permission to include Tables 1 and 2, Figures 1, 3, 4, 5B, 6, 11 of IS:16014, and Table 1 of IS:6512;
2. The Indian Road Congress, New Delhi;
3. Presses polytechniques et universitaires romandes, Lausanne, Switzerland;
4. Standard Publisher Distributors, New Delhi.

The authors also wish to acknowledge and extend thanks to the following:

1. Dr. Guillermo Cerrillo Tardio, Spain, Mr. Manoj Chandran$_{IFS}$, Government of Uttarakhand, Dr. Ram Suresh, Professor RPCAU, Dr. S. K. Jena, ICAR-Indian Institute of Water Management, Dr. Hamid R. Solaymani, Tehran;
2. The authors and publishers of various books and technical papers, referred at appropriate places;
3. U.S. Geological Survey, Food and Agriculture Organisation of the United Nations, Web site owners, and others for the use of illustrations and texts;
4. Researchers and design engineers working in the water conservation.

The authors extend special thanks to Dr. Gagandeep Singh, Ms. Jahnavi Vaid, and the staff of CRC Press Publishers – Taylor & Francis – for their cooperation at all the stages of production of this book and bringing out the book nicely.

The authors would like to express their heartfelt gratitude to each and every one who has helped us directly or indirectly in this endeavour.

We would like to apologize for any phrase or illustrations used in this book inadvertently without acknowledgement. The complete details of sources are presented in the references at the end of each chapter.

About the Authors

Rajendra Chalisgaonkar, former Engineer-in-Chief, Irrigation Department, Government of Uttarakhand, India, graduated from Madhav Institute of Technology & Science, Gwalior, India, in 1978 in civil engineering and then obtained an M.E. (structural engineering) in 1982 from the University of Roorkee, now IIT Roorkee. He was also Managing Director of the Uttarakhand Project Development and Construction Corporation Ltd., a government of Uttarakhand enterprise. He was responsible for the development of irrigation, policymaking at the apex level, planning, management, and construction of water resources and infrastructure projects. He has more than 40 years of varied experience, which includes investigation, planning, design, research, CAD applications, and construction, and has also worked on prestigious projects of Uttar Pradesh and Uttarakhand in India. He has authored two books, *Computer Aided Concrete Mix Design* and *Design of Breast Walls: A Practical Solution Approach*, and authored more than 125 research papers published in various journals and symposia. He has also delivered keynote/special lectures at various institutes and organisations. After superannuation, he has been engaged as Adjunct Professor at MITS, Gwalior, Water Resources and Hydropower Consultant, IIT Roorkee, Dam Safety Management Expert, National Dam Safety Authority, Consultant, DMR Hydroengineering & Infrastructures Ltd., Consultant, NJS Engineers India Ltd., Consultant, DHI, India, Water & Environment Pvt. Ltd. and other organisations. He is the recipient of many prestigious awards, including the Life Time Achievement Award in 2021 by the Indian Society for Hydraulics, CBIP-Young Engineers Award in 1992, IEI-HYDROSEM Award in 1996, IEI-Eminent Engineers Award in 2014, IEI-Engineering Excellence in Field Award in 2017, IEI-Eminent Authors Award in 2021, CBIP-D.B. Anand Medal in 1992, Gwalior Ratna in 1997, among other honours.

Mukesh Mohan, former Engineer-in-Chief, Irrigation Department, Government of Uttarakhand, and Managing Director, Uttarakhand Project Development and Construction Corporation Ltd., obtained his B.E. (civil engineering) and M.E. (hydraulics engineering) from the University of Roorkee, now IIT Roorkee, in 1985 and 1987. He has an experience of more than 35 years in the field of design, planning, management and construction of water resource structures, water supply projects and sewerage projects, sewerage network, and sewage treatment plant. Mr. Mukesh Mohan was involved in planning, survey, and construction work of barrages and allied structures along with the 4.50 km long head race tunnel of the 304 MW Maneri Bhali Stage-II Hydro Electric Project, Uttarkashi, India. He was also involved in the laying of water supply networks, sewer networks, and sewerage treatment plants in Roorkee and Dehradun cities. He has co-authored ten research papers, which has been published in journals and symposia. Mr. Mukesh Mohan is the recipient of the prestigious IEI-Eminent Engineers Award from Uttarakhand State Centre for his contributions in the field of water resources.

Notations

The notations have been explained, wherever they appear. The following notations have been more commonly used in the book:

α	Angle of Earth Face of Wall with Vertical
δ	Angle of Friction between the Masonry/Concrete Wall and Backfill
ϕ	Angle of Repose of Backfill
ι	Angle of Surcharge of Backfill
μ	Coefficient of Friction between the Masonry and Soil/Rock Mass
γ_{sat}	Unit Weight of Saturated Earthfill
γ_{sub}	Unit Weight of Submerged Earthfill
γ_{Box}	Unit Weight of Gabion Boxes with filling Material
$\gamma_{concrete}$	Unit Weight of Concrete
$\gamma_{masonry}$	Unit Weight of R. R. Masonry
γ_{water}	Unit Weight of Water
B	Base Width
B_{ABUT}	Base Width of Abutment
$B_{D/SWING}$	Base Width of D/S Wing Wall
B_{NOF}	Base Width of Non-Over Flow
B_{WEIR}	Base Width of Weir
f	Silt Factor
F_{ϕ}	Partial Factor of Safety in Respect of Friction
F_{C}	Partial Factor of Safety in Respect of Cohesion
K_{a}	Coefficient of Active Earth Pressure
K_{ah}	Horizontal Component of Coefficient of Active Earth Pressure
F_{SLD}	Factor of Safety against Sliding
F_{OVT}	Factor of Safety against Overturning
H_{DS}	Horizontal Component of Slope in Downstream in Non-Over Flow
$H_{DS'}$	Horizontal Component of Slope in Downstream in Weir
H_{US}	Horizontal Component of Slope in Upstream in Non-Over Flow
$H_{US'}$	Horizontal Component of Slope in Upstream in Weir
L	Width of River/Stream/Gadhera
$L_{D/SABUT}$	Length of Abutment in Downstream
$L_{D/SWING}$	Length of D/S Wing Wall
$L_{D/S\ APRON}$	Length of D/S Apron
$L_{U/SABUT}$	Length of Abutment in Upstream
P_{a}	Active Earth Pressure along the Length of the Wall
P_{p}	Passive Earth Pressure along the Length of the Wall
Q	Flood Discharge
q	Discharge Intensity

R	Depth of Scour
TW	Top Width
TW_{ABUT}	Top Width of Abutment
$TW_{D/SWING}$	Top Width of D/S Wing Wall
TW_{NOF}	Top Width of Non-Over Flow
TW_{WEIR}	Top Width of Weir
W	Width of Weir

1 Check Dams and Their Functions

1.1 WHAT IS A CHECK DAM?

Check dams are small linear barriers built across the width of rivers and streams in the direction of water flow to harvest water, decrease the velocity of flow, minimise scour and promote the deposition of eroded material. Check dams have been and are being built throughout the world for flood mitigation, sediment control, water supply, agricultural land development, watershed restoration, etc. However, it has been observed that the number of check dams constructed to control floods, regulate sediment transport, reduce upstream reach slopes, and stabilise river channel beds continues to increase, and they have experienced heavy damages or have been washed away during floods. Causes of failure include inappropriate check dam location, lack of adequate design criteria and/or poor construction quality. These failures have created a lack of confidence in adopting check dams as restoration tools.

Check dams are built in a range of sizes using a variety of materials, including clay, stone, and cement. Earthen check dams or embankments can easily be constructed by the farmers. Masonry and cement concrete structures, on the other hand, require some degree of advanced construction experience and monetary inputs. Earthen dams do not allow the overflow of water, in contrast to masonry and concrete structures which allow excess water to flow over the weir.

1.2 PURPOSE OF CHECK DAMS

The purpose of check dams is to retain water upstream so that the water percolates into the ground and recharges the ground water table. The resulting water can then be repurposed for irrigation, drinking water for cattle, or even pumped and filtered to provide water for household drinking and other purposes. As concerns arise around large dams and the environment, check dams are proving their worth over and over. Check dams built on seasonal rivers stop rain water from discharging into the sea or going to waste and provide natural freshwater bodies for local use as well as strengthening ecosystems. Check dams also serve to trap sediment and pollutants, strengthen local biodiversity, and help improve the overall quality of the water.

1.3 WHERE TO CONSTRUCT CHECK DAMS?

Check dams can be constructed by blocking the flow of water at a suitable location in the small stream/river. The stream or other flowing water should have clear banks on both sides so that the check dam can create a small reservoir. The banks should be higher than the proposed height of the check dam or the water will simply flow

DOI: 10.1201/9781003605621-1

around the check dam and divert the stream. Once a suitable location is identified, a check dam can be constructed out of any sturdy material that will withstand the force of the flowing water.

1.4 FUNCTIONS OF CHECK DAM

A check dam is generally designed to perform the following functions:

- store water for drinking, irrigation, etc., and to increase infiltration and percolation in the soil.
- reduce the velocity of water to a non-erodible one and trap the silt going downstream in the river.
- prevent erosion by not allowing water to run on long slopes.

1.5 TYPES OF CHECK DAM

1.5.1 Temporary Check Dam

The temporary check dams are constructed to maintain the stability of a stream/rivulet and to make possible the establishment of vegetative cover. These are not durable and need frequent maintenance, though these are inexpensive and easy to build. These structures are constructed using the locally available material. These check dams are only built for a short period of time, i.e. between 3–8 years by using sand bags, pine needles, brushwood, etc.

1.5.1.1 Sand Bag Check Dam

A sand bag is a sack made of burlap polypropylene, used jute or polyethylene bags, etc., filled with sand or soil and used for flood control, soils erosion, etc. The bags are piled up to a maximum of three to four layers to form a small check dam. This cheap technique is particularly useful in areas where stones are not available for building ordinary check dams. The expected or designed service life of such check dams is about one year.

1.5.1.2 Brushwood Check Dam

These type of check dams are constructed with wooden poles and brushwood available locally, across small streams/rivulets, where the side slope of the rivulet is less than 45°. Depending on the width of the stream and catchment area, wooden poles of small diameter are driven into the ground in a single or double row, at a right angle to the flow, and brushwood is packed on the upstream face of the check dam. Single-row brushwood check dams are used in small rivulets not deeper than 1.00 m, and double-row brushwood check dams are most suitable for rivulets, deep up to about 2 m. The expected service life of such check dams is about two years.

1.5.1.3 Pirul (Pine Needle) Check Dam

Pinus roxburghii, or the Himalayan longleaf pine, is the most important and widely spread forest type found in the Western Himalayas. Pirul check dams are built by packing the dry pine needles tightly into woven coir nets and rolled into logs of

FIGURE 1.1 Sand Bag Check Dam.

Source: Kumar et al. (2012)

FIGURE 1.2 Brushwood Check Dam.

Source: Jonga et al. (2021)

appropriate length and diameter. These logs are then plugged in to eroding rivulets and reinforced with cuttings of easy-to-sprout vegetation. These pine needle check dams, instead of decomposing after a few years, are found to get stronger by accumulation and trapping of clay and cemented with calcium getting precipitated within. Moreover, the pirul check dam is found to be more cost-effective and

FIGURE 1.3 Pirul Check Dam.

Source: Rawat and Chandran (2009)

environmentally friendly compared to traditional check dams and also contributes to the reduction of carbon dioxide emissions by reducing forest fires. The expected service life of such check dams is about one year.

1.5.2 Semi-Permanent Check Dam

Semi-permanent check dams, built for a short period of time using logs, stones, etc., are also constructed using local materials.

1.5.2.1 Log Check Dam

These type of check dams are constructed in an area where wooden logs are available in abundance. White ants are the greatest enemy of log check dams, and therefore, appropriate measures should be taken before the construction of such check dams. The expected service life of such check dams is about two years.

1.5.2.2 Loose Stone Check Dam

These types of check dams are constructed across steep and broad streams, where good-sized stones are available in abundance. Dry stone/loose rock check dams are made up of relatively small rocks/stones, usually constructed up to about 1.25 m in height, have longer life and usually require less maintenance. The expected service life of such check dams is between two and five years.

FIGURE 1.4 Log Check Dam.

Source: Tardio and Rodriguez (2016)

FIGURE 1.5 Loose Stone Check Dam.

Source: Tardio and Rodriguez (2016)

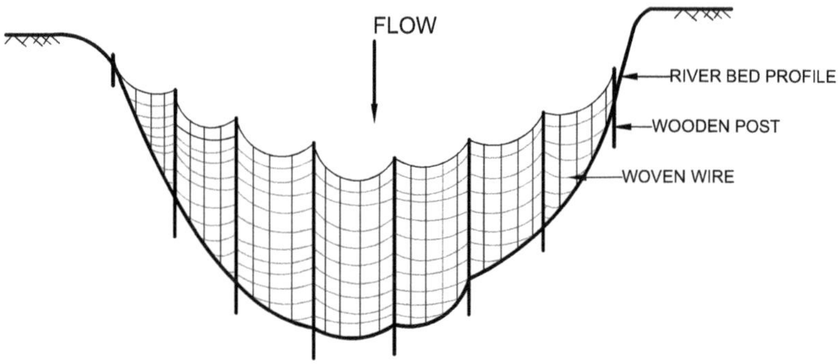

FIGURE 1.6 Netting Check Dam.

Source: Suresh (2023)

1.5.2.3 Netting Check Dam

Netting check dam consists of wooden posts which are driven into the stream bed to support a strip of wire netting (Figure 1.6) and, thus, form a low wall across the stream. Such check dams are constructed up to 60 cm in height. The lower edge of the netting is buried into the stream bed. The brush or straw, which is porous in nature, is also piled loosely towards the upstream side of the netting wall to form a barrier. The netting dams are usually located near the top end of the streams/rivulets.

1.5.3 Permanent Check Dam

Permanent check dams are constructed when the benefits from such structures are justifiable compared to the cost of construction. General requirements of the permanent check dams are as follows:

i they should be constructed with permanent materials,
ii they should have adequate capacity to handle the runoff and
iii they should help in stabilising the stream/rivulet and store water wherever necessary.

Permanent check dams are often built of boulder, gabion, masonry, concrete, etc.

1.5.3.1 Boulder Check Dam

Boulder check dams are constructed across streams, where big boulders are available. When properly made, they are almost like permanent structures, such as gabions, masonry, and concrete structures. The expected service life of boulder check dam is between two and five years.

1.5.3.2 Gabion Check Dam

Gabion check dams are constructed by the filling of stones in wire mesh cage and are used for retaining water and debris in the stream/rivulet. These structures are widely adopted for the treatment of drainage lines because they are flexible, porous, and economical as compared to masonry structures. These structures are

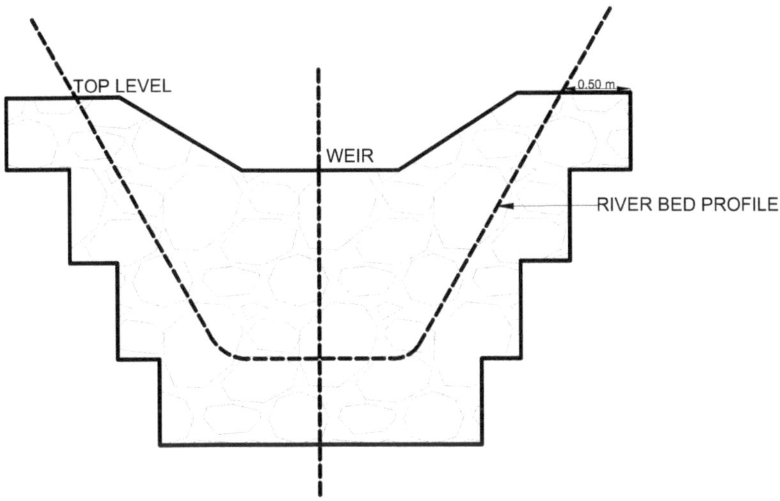

FIGURE 1.7 Boulder Check Dam.

Source: Geyik (1986)

FIGURE 1.8 Gabion Check Dam.

Source: Solaymani and Gosain (2013)

constructed up to a height of about 1.00 m to 3.00 m. The foundation of these check dams is generally placed below the ground level by 0.30 m to 0.50 m and embedded in to the banks from 0.30 m to 0.60 m. The expected service life of such check dams is between two and five years or more depending on the type and quality of construction.

FIGURE 1.9 Masonry Check Dam.

Source: Sreedevi et al. (2004)

1.5.3.3 Masonry Check Dam

Masonry check dams, designed on the principle of gravity dams and constructed with cement mortar and non-disintegrating stones, are generally used (Figure 1.9). The stones used in constructing masonry check dams must be hard enough to withstand abrasion, non-disintegrating, and resistant to weathering. The main objective of the masonry check dam is to hold fine and coarse material carried by flowing water in the stream/rivulet. Wing walls are also built to protect them against flash water. The angle between the wing and wing wall is kept from 30° to 45°. The expected service life of such check dams is more than five years.

1.5.3.4 Concrete Check Dam

Concrete check dams, designed as gravity dams, are usually constructed when adequate materials are not available for constructing the masonry check dams. The specifications used for construction of masonry dams can be suitably used for concrete dams. The disadvantage of this structure is that once it is damaged, it cannot be repaired easily. The expected service life of such check dams is more than five years.

1.5.4 Flexible Check Dam

A flexible check dam is an inflatable structure or rubber dam built across a stream for water conservation, flood control, and regulating the flow of water in the stream. When it is inflated, it serves as a check dam/weir, and when it is deflated, it functions as a flood mitigation device and allows the rubber dam to be routinely cleaned from deposited silt. In the recent past, incidents of check dams being washed away

FIGURE 1.10 Concrete Check Dam.

during floods in India are common, and therefore, rubber dams have been increasingly accepted and are widely used worldwide. The expected service life of such check dams is about 25 to 30 years.

1.5.5 Nonporous Check Dam

Nonporous check dams, such as the dams built from earth, concrete, etc., receive heavy impact from the hydrostatic forces of stream flow. These forces require effective anchoring of the check dam into the stream banks, where most of the pressure is transmitted. In exceptional cases only, the earthen check dams be constructed to control stream erosion due to the likely failure of earth materials.

1.5.6 Porous Check Dam

Porous check dams have holes in an otherwise impermeable structure that allow water to seep through and drain the structure, creating less pressure on the banks of streams than nonporous check dams. Because streams generally form on erodible soils, constructing porous check dams is easier, cheaper and often more effective than nonporous check dams. Loose rocks, rough stone masonry, gabions, etc., have been used successfully to construct porous dams.

1.6 BENEFITS OF A CHECK DAM

- The check dams retain excess water flow during monsoon rains in a small catchment area behind the structure, which is primarily intended for use in irrigation during the monsoon, and later, during the dry season, it can also be used for livestock and domestic needs.

FIGURE 1.11 Flexible/Rubber Check Dam.

Source: **Indian Institute of Water Management, Bhubaneswar**

- It helps in preventing flooding and erosion.
- The silt deposited can be used in farms all throughout the year as a possible substitute for chemical fertilizers and can be a better organic alternative to the fertiliser.
- The major environmental benefit is the replenishment of nearby ground water reserves and wells.
- It slows the flow rate and aids in protecting river banks.
- Check dams are inexpensive, cheaper than other types of dams and easy to construct.
- The construction of check dams is straightforward and very simple and therefore, there is no need for cutting-edge equipment and technology.

1.7 NEGATIVE ASPECTS OF CHECK DAM

- Check dams may only be used in a small drainage basin, and their strength and life span are constrained.
- Upkeep/maintenance is necessary after heavy rain/flood.
- The water table below the ground rises as a result.

- Aquatic life is affected.
- Removal could be expensive, depending on the type of check dam placed.

1.8 CAUSES OF FAILURES OF CHECK DAMS

In the case of check dams, the main types of failure which may occur are similar to dam structures and can be grouped into the following categories:

1.8.1 Sliding Failure

When the check dam slides over its foundation or one part of the check dam slides over a part of itself, it is called a sliding failure. It occurs when the net horizontal forces acting on the gravity dam exceed the frictional resistance produced between the body of the dam and the foundation. The resistance may be due to friction alone, or it may be due to a combination of friction and shear strength.

1.8.2 Overturning Failure

Horizontal and vertical forces, such as water pressure and uplift pressure, can act against a check dam, creating overturning force or rotation of the structure.

1.8.3 Scour Failure

If the foundation of the check dam is not provided up to the depth of scour or the cutoff is not provided in the downstream up to the scour depth, the structure may fail due to scour.

1.8.4 Structural Failure

Structural failure occurs when the tensile or compressive stresses in the dam exceed the strength of the materials used in the construction of check dam. Masonry and concrete are weak in tension, and therefore, they are so designed that there is no tension in the structure. If the compressive stresses exceed the allowable stresses of the masonry or concrete, the materials may be crushed, leading to dam failure.

The structure failure is caused mainly due to faulty hydrologic, hydraulic, or structural design either alone or in combination. However, it has also been seen that there are cases when check dams have failed because of the following:

- Insufficient flood passing capacity of the check dam weir
- Insufficient provision for the dissipation of kinetic energy
- Unprotected banks in upstream
- Uneven settlement of the foundation leading to deformation or collapse
- Improper foundation causing uplift pressure to prevail over the body of the structure

1.8.5 Some Other Causes of Failure

There have been some causes of failure of the check dams due to improper/wrong data, improper design, poor workmanship at the site and/or material failure. Some of such failures include the following:

1) Poor design
   The main reason for the failure could be the improper design of the check dams and supporting system.
2) Poor foundation preparation
   Inadequate site preparation, such as improper grading or compacting the base soil, can lead to an unstable foundation.
3) Poor workmanship
   Due to the lack of proper knowledge and know-how of construction techniques, often, the check dams are not constructed as monolithic structures at the site and thus fail due to poor workmanship.

1.9 PREVENTIVE MEASURES TO AVOID FAILURE

1.9.1 Proper Site Preparation

Ensure thorough site preparation, including adequate grading, compaction, and soil stabilisation, to create a stable foundation for the check dam.

1.9.2 Geotechnical Analysis

Conduct a geotechnical analysis of the site to assess soil conditions, evaluate load-bearing capacity, and choose the appropriate design for check dam.

1.9.3 Engineered Solutions

In some cases, check dams may need to be engineered with additional features, such as reinforced concrete foundations, tiebacks or geogrid reinforcement, etc.

1.9.4 Consult Professionals/Experts

Seek the expertise of geotechnical/structural engineers/water resources specialists when designing, installing, or constructing check dams.

1.9.5 Regular Inspection

Periodic inspections and maintenance can identify issues early and prevent extensive damages during floods. Replacing deteriorated stones/gabion boxes is essential in case of masonry and gabion check dams.

1.9.6 Proper Filling of Gabions

Ensure that gabion boxes are filled to the required density and maintain the structural integrity of the wall.

REFERENCES

Geyik, M. P. (1986) *FAO Watershed Management Field Manual – Gully Control*, FAO Conservation Guide 13/2, Food and Agriculture Organization of United Nations, Rome.

Jonga A., E. Meilianda and Nizamuddin (2021) *Community-based Intervention in Reducing Flood Impacts in Gambia*, The 10th AIC-ELS 2020, IOP Conference Series: Earth and Environmental Science 711, 012020, March.

Kumar, Sanjay, A. K. Singh, K. A. Gopinath, Sreenath Dixit and B. Venkateswarlu (2012) *Sand Bag Check Dam – A Case of Successful Mobilization of Community Resources against Climate Variability*, Zonal Project Directorate, Zone – II, Indian Council of Agricultural Research, Kolkata.

Rawat, R. B. S. and Manoj Chandran (2009) *Pine Needle Check Dams for Soil and Water Conservation*, World Forestry Congress 2009, Buenos Aires.

Solaymani, Hamid R. and A. K. Gosain (2013) *Exploring Adaptation Options to Climate Change in Semi-Arid Watershed Using Choice of BMPs*, SWAT 2013 Conference Proceedings, Texas Water Resources Institute Report – TR-471, Paul Sabatier University, Toulouse, July 17–19.

Sreedevi, T. K., B. Shiferaw and S. P. Wani (2004) *Adarsha Watershed in Kothapally Understanding the Drivers of Higher Impact*, Global Theme on Agroecosystems, Report No. 10, International Crops Research Institute for the Semi-Arid Tropics, Patancheru, Andhra Pradesh.

Suresh, R. (2023) *Soil and Water Conservation Engineering*, Fifth edition, Standard Publisher Distributors, New Delhi.

Tardio, Guillermo Cerrillo and Jose Luis Garcia Rodriguez (2016) *Monitoring of Erosion Preventive Structures Based on Eco-Engineering Approaches: The Case of the Mixed Check Dams of Masonry and Forest Residues*, Journal of Engineering Science and Technology Review 9 (1) 103–107.

2 Investigations and Hydrological Aspects

2.1 INTRODUCTION

Surveys, investigations, and hydrological analyses provide data required for planning and designing a check dam project. A project can be designed optimally and economically, only when the necessary field data is available. Detailed surveys and investigations, such as topographical surveys, hydrological data, geological and geotechnical investigations, construction material investigations, socio-economic, ecological and environmental impact assessment studies, etc., should be undertaken to examine the feasibility of the project.

2.2 TOPOGRAPHICAL SURVEY

Topographical surveys required for any water conservation project may include river survey, reservoir survey, site survey, command area survey, etc. Methods of topographical survey vary depending on the degree of accuracy to be attained, extent/ size of the water conservation project and topography of the project area. Some of the topographical surveys to be carried out for any check dam project include the following:

2.2.1 River Surveys

River surveys cover the longitudinal section (L-Section) and cross section (X-Section) of the river upstream and downstream of the proposed check dam project.

i The L-Section on the upstream side shall extend from the axis of the check dam to the point up to which the back water effect is likely to extend or up to the maximum water level (MWL) + 0.50 m whichever is more and about 300 m length in the downstream of the proposed axis of the check dam.

The levelling of the L-Section shall be done at 50 m or fewer intervals, along the deep channel, and the following details shall also be indicated in the L-Section:

a) Date of survey of the particular reach and average water level on that day; and
b) Maximum historical observed highest flood level (H.F.L.).

ii The cross section on the upstream side shall be taken at 50 m intervals up to the point to which the back water effect is likely to extend or up to the

DOI: 10.1201/9781003605621-2

maximum water level (MWL) + 0.50 m whichever is more and about 300 m length in the downstream of the proposed axis of the check dam. A cross section shall also be taken along the axis of the check dam.

The levelling of the cross section shall be done at 25 m or fewer intervals, depending on the width of the stream/rivulet, and the following details shall also be indicated on the cross section:

a) Date of survey of the particular reach and average water level on that day; and
b) Maximum historical observed highest flood level (H.F.L.).

iii The longitudinal section and cross section shall be plotted on a suitable scale depending on the width of the stream/rivulet.

2.2.2 Reservoir Survey

A survey of reservoir, formed due to construction of the check dam, is necessary for developing the submergence area plan and assessing the reservoir storage capacity and affected population, if any. The scale and contour interval for the reservoir survey depend on the extent of the area and the topography of the site. Reservoir survey maps shall be made on scales ranging from 1:500 to 1:1000 with a contour interval of 0.50 m or more corresponding to natural ground slopes.

2.2.3 Command Area Survey

The command area is the area that is likely to be benefited from the check dam project. The command area survey helps in estimating the capability of the command area to make maximum use of the benefits from the check dam project.

2.3 SOCIO-ECONOMIC AND ENVIRONMENTAL SURVEY

As the project construction is likely to make significant environmental and economic impact in the region, it is necessary to carry out a detailed study to assess the likely effects on various aspects, including socio-economic factors and the environment at the time of the preparation of the detailed project report.

2.4 HYDROLOGICAL AND METEOROLOGICAL INVESTIGATION

This investigation helps in deciding the height of the check dam, weir/spillway capacity and the size of the reservoir. In this investigation, data is mainly collected regarding the catchment, such as the size, shape, density of vegetation, slope, permeability, etc., and the rainfall pattern, such as total monsoon rainfall, intensity of rainfall, distribution, etc.

The catchment area of the project is one of the most important aspects, as the capacity of the reservoir is assessed on the basis of yield from the catchment. The shape of the catchment also plays an important role in the occurrence of floods. In the case of a long and narrow catchment (leaf shaped), the time of concentration is longer, and hence, the chance of occurrence of heavy flood is much less. While in the case of compact (circular or fan shaped) catchments, the time of concentration is shorter, and hence, the chance of occurrence of heavy flood is more. The climatic factors, such as rainfall, temperature, humidity, evaporation, evapo-transpiration, wind velocity, etc., also affect the water yield and peak flood from the catchment. The essential minimum data required for the planning and design of the check dams include the following:

- Catchment area of the stream/rivulet
- Rainfall data from the nearby rain gauge stations
- The discharge data of the stream/rivulet at or near the proposed site for at least three years

2.5 GEOTECHNICAL INVESTIGATIONS

The basic purpose of geotechnical investigation is to find out the type of rock/soil in the area, the properties of rock/soil, including the bearing capacity at the check dam axis, and the permeability of the subsoil at the site.

2.6 HYDROLOGICAL ANALYSIS

When a storm occurs, some portion of rainfall may evaporate, some may infiltrate into the ground and the remaining portion flows over the land surface which is termed *surface flow*. If there is a relatively impermeable stratum in the subsoil, the infiltrating water moves laterally in the surface soil and joins the stream flow, which is termed *sub-surface flow*, or interflow (Figure 2.1). If there is no impeding layer in the subsoil, the infiltrating water percolates into the ground as deep seepage and builds up the ground water table or phreatic surface. The ground water may also contribute to the stream flow, if the ground water table is higher than the water surface level of the stream, creating a hydraulic gradient towards the stream. Low soil permeability favours surface flow. While all the three types of flow contribute to the stream flow, it is the surface flow that reaches the stream first, the subsurface flow, being slower, reaches after a few hours, and the ground water flow, being the slowest, reaches the stream after some days. The term *direct runoff* is used to include the surface flow and the subsurface flow.

When the surface flow starts, some flowing water is held in puddles, pits, and small ponds; this water stored is called *depression storage*. The volume of water in transit in the surface flow which has not yet reached the stream channel is called *surface detention* or *detention storage*. The portion of runoff in a rising flood in a stream, which is absorbed by the permeable boundaries of the stream above the normal phreatic surface, is called bank storage (Figure 2.2).

2.6.1 Catchment Characteristics

The number of streams, length of streams, stream density, and drainage density define the characteristics of the drainage, and the entire area of a river basin whose

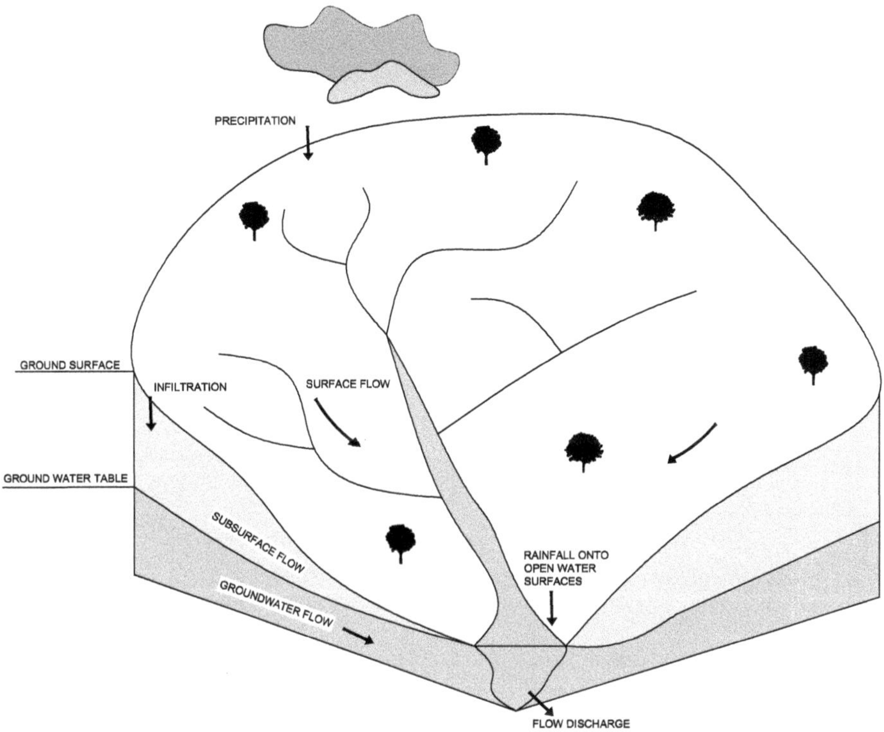

FIGURE 2.1 Rain Water Flow in Catchment.

Source: **Musy and Higy (2004)**

surface runoff drains into the river in the basin is considered as a hydrologic unit and is called *drainage basin*, watershed, or catchment area of the river. The characteristics of the catchment area are generally described by the stream order, hydrologic length, form factor, compactness coefficient, circulatory ratio, elongation ratio, stream density, drainage density, bed slope, etc.

2.6.1.1 Stream Order

Stream order is an important aspect of a drainage basin. It is defined as the measure of the position of a stream in the hierarchy of streams. The first-order streams are defined as those streams that have no tributaries. The junction of two first-order streams forms a second-order stream. A third-order stream is formed by the junction of two second-order streams. Thus, a stream of any order has two or more tributaries of the previous lower order. However, the stream order does not increase if two streams of different orders join. The highest stream order becomes the catchment order. First- through third-order streams are usually called *headwater streams.* Streams classified as fourth- through sixth-order are considered *medium streams.* A stream that is seventh-order or larger constitutes a *river.* The most common method used to define stream order is the Strahler method (Figure 2.3).

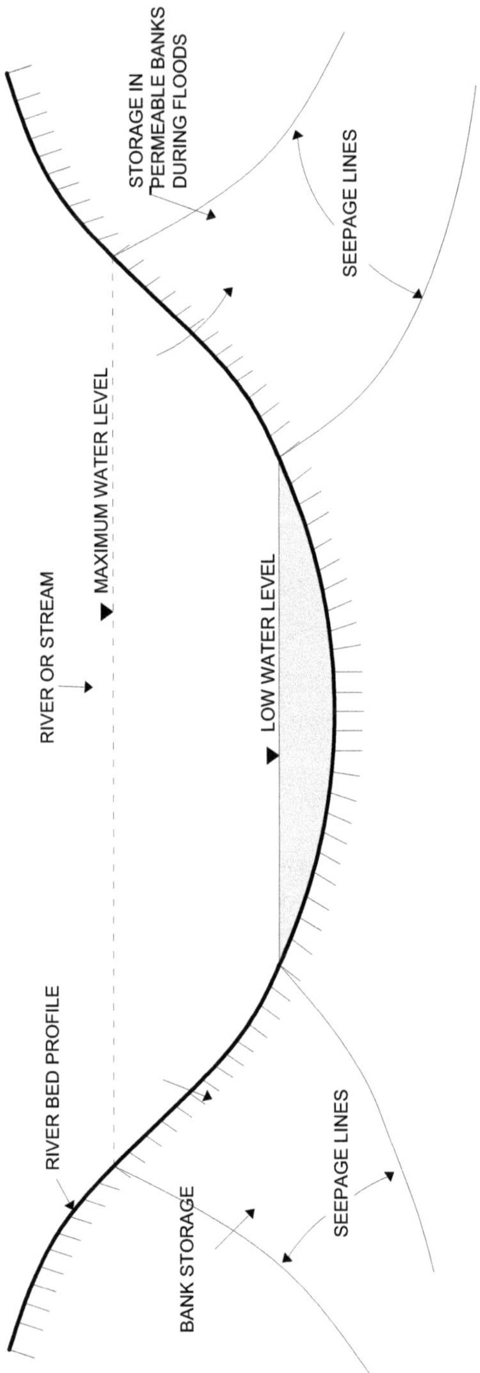

FIGURE 2.2 Bank Storage.

Source: **Winter et al. (1998)**

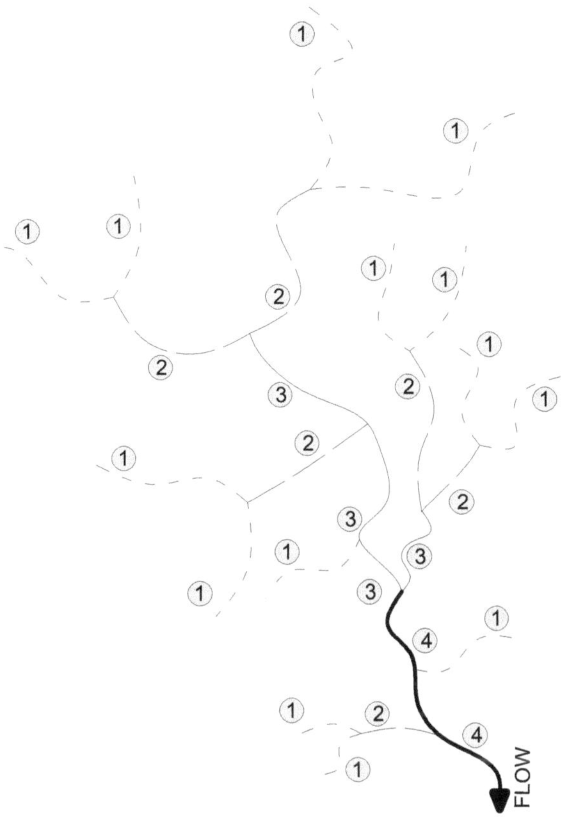

FIGURE 2.3 System of Stream Ordering.

Source: Strahler (1957)

2.6.1.2 Hydrologic Length

Conceptually, the basin length is the distance travelled by the surface drainage and sometimes more appropriately labeled as hydrologic length. This length is generally used in computing a time parameter, which is a measure of the travel time of water through a watershed. The hydrologic length is, therefore, measured along the principal flow path from the outlet to the catchment/basin boundary.

Basin length, L_b, is the longest dimension of a catchment/basin parallel to its principal drain, and basin width can be measured in a direction approximately perpendicular to the length measurement. The relation between mainstream length and drainage-basin area for small watershed is given by the following:

$$L_b = 1.312\ A^{0.568} \tag{2.1}$$

where

A = Area of catchment/basin, km^2

2.6.1.3 Form Factor

Form factor is the ratio of the area of the basin and the square of the axial length of the basin. It is commonly used to represent different basin shapes (Horton, 1932).

$$\text{Form Factor}, F_f = \frac{W_b}{L_b}$$

$$= \frac{A}{L_b^{\,2}} \tag{2.2}$$

where

W_b = Axial width of basin

L_b = Axial length of basin, *i.e.* the distance from the measuring point to the most remote point on the basin.

The value of form factor is in between 0.1–0.8. The value of form factor would always be greater than 0.78 for a perfectly circular basin. The smaller the value of the form factor, the more elongated the basin will be. The basins with high form factors (0.8) have high peak flows of shorter duration, whereas elongated drainage basins with low form factors have lower peak flow of longer duration.

2.6.1.4 Compactness Coefficient

Compactness coefficient is the perimeter of the basin divided by circumference of equivalent circular area.

$$\text{Compactness Coefficient, } C_c = \frac{P_b}{2\sqrt{(\pi \mathrm{A})}} \tag{2.3}$$

where

P_b = Perimeter of the basin

$2\sqrt{(\pi \times \mathrm{A})}$ = circumference of circular area, which equals the area of the basin.

If R is the radius of an equivalent circular area,

$$A = \pi R^2,$$

$$R = \sqrt{(\mathrm{A}/\pi)}$$

Circumference of the equivalent circular area $= 2\pi\mathrm{R} = 2\sqrt{(\pi \times \mathrm{A})}$

The compactness coefficient is independent of the size of the catchment and is dependent only on the slope.

2.6.1.5 Circularity Ratio

Circularity ratio is the ratio of basin area to the area of a circle having the same perimeter as the basin.

$$\text{Circularity Ratio } \mathrm{R_c} = 12.57\mathrm{A}/\mathrm{P_b^2} \tag{2.4}$$

2.6.1.6 Elongation Ratio

Elongation ratio is the ratio of the diameter of a circle of the same area as the basin to maximum basin length.

$$\text{Elongation Ratio } R_e = 1.128A^{0.5} / L_b \tag{2.5}$$

2.6.1.7 Stream Density

The stream density of a drainage basin is expressed as the number of streams per km².

$$\text{Stream Density, } D_s = \frac{N_s}{A} \tag{2.6}$$

where

N_s = Number of streams

2.6.1.8 Drainage Density

Drainage density is expressed as the total length of all stream channels (perennial and intermittent) per unit area of the basin and serves as an index of the areal channel development of the basin.

$$\text{Drainage Density, } D_d = \frac{L_s}{A} \tag{2.7}$$

where

L_s = Total length of all stream channels in the basin.

Drainage density varies inversely with the length of overland flow and indicates the drainage efficiency of the basin. A high value indicates a well-developed network and torrential runoff causing intense floods while a low value indicates moderate runoff and high permeability of the terrain.

2.6.1.9 Average Stream Slope

Stream slope affects the momentum of runoff. It reflects the rate of change of elevation with respect to distance along the principal flow path. The average stream slope is defined by

$$\text{Average Stream Slope} = \frac{\text{Total Fall of Longest Water Course}}{\text{Length of Longest Water Course}} \tag{2.8}$$

Horton (1932) has suggested a method of determining the slope of large drainage areas, *i.e.* the area is subdivided into a number of square grids of equal size. The number of contours crossed by each subdividing line is counted, and the lengths of the grid lines are scaled. Then the slope of the basin is given by the following:

$$S = \frac{1.5 \times CI \times N_c}{\sum L} \tag{2.9}$$

where

S = Slope of the basin

CI = Contour interval

N_c = Number of contours crossed by all the subdividing lines

ΣL = Total length of the subdividing lines

2.6.2 Catchment Shapes

The shape of a catchment is an essential element because it influences directly the shape of the hydrograph at the outlet. For example, for the same rain event, a catchment with a long narrow shape will result in lower runoff at the discharge site because it requires more time for the water to reach the outlet. Figure 2.4 show the catchment shapes and their hypothetical effect on discharge response to a rainfall event. A broad-shaped catchment (Figure 2.4 A) produces high runoff, as runoff from different tributaries meets the main stream at the middle at the same time resulting in high flood. Whereas in the fern-shaped (leaf or elongated) catchments (Figure 2.4 C), the time of concentration is more and the discharge is distributed over a long period, as the runoff from the tributaries nearest to discharge site is already drained out before the arrival of the flood from the farthest tributaries. In a fan-shaped catchment (Figure 2.4 B), the runoff peak is between broad- and fern-shaped catchments.

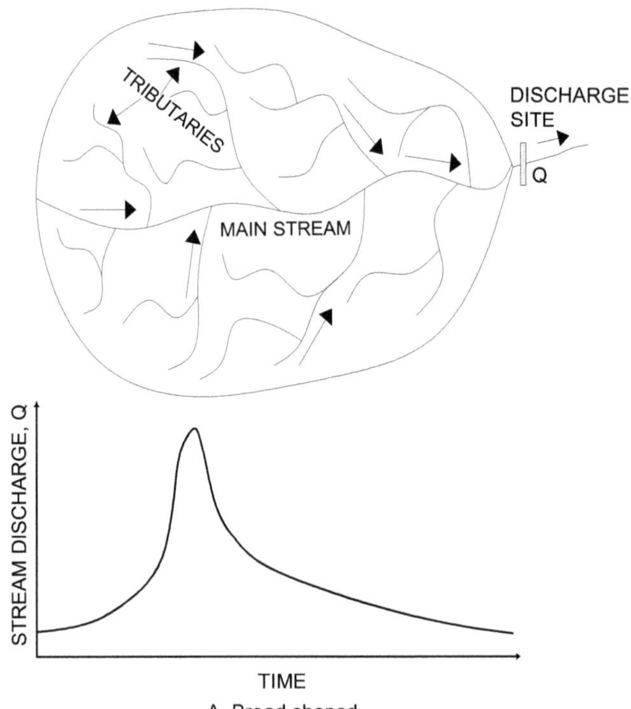

FIGURE 2.4A, 2.4B and 2.4C Broad-, Fan-, and Fern-Shaped Catchments and Their Hypothetical Effect on Discharge.

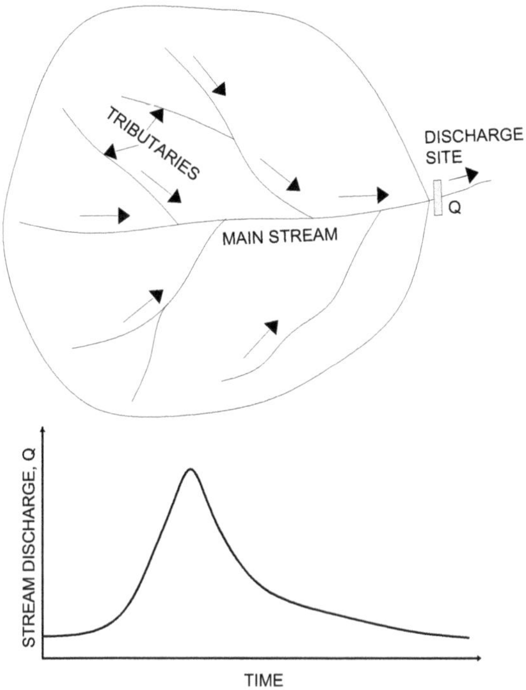

B. Fan shaped

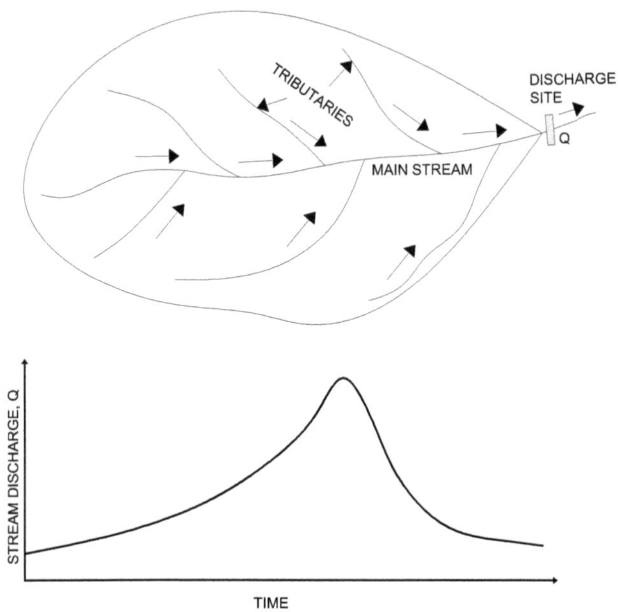

C. Fern (leaf) shaped

FIGURE 2.4A, 2.4B and 2.4C (Continued)

2.6.3 Classification of Streams

Streams may be classified as follows:

(i) Influent and effluent streams
(ii) Intermittent and perennial streams

2.6.3.1 Influent and Effluent streams

If the ground water table is below the bed of the stream, the seepage from the stream feeds the ground water, resulting in the build-up of water mound (Figure 2.5). Such streams are called influent streams, and many rivers which cross desert areas do so. Irrigation channels function as influent streams, and such streams dry up completely in the dry period and are called ephemeral streams. The ephemeral streams are generally seen in arid regions, which flow only for a few hours after the rainfall.

When the ground water table is above water surface elevation in the stream, the ground water feeds the stream (Figure 2.6). Such streams are called effluent streams. Most of the perennial streams are mainly effluent streams.

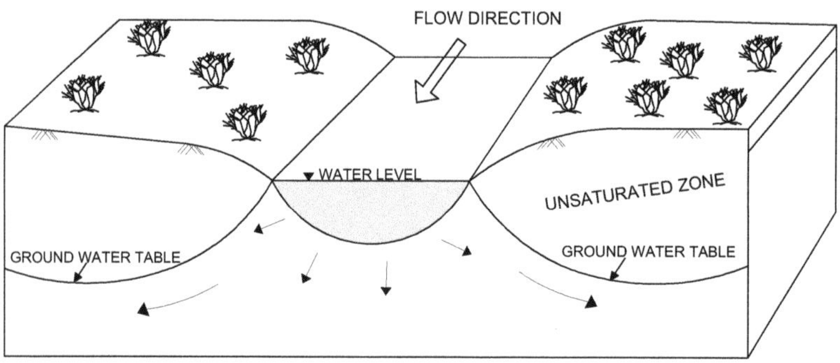

FIGURE 2.5 Influent Streams.

***Source*: Winter et al (1998)**

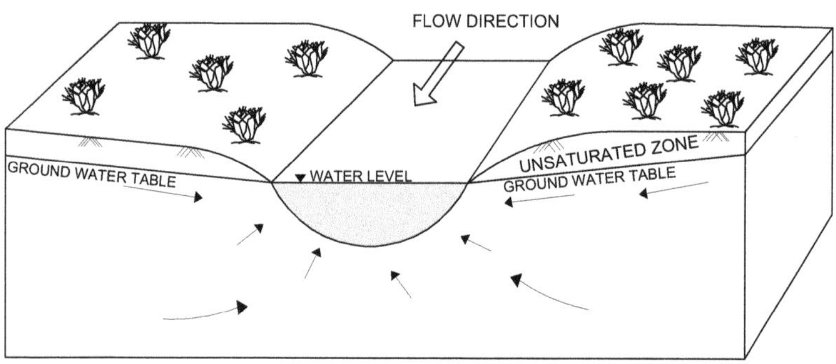

FIGURE 2.6 Effluent Streams

***Source*: Winter et al (1998).**

2.6.3.2 Intermittent and Perennial Streams

If the ground water table lies above the bed of the stream during the wet season but drops below the river bed during the dry season, the stream flows during wet season but becomes dry during dry season. Such streams are called intermittent streams.

Whereas in the case of perennial streams, even in the most severe droughts, the ground water table never drops below the bed of the stream, and therefore, they flow throughout the year.

2.6.4 Factors Affecting Runoff

The factors that affect the runoff from a drainage basin are shown in Table 2.1 (Rajab, 2021).

TABLE 2.1
Factors Affecting Runoff from a Drainage Basin

	Characteristics	Factors
i	Storm characteristics	• Type or nature of storm and season • Intensity • Duration • A real extent (distribution) • Frequency • Antecedent precipitation • Direction of storm movement
ii	Meteorological characteristics	• Temperature • Humidity • Wind velocity • Pressure variation
iii	Basin characteristics	• Size • Shape • Slope • Altitude • Topography • Type of soil • Land use/vegetation • Orientation • Type of drainage net • Proximity to ocean and mountain ranges
iv	Storage characteristics	• Depressions • Pools, ponds, lakes, streams, channels • Check dams • Upstream reservoir • Flood plains, swamps • Ground water storage in aquifers

Low-intensity storms over longer spells contribute to ground water storage and produce relatively less runoff. A high-intensity storm or smaller area covered by it increases the runoff since the losses, like infiltration and evaporation, are less. If there is a succession of storms, runoff will increase due to the soil's initial wetness from antecedent rainfall. Rain during the summer season will produce less runoff, while that during winter will produce more.

Peak runoff decreases as the catchment area increases due to the higher time of concentration. A fan-shaped catchment produces greater flood intensity than a fern-shaped catchment.

Steep rocky catchments with less vegetation will produce more runoff compared to flat tracts with more vegetation. If the vegetation is thick, greater is the absorption of water, so less runoff. If the direction of the storm producing rain is down the stream receiving the surface flow, it will produce greater flood discharge than when it is up the stream. The land use pattern – arable land, grass land, forest, or cultivated area – also affect runoff.

2.6.5 Design Flood Estimation

A flood is an unusually high stage in a river, usually the level at which the river overflows its banks and inundates the adjoining area, and the estimation of the design flood for check dams is a very significant component of hydrological studies, as a higher value will result in the increase of the cost of the check dam and an underestimated value is likely to place the check dam at risk. Also, flood estimation is inherently more difficult on smaller catchments than larger ones. Catchment characteristics, used in the estimation of flood parameters at ungauged sites, are more difficult to extract from smaller catchments; errors that escape detection will have a proportionally greater effect on the final estimate.

In order to design check dams, it is important to make an estimate of peak flood or runoff from small catchments. To estimate the magnitude of flood peaks from small catchments, the following methods are available:

1. Slope area method
2. Velocity area method
3. Rational method
4. Empirical method

Since the check dam projects are mainly constructed across ungauged streams/rivulets, the rational and empirical methods are generally used for estimating the peak flood.

2.6.5.1 Slope Area Method

In this method (Hiranandani and Chitale, 1960), the flow area, corresponding to the ever-maximum recorded flood, is obtained from the survey, and the velocity is calculated using the Mannings equation for rugosity coefficient and bed slope of the river/stream.

$$V = \frac{1}{n} R^{2/3} S^{1/2} \tag{2.10}$$

where

V = Velocity of flow, m/s

n = Roughness coefficient, specially known as Manning's n

R = Hydraulic radius = A/P

S = Bed slope

and

$$Q = VA \tag{2.11}$$

where

Q = Estimated flood, m^3/s

A = Cross-sectional area of the river/stream, m^2

The changes occurring in the cross-sectional profile during floods are generally ignored, and quite often, the values of n and S are assumed arbitrarily. Besides, the formula is not applicable to rivers with a lot of bed load or mudflow.

2.6.5.2 Velocity Area Method

The velocity area method (Hiranandani and Chitale, 1960) uses stream velocity and the stream cross-sectional area to obtain stream discharge as multiplication of both.

In practice, this is carried out by dividing the stream into small sections and measuring the velocity of flow going through each cross-sectional area. The number of cross-sectional areas that are used in a discharge measurement depend upon the width and smoothness of the stream bed. If the bed is particularly rough, it is necessary to use more cross-sectional areas so that the estimates are as close to reality as possible. The water velocity measurement is usually taken with a current meter and inserted into the stream which records the number of propeller turns with time. This reading can be easily converted into a stream velocity using the calibration equation supplied with the flow meter.

Where there is no current meter available, it may be possible to make a very rough estimate of stream velocity using a float in the stream (i.e. the time it takes to cover a measured distance). However, an allowance must be made for the fact that the float is travelling on the surface of the stream at a faster rate than water closer to the stream bed. The velocity-area method is an effective technique for measuring stream flow in small rivers, but its reliability is heavily dependent on the sampling strategy. The technique is also less reliable in small turbid streams with a rough bed (e.g. mountain streams).

2.6.5.3 Rational Method

Rational method, proposed in 1851, is most widely used for the analysis of runoff response from small catchment. The structure of the formula has remained the same all over these years, though the concepts in evaluating the parameters, especially the runoff coefficient and the rainfall intensity, have undergone considerable change and development. Initially, the formula was developed for urban areas, but due to its simplicity, it has now been used in rural areas also. The structure of the formula also provides a means of estimating the design flood of a certain return period, with the rainfall duration equal to the concentration time. The method takes into account the following hydrological characteristics:

- Rainfall intensity, duration, and frequency
- Catchment area
- Hydrologic abstractions
- Runoff concentration and diffusion

When the rainfall continues beyond the time of concentration t_c i.e. $t_p \geq t_c$, the runoff will be constant and at the peak value. The peak value of runoff is given by

$$Q_p = \frac{1}{36} K \times P_c \times A \tag{2.12}$$

where

Q_p = Peak discharge, m^3/sec
P_c = Critical intensity of rainfall corresponding to the time of concentration t_c, cm/hr
A = Catchment area, hectare
t_c = Time of concentration
K = Coefficient of runoff (Table 2.2)

The proportion of rain which becomes runoff depends on many factors like topography, vegetation, infiltration rate, soil storage capacity, drainage pattern, etc. It is at the same time the virtue of the rational formula and its weakness that all these factors are combined into a single runoff coefficient K.

2.6.5.4 Empirical Method

The flood discharge can be evaluated by using various empirical formulae involving the area of the catchment and some coefficient depending on the location of the catchment.

TABLE 2.2
Runoff Coefficient for Catchments

Catchment Characteristics	Runoff Coefficient, K
Steep, bore rock	0.9
Rock, steep but wooded	0.8
Plateau slightly covered, ordinary ground, bore	0.7
Clayey soils, stiff and bore	0.6
Clayey soils, lightly covered	0.5
Loom, lightly cultivated or covered	0.4
Loom, largely cultivated	0.3
Sandy soil, light growth	0.2
Sandy soil, covered, heavy bush	0.1
Jungle areas	0.1–0.2

a. Dicken's formula

Dicken in 1885 made the first attempt in India to derive a general formula for estimating the maximum flood on the basis of studies conducted for Northern India but can now be used in most of the States of India with the modification of the value of the coefficient C is given by the following:

$$Q = CA^{3/4} \tag{2.13}$$

where

Q = Discharge, m^3/s
C = 11.4 for North India
= 13.9 to 19.5 for Central India
= 22.5 to 25 for Western Ghats
A = Catchment area, km^2

b. Ryve's formula

Ryve in 1884 modified Dicken's formula to suit South Indian conditions and was originally devised for the Madras (Now Tamilnadu, India) State and is expressed as follows:

$$Q = CA^{2/3} \tag{2.14}$$

where

Q = Discharge, m^3/s
C = 6.75 if area is within 24 km from the coast
= 8.45, if area from 24 km–161 km from the coast
= 10.1, if area is near hills
A = Catchment area, km^2

c. Inglis formula

Inglis (1930) formula is based on the flood data of catchments in Western Ghats in Maharashtra, India, and the flood peak is expressed as follows:

$$Q = \frac{123A}{\sqrt{(A + 10.4)}} \tag{2.15}$$

where

Q = Discharge, m^3/s

d. Ali Nawaj Jung Bahadur's formula

This formula is applicable mainly in Deccan plains and is expressed as follows:

$$Q = CA^{(0.993 - 1/14 \log A)} \tag{2.16}$$

where

Q = Discharge, m^3/s

A = Catchment area, km^2

C = Coefficient, which varies from 48 to 60.

REFERENCES

Hiranandani, M. G. and S. V. Chitale (1960) *Stream Gauging – A Manual*, Central Water and Power Research Station, Pune.

Horton, R. E. (1932) *Drainage-basin Characteristics*, Transactions, American Geophysical Union 13 350–361.

Inglis, C. C. (1930) *The Behaviour and Control of Rivers and Canals*, Research Publication No. 13, Central Water and Power Research Station, Pune.

Musy, André and Christophe Higy (2004) *Hydrologie, une science de la nature*, Presses polytechniques et universitaires romandes, Lausanne.

Rajab, Jasim Mohammed Al (2021) *Hydrometeorology 2021 – Hydrology – Chapter 4 Runoff*, https://uomustansiriyah.edu.iq/media/lectures/6/6_2021_09_16!09_47_30_PM.pdf.

Strahler, A. N. (1957) *Quantitative Analysis of Watershed Geomorphology*, Transactions of the American Geophysical Union 38 (6) 913–920.

Winter, Thomas C., Judson W. Harvey, O. Lehn Franke and William M. Alley (1998) *Groundwater and Surface Water, A Single Resource*, 1139 U.S. Geological Survey Circular 9, Denver, CO.

3 Design Philosophy

3.1 COMPONENTS OF CHECK DAMS

Permanent check dams are small linear barriers and mainly consist of following components:

- Non-over flow portion
- Weir (overflow) portion
- Wing walls/abutments
- Return wings
- Apron
- Cutoff walls

The typical plan and section depicting the various components of permanent check dams have been illustrated in Figure 3.1.

3.1.1 Non-Over Flow Portion

The non-over flow portion in the permanent check dams is provided to check the flow of water from the sides. Its top level is decided on the basis of afflux and free board and is designed as a gravity dam.

3.1.2 Weir (Overflow Portion)

Weirs are provided to release surplus or flood water which cannot be contained in the storage space. Ordinarily, the excess water released through a weir is passed into the river or to some natural drainage channel. The check dam is constructed across the stream width for making easy water conveyance, and its width should be sufficient to pass the designed flood water very safely. The overflow or weir section is also designed as a gravity dam.

3.1.3 Wing Walls/Abutments

These walls are constructed in the side along the stream. The function of side walls is to prevent the splashing of water flow over the stream banks and also to confine the water flow within the banks. The abutment/wing walls are designed as gravity earth-retaining walls, and their top level is fixed on the basis of afflux and free board.

3.1.4 Return Wings

These walls are constructed at rear end of the structure to prevent the back flow.

DOI: 10.1201/9781003605621-3

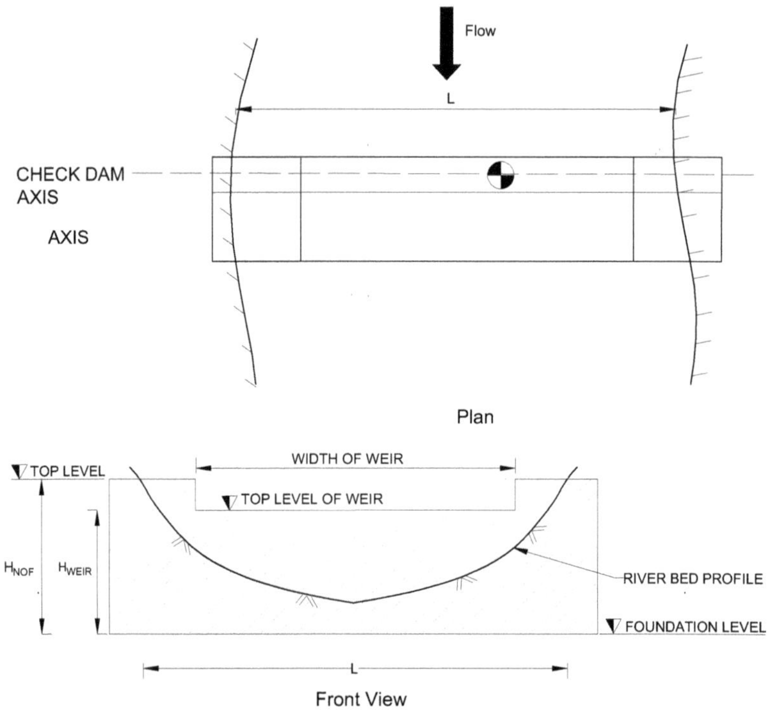

FIGURE 3.1 Components of Permanent Check Dams.

3.1.5 Apron

It is one of the main downstream components of the weir and is provided to dissipate the energy of falling water, and the structure is prevented from undercut and downstream scouring. Many times, some blocks, end sill, etc., are provided to dissipate more kinetic energy. In general, the apron should have a minimum length of 1.5 times the height of the check dam, and for streams with a slope of more than 15%, the apron length should be at least 1.8 times the height of the check dam. It is generally placed on the ground level, and the thickness of apron is kept about 0.50 m.

3.1.6 Cutoff Walls

These walls are constructed to provide structural strength against the sliding of the structure. They increase frictional resistance of the structure, which opposes the force causing to slide. In other words, cutoff walls act as key to the structure. The undermining of the floor starts from the downstream end of the D/S apron, and if not checked, it travels upstream towards the weir. The undermining starts only when the exit gradient is unsafe for the subsoil on which the weir is founded. It is, therefore, absolutely necessary to have a reasonably deep vertical cutoff at the downstream end of the apron to prevent undermining. The depth of this D/S vertical cutoff is governed by two considerations, i.e.

(i) maximum depth of scour; (ii) safe exit gradient.

Downstream cutoff is designed from the maximum scour depth considerations and is checked for exit gradient. If a safe value of exit gradient is not obtained, then the depth of cutoff is increased accordingly.

The upstream and downstream cutoffs are generally provided to cater for scours up to 1.0 R and 1.25 R, respectively, where R is the depth of scour. The concentration factor shall be taken into account in fixing depth of cutoffs.

3.2 SCOUR DEPTH ESTIMATION

3.2.1 Alluvial Rivers

3.2.1.1 Lacey's Method

Lacey (1929) analysed the data for stable channels carrying loose sandy bed material in the Indo-Gangetic plain. Due to the provision of silt excluder/ejector in the upstream part of the channel, the channel has low bed load carrying capacity. Lacey's field data analysis provides two relations for computation of scour depth R and perimeter (or width) P (IS: 6966 Part 1, 1989).

$$R = 0.473\left(\frac{Q}{f}\right)^{1/3} \text{ (applicable when looseness factor is more than 1)} \quad (3.1)$$

$$R = 1.35\left(\frac{q^2}{f}\right)^{1/3} \text{ (applicable when looseness factor is less than 1)} \quad (3.2)$$

where

R = Scour depth below the high flood level (HFL), m;
Q = High flood discharge, m^3/s;
q = Intensity of flood discharge, m^3/s per m width; and
f = Silt factor, which may be computed from the equation

$$f = 1.76\sqrt{d_{50}} \quad (3.3)$$

where

d_{50} = Median size of bed material, mm.

Equations 3.1 and 3.2 were meant for alluvial channels having constant discharge with the sediment size in the range 0.13 to 0.43 mm.

3.2.2 Non-Alluvial–Gravelly Rivers

Lacey's method which was developed for Indo-Gangetic plain alluvial rivers flowing through loose alluvium is still being used for computing scour depth, as not much work has been done for non-alluvial rivers. Most of the formulae developed from the studies have been carried out in European rivers, which are relatively small and have considerable variation. For calculating the regime depth of flow in gravelly or bouldery rivers, different formulae have been developed. For the average diameter of bed material up to 400 mm, the formulae developed by Hey and Thorne (1986) can be used to estimate the scour depth for design discharge up to 500 m^3/s.

$$R = 0.22Q^{0.37}d^{-0.11} \tag{3.4}$$

where

R = Scour depth below the high flood level (HFL), m,
Q = Total discharge in the river, m^3/s,
d = Median size of bed material, mm.

3.3 DESIGN CONSIDERATIONS

The following points be considered at the time of designing check dams:

(i) Proper assessment of flood discharge;
(ii) Evaluation of foundation conditions and foundation treatment;
(iii) Incorporation of weir so that the surplus water can be released;
(iv) Check dam structure should have sufficient strength to withstand the pressure exerted by flowing water; and
(v) Structure should be protected from scouring/erosion.

These points refer to the hydrologic design, hydraulic design, and structural design of structures. The design vigour depends on the type of structures, such as permanent check dam structures, which require more vigorous design than temporary structures. However, no complex engineering design is involved in the design of permanent check dams, yet proper design of permanent check dams is very crucial to fulfill their objective and service life. At the time of designing the check dams, the following points need to be considered:

- Low check dams are less likely to fail than high dams, as the high check dams will impound more water and thus bear more pressure, which may lead to seepage, scouring and undermining.
- A check dam should have a properly designed weir in the centre to discharge surplus water and shoulders on either side to prevent water cutting around.
- A check dam should have apron in the downstream side to protect it from scouring/undercutting.
- A check dam should be properly keyed to the floor and embedded in the sides of the stream from the stability and safety point of view.
- Construction of check dams should start from the upper end to reduce the risk of failure.

3.4 FORCES ACTING ON CHECK DAMS

In general, the following forces are considered in the design of check dams:

- Self weight of the check dam;
- Earth pressure acting on the wall;
- Surcharge;
- Water pressure;
- Allowable soil-bearing pressure; and
- Frictional forces on base against sliding.

In general, the forces acting on the check dam weir are depicted in Figure 3.2.

3.4.1 Self Weight of the Check Dam

The self weight of a unit length of a check dam is simply the product of the dam cross section and the unit weight of material. The weight of the dam acts vertically through the centroid of its cross-section area.

3.4.2 Water Pressure

When water accumulates behind a check dam, lateral forces are exerted on the check dam, and the greater the depth of water behind a check dam, the higher the hydrostatic pressure on the dam.

3.4.3 Uplift Pressure

The water has a tendency to seep through voids and through the joints between the body of the dam and the foundation, and this seeping water exerts pressure in an upward direction, which has a tendency to uplift the dam. It is exerted in all situations, when there is water storage in the upstream and/or downstream. Therefore, it is necessary to check the stability of the check dam against hydraulic uplift and ensure that the uplift force is not greater than the total weight of the check dam. The general flow net diagram below the base of the check dam is depicted in Figure 3.3.

3.4.4 Earthquake Forces

Since the height of the check dam is very low, generally, the check dams are not designed for earthquake forces, as *it will always be economical to repair damaged portion after earthquake.*

3.4.5 Allowable Soil-Bearing Pressure

The foundations shall be designed to take care of the shear capacity of the soil/rock formation and the permissible settlements which structure can tolerate. It is generally the

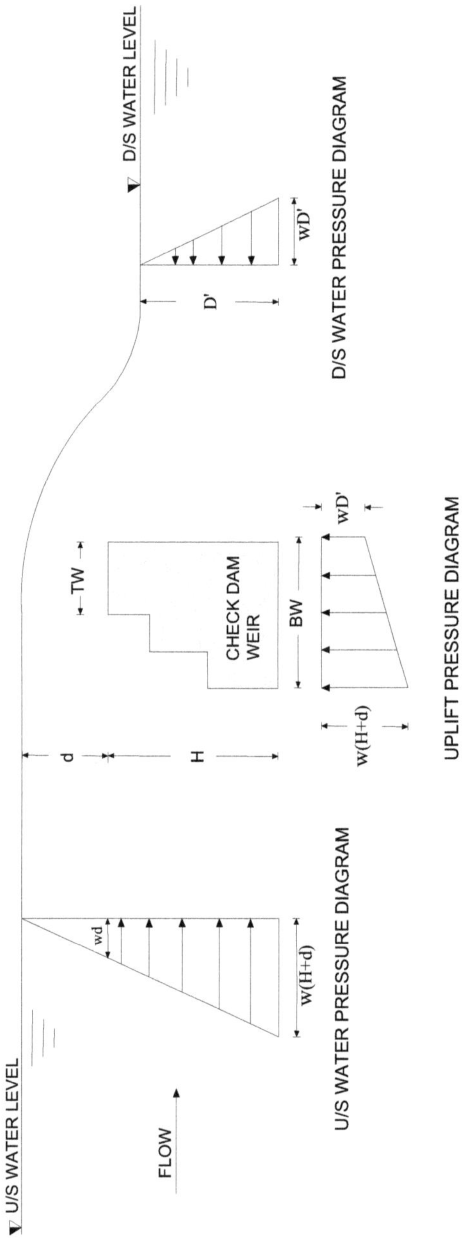

FIGURE 3.2 Forces Acting on the Check Dam Weir.

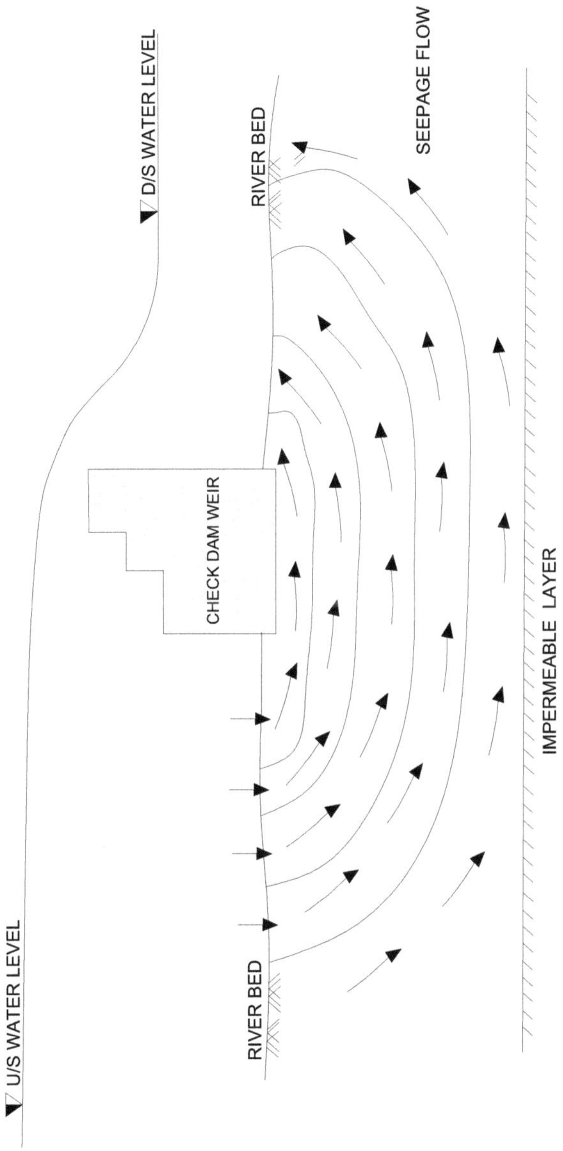

FIGURE 3.3 Flow Net for a Check Dam Weir.

settlement criterion that usually governs the value of the allowable bearing pressure and shall be used for design of foundations. The foundations of check dams are normally so proportioned that no tension is created at the foundation plane under static loading.

3.4.6 Frictional Forces on Base against Sliding

Para 706.3.4 of the Standard Specifications and Code of Practice for Road Bridges Section: VII Foundations and Substructure IRC:78 (2014) states, "Frictional co-efficient between concrete and soil/rock will be Tan ϕ, ϕ being angle of friction. Founding soil in foundation of bridge being generally properly consolidated, following values may be adopted":

Friction co-efficient between soil and concrete, $\mu = 0.50$
Friction co-efficient between rock and concrete, $\mu = 0.80$ for good rock
0.70 for fissured rock

3.4.7 Active Earth Pressure on Abutments/Wing Walls

Coulomb (1776) developed a method for the determination of the earth pressure considering the equilibrium of the sliding wedge formed due to the movement of the retaining wall. In the Active Earth Pressure case, the sliding wedge moves downwards and outwards on a slip surface relative to the intact backfill, and in case of Passive Earth Pressure, the sliding wedge moves upward and inwards. The pressure on the wall is, in fact, a force of reaction which it has to exert to keep the sliding wedge in equilibrium. The lateral pressure on the wall is equal and opposite to the reactive force exerted by the wall to keep the sliding wedge in equilibrium. Assumptions made in Coulomb's theory include the following:

i) The backfill is cohesionless, dry, homogenous, isotropic, and ideally plastic material.
ii) The slip surface is a plane surface which passes through the heel of the wall.
iii) The sliding wedge behaves like a rigid body, and the magnitude of the earth pressure is obtained by considering the equilibrium of the wedge as a whole.
iv) The back of the wall is rough.
v) The position and direction of the resultant earth pressure are known. It acts at a distance of one-third the height of the wall above the base and is inclined at an angle δ to the normal of the back of the wall, where δ is the angle of friction between the wall and the backfill.

Active Pressure Due to Backfill

The active pressure exerted against the retaining wall shall be the following:

$$P_a = \frac{1}{2}\gamma H^2 K_a \tag{3.5}$$

where

P_a = Active earth pressure acting at an angle ($\alpha + \delta$) with horizontal
γ = Unit weight of the backfill
H = Height of the wall/backfill
K_a = Coefficient of active pressure, which takes into account backfill slope, friction angle at wall face, angle of repose of backfill and angle of wall face with vertical:

$$K_a = \frac{\cos^2(\phi - \alpha)}{\cos^2 \alpha \cdot \cos(\delta + \alpha)\left[1 + \sqrt{\frac{\sin(\phi + \delta)\sin(\phi - \iota)}{\cos(\alpha - \iota)\cos(\delta + \alpha)}}\right]^2} \tag{3.6}$$

where

ϕ = Angle of the internal friction of the soil
δ = Angle of the friction between the wall and the backfill
ι = Angle of the earth surface with the horizontal behind the earth retaining structure
α = Angle of the earth face of the wall with vertical

All these design parameters influence on the estimation of active earth pressures and design of retaining walls and therefore appropriate parameters be selected for the techno-economic design of retaining walls (Chalisgaonkar, 2018).

If the backfill is levelled, the earth face of the wall is vertical, and if zero friction is assumed between the soil and the wall, then the Coulomb equation reduces to the familiar Rankine's equation (1857):

$$K_a = \frac{1 - \sin\phi}{1 + \sin\phi} \tag{3.7}$$

3.5 STABILITY OF CHECK DAMS

The basic principles of check dam design include safety against failure criterion, safety against adverse meteorological condition and geological feature of the location. The following stability conditions should be considered at the time of designing the check dam:

- It should be safe against overturning.
- There must be no tension in the masonry/concrete or in the contact plane between the check dam and the foundation.
- There must be no tendency to slide on the joint with foundation or any horizontal plane above the base.
- The maximum toe and heel pressures on the foundations should not exceed the safe bearing capacity of the soil.

3.5.1 Stability against Overturning

If the check dam structure is to overturn, it would do so with the toe acting as the centre of rotation. Thus, the factor of safety against overturning is defined as the ratio between the sum of resisting moments and the sum of overturning moments about the toe.

$$F.S._{overturn} = \frac{\sum M_{resisting}}{\sum M_{overturn}} \tag{3.8}$$

3.5.2 Stability against Sliding

The stability of a check dam against sliding is evaluated by comparing the minimum total available resistance along the critical path of sliding (that is, along that plane or combination of planes which mobilises the least resistance to sliding) to the total magnitude of the forces tending to induce sliding.

Sliding resistance is also a function of the cohesion inherent in the materials at their contact and the angle of internal friction of the material at the surface of sliding. The factor of safety against sliding (F) along a plane may be computed from the following equation and shall not be less than 1.0:

$$F = \frac{\left\{\frac{(W-U)tan\varnothing}{F\varnothing} + \frac{CA}{Fc}\right\}}{P} \tag{3.9}$$

where

F = Factor of safety against sliding
W = Total weight
P = Total horizontal force
U = Total uplift force
C = Cohesion of material at the plane considered
A = Area under consideration for cohesion
tan ϕ = Coefficient of internal friction of the material
Fϕ = Partial factor of safety in respect of friction
F_c = Partial factor of safety in respect of cohesion

The partial factors of safety in respect of friction and cohesion for different load combinations in gravity dam design are given in Table 3.1.

The value of cohesion and internal friction may be estimated for the purpose of preliminary designs on the basis of available data on similar or comparable materials. The check dam, being a low-height structure, values of Fϕ and F_c have been adopted as 1.50 and 3.60, respectively, for evaluating the factor of safety against sliding.

TABLE 3.1
Partial Factors of Safety against Sliding

	Loading Combinations	$F\phi$	F_c		
			For Dams and the Contact Plane with Foundation	For Foundation	
				Thoroughly Investigated	Others
i	A, B, C	1.50	3.60	4.00	4.50
ii	D, E	1.20	2.40	2.70	3.00
iii	F, G	1.00	1.20	1.35	1.50

Source: IS:6512 (2019)

3.5.3 Bearing Pressure

Similar to any footing design, the width of the base of a check dam must be adequate to distribute the vertical reaction to the foundation soil without causing excessive settlement or rotation. When the base pressure exerted exceeds the safe bearing capacity of the soil, the wall fails due to the bearing pressure.

These steps are repeated iteratively until a suitable design that meets all criteria is achieved.

3.6 LENGTH OF CHECK DAM

Depending on the condition, the length of the check dam should be fixed in such a way that the afflux caused by the obstruction is minimal. The afflux is defined as the rise in water level upstream of the check dam as a result of the obstruction to the natural flow caused by the construction of the check dam. This afflux is confined only to a short reach, equal to the length of the back water curve.

3.6.1 Afflux in Streams with Non-Erodible Beds

For streams with non-erodible beds, the afflux may be worked out by the Molesworth formula (IRC:89, 2019) given next:

$$h = (V^2/17.88 + 0.015) \times ((A/a)^2 - 1) \quad (3.10)$$

where

h = Afflux, m
V = Average velocity of the water in the river prior to constriction, m/s
A = Unobstructed sectional area of the river at proposed site, m^2
a = Obstructed waterway area of the river, m^2

3.6.2 Afflux in Streams with Erodible Beds

However, as per Para 4.7.2 of the Indian Railway Standard Code of Practice for the Design of Sub-structures and Foundation (2003), the full afflux, as calculated by the Molesworth formula, may not occur in case of rivers with erodible beds.

It has been observed that the afflux and discharge intensity are dependent upon the crest levels of weir. If the crest level of weir is low, afflux shall be less, and since the depth of water over the crest will be more, it shall lead to a higher discharge intensity. The amount of afflux determines the top levels of abutments and upstream wing walls. By providing a higher afflux, the waterway and, therefore, the length of the weir can be reduced, but it will increase the cost of training works and the risk of failure by outflanking. At the same time, the discharge intensity and the consequent scour shall go up, and hence, the sections of loose protection upstream and downstream will become larger, thereby making it costly. It is, therefore, always desirable to limit afflux to a safe value of 1.0 m to 1.2 m. However, in steep reaches with rocky bed, a higher value of afflux may be permitted.

3.7 HEIGHT OF A CHECK DAM

The effective height of a check dam is the elevation of the crest of the weir above the original bed level of the stream. The height not only influences structural spacing but also volume of sediment deposits.

Heede and Mufich (1973) developed an equation that relates the volume of sediment deposits to the spacing and effective height of a dam:

$$V_S = 1/2\ H_E\ S \cos\alpha\ L_{HE} \tag{3.11}$$

where

V_S = Sediment volume

S = Spacing between two check dams

= $H_E/(KG\cos\alpha)$

H_E = Effective height of the check dam (weir height)

α = Angle corresponding to the stream gradient

L_{HE} = average length of a dam, considered for effective dam height

= $L_B + (L_U - L_B) \times H_E/2D$

L_B = Bottom width

L_U = Bank width of stream, measured from brink to brink

D = Depth of stream

K = Constant

= for G or $\tan\alpha$ = < 0.20 K = 0.30 and

= for G or $\tan\alpha$ > 0.20 K = 0.50

If the value of S is substituted in equation 3.11, it becomes

$$V_S = H_E^{\,2}\,L_{HE}\,/2KG \tag{3.12}$$

The equation 3.12 indicates that sediment deposits increase with the square of effective dam height. Thus, it can be said that a 6 m high check dam accumulates about nine times as much sediment as a 2 m high check dam.

In most cases, however, the height of a check dam is restricted by one or all of the following criteria:

i cost;
ii stability; and
iii channel geometry in relation to weir requirements.

3.8 WEIR DESIGN

The design of the weir section of a check dam is difficult because there are no accurate formulas to estimate high water, bed load, flow velocity, and a possible mudflow. Weir has to be designed to carry the maximum flow without overtopping or breaching the check dam. Therefore, it should be wide enough to accommodate the maximum flow expected once in ten years. The discharge passing capacity of a weir can be computed by using the following formula for different types of weirs:

3.8.1 Broad-Crested Weir

$$Q = 1.70\,L\,H^{3/2} \tag{3.13}$$

where
Q = Discharge, m^3/s
H = Total head, including velocity head, m
L = Clear span of weir, m

3.8.2 Sharp-Crested Weir

$$Q = 1.84\,L\,H^{3/2} \tag{3.14}$$

If the head over the weir crest is more than 1.5 times the width of the weir, the weir behaves as a *sharp-crested weir.*

3.9 SHAPE OF CROSS SECTIONS OF WEIR

Figure 3.4 shows the typical basic forms of cross sections of weir, adopted in check dam constructions.

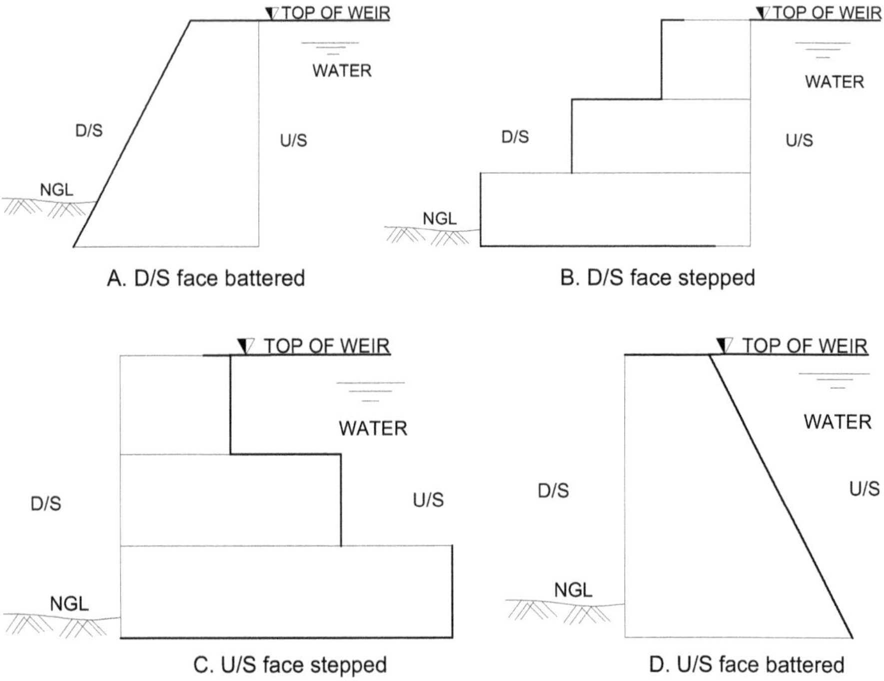

FIGURE 3.4A, 3.4B, 3.4C, and 3.4D Typical Basic Forms of Cross Sections.

The shapes of the cross sections shown in Figure 3.4B and Figure 3.4C are generally used in gabion check dams. Types A and D basic forms are used in masonry/concrete check dams. Figure 3.4A and Figure 3.4B show the downstream inclined wall and vertical upstream wall, where water with bed load falls over the inclined or terraced wall and damages the wall. The forms Figure 3.4C and Figure 3.4D are suitable for the bed load fall. The shape of the downstream side of a weir is chosen considering the following factors:

- drop height,
- hydraulic charge,
- characteristics of the materials used in construction, and
- characteristics of local natural soils.

There are no general rules for the choice of a particular shape. However, if the drop's height does not exceed 3.0 m to 5.0 m, the weir with downstream vertical shape is always the most appropriate. If the drop height is more than 5 m, the weir must be designed with a stepped or battered shape in the downstream. However, the stepped and battered downstream-shaped weirs are not well fitted to natural streams with a significant solid transportation, especially when gravel and boulder are transported by the flow. In fact, the continuous dropping and sliding of particles can damage the downstream faces.

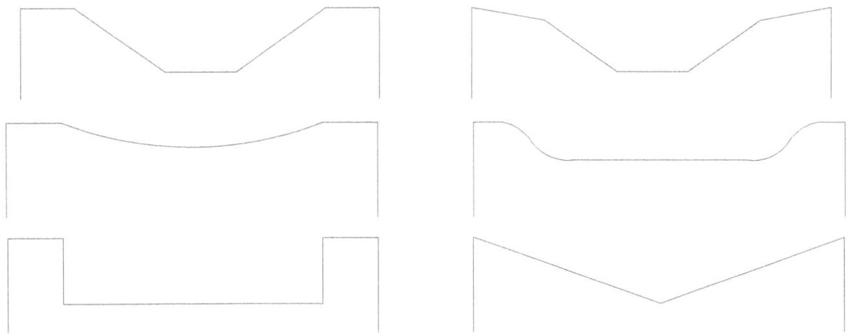

FIGURE 3.5 Symmetrical Weir Sections.

3.10 LENGTH OF WEIR SECTION

The length of the weir section varies according to the width of the river bed and shall be designed in such a way that a stilling basin is formed without endangering the banks. The symmetrical weir sections have different forms, such as a circle segment, a trapezium, a trapezium with rounded edges, a rectangle, or a triangle (Figure 3.5).

The circle segment with a big centre angle holds the water together best and does not favour the deposition of stones in the weir section. This form has the disadvantage that not only the power of the water is concentrated at the lowest part but also the bed load, which results in a big abrasion. Rectangular forms should be avoided unless guide walls are provided to lead the water and bed load to the weir section; otherwise, the crown shoulders shall be fully exposed to a mudflow. The triangle form should be avoided, too, because water is concentrated in one place.

Thus, the form of the weir section is selected according to the local situation and according to the construction material used.

3.11 COMMON TYPE OF WEIRS USED IN CHECK DAMS

3.11.1 Straight Drop Weir

It is one of the most suitable and common type of weir used in check dams. Several such drop structures are constructed across the stream width throughout the length at fixed intervals. The series of such structures develop a continuous break to the flow of water, causing deposition of sediments, thus filling the stream section. These drop weirs are also used at the stream head to pass the flow safely. These weirs are less susceptible to get structural damage compared to the other structures. This type of weir consists of a raised crest with vertical or slightly sloped downstream face, a horizontal floor, and an energy-dissipation mechanism. Such types of structure have the following advantages over the other types:

- The construction of such weirs is very simple
- Suitable for hard clay and consolidated foundations

Their design and analysis can be done with the simple methods.

3.11.2 Stepped Weirs

This kind of weir is generally used in relation to low specific flows and significant drop heights. The construction of steps on the weir contributes to the dissipation of the energy, subsequently diminishing 10%–30% of the length of the stilling basin compared to the length obtained using traditional methods.

3.11.3 Battered Weirs

This type of weir is generally well suited to significant drop heights and low specific flows. The specific flow has to be limited only if the weirs are built with gabions, which easily gets damaged. In fact, violent water flows can provoke stones rubbing the downstream face, consequently leading to breakage/damage. The lining on the downstream face of the battered weir is suggested when the specific flow is high.

3.12 ENERGY DISSIPATION BELOW WEIRS

After having fixed the weir's main dimensions (e.g. height, width, etc.), it is necessary to verify what happens downstream of the weir, and it should be ensured that the energy is dissipated immediately downstream of the weir. This condition is satisfied when a subcritical flow takes place in the downstream of structure. The choice of the most appropriate energy dissipation method depends on the weir type. As the height of a check dam is not much, the energy of falling water passing through the weir to be dissipated will also be less. In general, the following types of arrangements are proposed in the downstream of the weir to dissipate the energy.

3.12.1 Simple Weir

For a very small weir, with limited specific flow, the conventional stilling basin can be omitted for energy dissipation. This solution is likely to be chosen especially if the stream bed material has sufficient resistance to erosion. Otherwise, the dropping water could dig a scour hole in the downstream of the weir.

3.12.1.1 Scour Hole

Scouring is the biggest enemy of check dams, which is caused by the energy of the falling water and bed load. The energy is destroyed in the scour hole or on the apron. If the check dams are not based on rock, big scour holes will be created. The foundation of gravity structures is especially more prone to the danger of collapsing.

Scour occurs when the bed velocity of the stream exceeds the velocity, which can move the particles of bed material. When the velocity is retarded, aggradation takes place, and when the velocity is increased, scouring takes place and silt is picked up.

3.12.2 Simple Weir with Apron

An apron is designed to dissipate the energy of falling water passing through the weir. The apron should be strong enough to withstand pressure of falling water and a surface wash and can be made from stone riprap, gabions, concrete, etc.

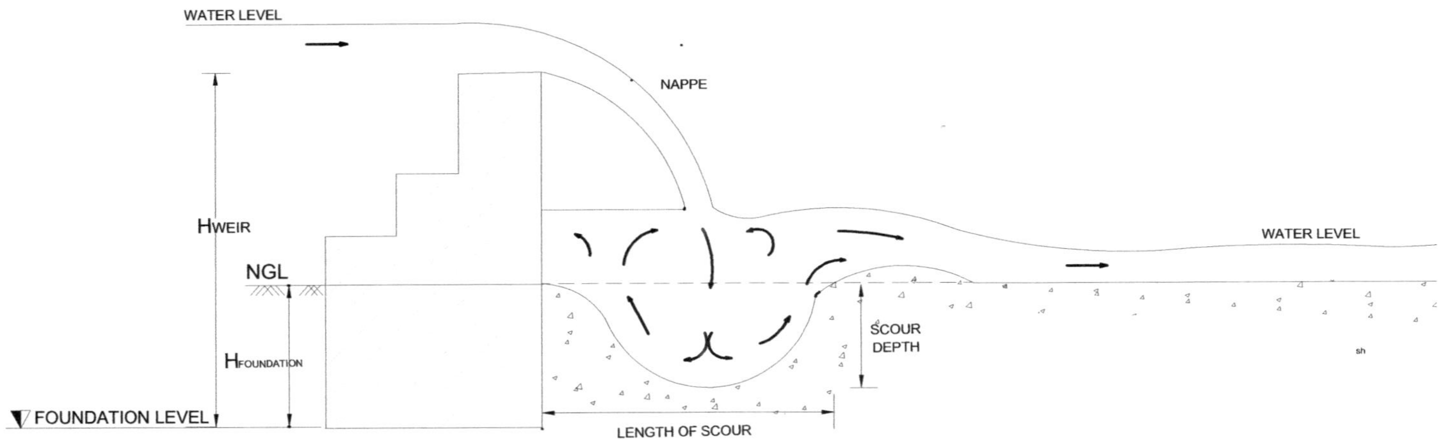

FIGURE 3.6 Scour Depth in Check Dam.

An apron is generally built below the ground surface to form a basin which, when filled with water, will function as a water cushion and dissipate the energy of water falling on it. Bligh's creep theory is used to design the aprons.

3.12.2.1 Bligh's Creep Theory

Bligh assumed that the water which percolates into the foundation creeps through the joint between the base of hydraulic structure and the subsoil (Figure 3.7). The total length covered by the percolating water till it emerges out at the downstream end is called a creep length, and the head loss of water is the difference of water levels on the upstream and the downstream ends. Also, an imaginary line which joins the water levels on the upstream and the downstream end is called a hydraulic gradient (HG) line.

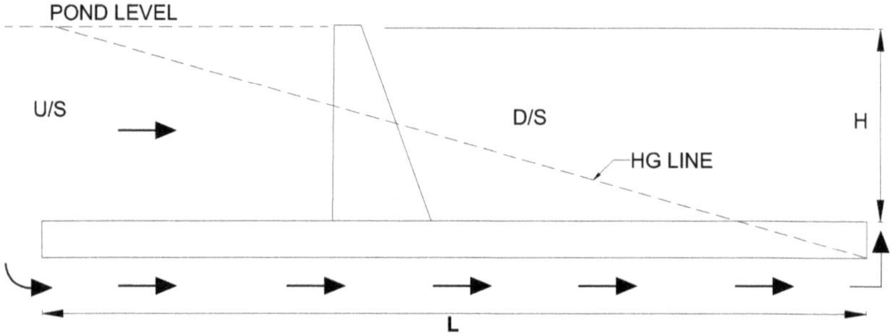

FIGURE 3.7 Bligh's Creep Theory.

In Figure 3.7, arrows show the path followed by the creeping water and

L = total creep length and

Loss of head per unit creep length will be H/L and is called hydraulic gradient.

3.12.2.1.1 Bligh's Criteria for the Safety of a Weir against Piping

The structure is safe against piping when the percolating water retains negligible upward pressure and emerges out at the downstream end of the weir. The path of percolation or creep length depends on the soil type and safe hydraulic gradient and is obtained from the equation

$$L = CH \tag{3.15}$$

where

L = Creep length or path of percolation;
C = Bligh's creep co-efficient for soil; and
H = Head of water against the weir.

Values of coefficient C for various soil types are given in Table 3.2.

TABLE 3.2
Values of Coefficient C for Different Types of Soils

Soil Type	Value of C
Silt and Sandy	18
Micaceous Sand (North Indian River)	15
Coarse Grained Sand	12
Sand Mixed with Boulder, Gravel, and Shingle	5 to 9
Loamy Soil	5 to 9
Gravel	5

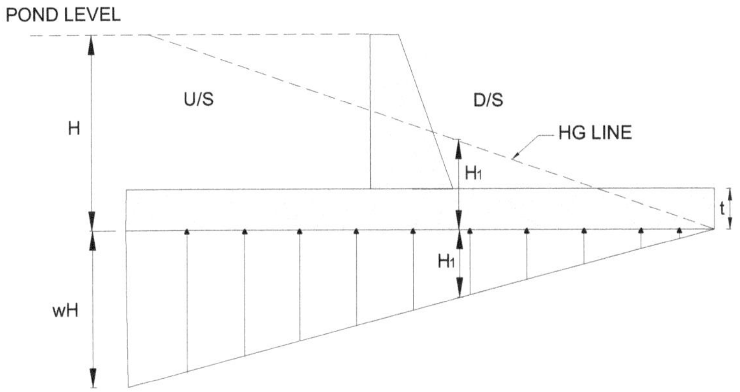

FIGURE 3.8 Uplift Pressure Diagram.

Obviously for no piping

$$H/L \leq 1/C \quad (3.16) \text{ or}$$

$$\text{Hydraulic Gradient} \leq 1/C \quad (3.17)$$

3.12.2.1.2 Bligh's Criteria for the Safety of a Weir against Uplift

The uplift pressure at any point, assumed by Bligh, is represented by the ordinate between the bottom of the apron floor and the hydraulic gradient line and is shown in Figure 3.8.

The thickness of the apron can be calculated on the basis of the residual head in the downstream and the unit weight of the downstream floor material. It is always better to provide greater apron length on the upstream side which requires minimum practical thickness to minimise the cost. However, it is necessary to provide some minimum length of apron on the downstream side to protect the bed of the channel.

Bligh has given some empirical formula for determining the length of the downstream floor. The cutoffs may be provided as per the provisions of Khosla's theory. The various empirical formulas put forward by Bligh are as follows:

$$\text{Length of Pucca Downstream Floor } L_d = 2.21\, C \sqrt{(h/10)} \tag{3.18}$$

$$\text{Length of D/S Protection Work } L_{D/S} = C \sqrt{(h/10)} \times \{18 \sqrt{(q/75)} - 2.21\} \tag{3.19}$$

$$\text{Length of U/S Protection Work } L_{U/S} = 0.5 \times L_{D/S} \tag{3.20}$$

However, with the check dams being small structures, an apron is provided only in the downstream of the check dam, and the length of the pucca downstream floor or apron is kept at least 1.5 times the height of the check dam.

3.13 DEPTH OF FOUNDATION FOR CHECK DAMS

The type of foundation soil and its geotechnical parameters not only affects the design but it also guides to decide the type of dam that would be suited at a site. The foundation depth of check dams is kept at least 0.50 m below the bed level for a flat ground condition of up to 6.00 m in height. Larger values may be required, depending on the depth of frost penetration, shrinkage and swelling of foundation soil, seismic activity, and scour. Foundation depth of check dams resting on rock can be kept between 0.30 m–0.50 m below the bed level.

3.14 SPACING BETWEEN CHECK DAMS

Spacing between check dams is very important as too distant check dams will practically be ineffective in preventing the further scouring of a nallah bed, whereas too closely placed check dams will lead to more number of check dams, thereby making the cost of project very expensive. Therefore, the distance between two check dams is so kept that at no point the flowing water should acquire an erosive velocity, which is achieved by keeping the maximum spacing between the check dams so that the toe of the upstream dam is at the same elevation as the top of the downstream dam.

3.14.1 Using Compensation Gradient

In order to reduce number of check dams and thereby cost, a small gradient called *compensation gradient*, which would limit the flow velocities within permissible limits, is usually allowed between the two consecutive check dams. The gradient between the top of the lower check dam and the bottom of the upper one is called *compensation gradient*, which is the future or final gradient of the stream. It is formed when material carried by flowing water fills the check dams to the weir level. The following formula is used to compute the spacing between check dams on the basis of compensation gradient and height of the check dams:

$$b = h \times 100/(S_o - S_e) \tag{3.21}$$

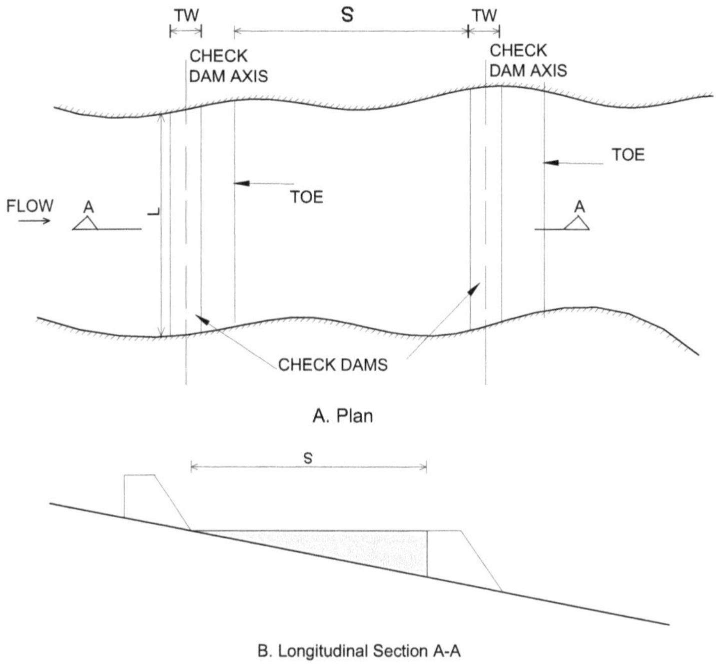

FIGURE 3.9A, 3.9B Spacing between Check Dams.

where

b = Spacing between two check dams, m

h = Effective height of the check dam (weir height), m

S_o = Existing bed slope, %

S_e = New slope (compensation gradient), %

3.14.2 Heede's Formula

Heede and Mufich (1973) developed an equation to calculate spacing between check dams:

$$S = H/(KG\cos\alpha)$$

or

$$S = H/(K \tan\alpha \cos\alpha) \tag{3.22}$$

The equation is based on the assumption that the gradient of the sediment deposits is (1 – K)G.

where

S = Spacing between two check dams, m

H = Effective height of the check dam (weir height), m

G = Stream gradient as a ratio
= tanα
K = Constant;
= for G or tanα = < 0.20 K = 0.30 and
= for G or tanα > 0.20 K = 0.50
α = Angle of stream bed with the horizontal

Based on equation 3.22, the minimum spacing between two consecutive check dams for different bed slopes and check dam height is given in Table 3.3.

3.14.3 Empirical Formula

Empirical formula based on *A Field Guide on Gully Prevention and Control* (Addis Ababa, 2012):

$$S = 1.20 \times H/G \tag{3.23}$$

TABLE 3.3
Minimum Spacing between Check Dams for Different Bed Slopes and Height

	Height of Check Dams, m				
Bed Slope	**1.00**	**1.50**	**2.00**	**2.50**	**3.00**
1%	333	500	667	833	1000
2%	167	250	333	417	500
3%	111	167	222	278	333
4%	83	125	167	208	250
5%	67	100	133	167	200
6%	56	83	111	139	167
7%	48	71	95	119	143
8%	42	63	83	104	125
9%	37	56	74	93	111
10%	33	50	67	83	100
11%	30	46	61	76	91
12%	28	42	56	70	84
13%	26	39	51	64	77
14%	24	36	48	60	72
15%	22	33	45	56	67
16%	21	31	42	52	63
17%	20	30	39	49	59
18%	19	28	37	47	56
19%	18	26	35	44	53
20%	17	25	34	42	50
21%	10	14	19	24	29
22%	9	14	18	23	27
23%	9	13	18	22	26
24%	8	13	17	21	25
25%	8	12	16	20	24

where

S = Spacing between check dams, m
H = Effective height of the check dam (weir height), m
G = Gradient of stream

3.15 NUMBER OF CHECK DAMS

The number of check dams required is determined from the following formula:

$$\text{Number of check dams} = (a - b)/H \quad (3.24)$$

where

a = Total vertical distance between the first and last check dam
b = The total vertical distance is calculated according to the compensation gradient and horizontal distance between the first and last check dam in that portion of the stream (compensation gradient)
= $S_e \times d'/100$
d' = Horizontal distance between first and last check dam, m
S_e = New Slope (compensation gradient), %
H = Average height, m

REFERENCES

Addis Ababa (2012) *A Field Guide on Gully Prevention and Control*, Nile Basin Initiative, Eastern Nile Watershed Management Project, Addis Ababa, March.

Chalisgaonkar, Rajendra (2018) *Influence of Design Parameters on Earth Pressures behind Retaining Walls*, Journal of Indian Highways, (Indian Road Congress, New Delhi) 46 (11) November.

Coulomb, C. A. (1776) *Essai sur une application des regles des maximis et minimis a quelques problems de statique relatifis a Parchitecture*, Mem. Acad. Roy. Pres divers savants, Vol. 7, Paris.

Heede, Burchard H. and John G. Mufich (1973) *Functional Relationships and a Computer Program for Structural Gully Control*, Journal of Environmental Management 1 321–344.

Hey, R. D. and C. R. Thorne (1986) *Stable Channels with Mobile Gravel Beds*, Journal of Hydraulics Division 112 (8) 671–689.

Indian Railway Standard Code of Practice for the Design of Sub-structures and Foundation (2003) *Indian Railway Standard Code of Practice for the Design of Sub Structure and Foundation of Bridges*, Research Designs and Standards Organisation, Ministry of Railways, Government of India, Lucknow.

IRC:78 (2014) *Standard Specifications and Code of Practice for Road Bridges Section: VII Foundations and Substructure IRC:78*, Indian Road Congress, New Delhi.

IRC:89 (2019) *Guidelines for Design and Construction of River Training and Control Works for road Bridges IRC:89*, Indian Road Congress, New Delhi, December.

IS:6512 (2019) *Indian Standard Criteria for Design of Solid Gravity Dams*, IS:6512, Bureau of Indian Standards, New Delhi.

Lacey, G. (1929) *Stable Channels in Alluviums*, Journal Institution of Engineers, Paper No. 4736 229.

Rankine, W. J. M. (1857) *On the Stability of Loose Earth*, Philosophical Transactions of the Royal Society A 147.

4 Types of Check Dams

Based on the construction materials, the check dams can be broadly categorised as follows:

- Gabion check dams
- Masonry check dams
- Concrete check dams
- Rubber check dams

4.1 GABION CHECK DAMS

A gabion (Italian word "gabbione" meaning *big cage*) is a rectangular cage made of steel wire mesh filled with stones/rocks. Individual gabion units are stacked, paired, and tied to each other with steel wire or fasteners to form the continuum. It is said that, for the first time, about 2,000 years ago, Egyptians used cylindrical willow baskets filled with small stones to protect the banks of the River Nile from erosion. The conventional gabion possesses the following peculiar technical and functional advantages:

- They are reinforced structures, capable of resisting tension and shear. The mesh not only acts to contain the stone fill but also provides a comprehensive reinforcement throughout to structure.
- Gabions are flexible structures, which do not diminish the structure but facilitate load redistribution.
- They are permeable structures.
- They are durable structures and require less maintenance.
- Their characteristics over time tend to gravitate towards establishing a natural state of equilibrium.
- They are easily installed, i.e. deployment is possible without the aid of special equipment or highly trained personnel. This aspect is notably important on highways, rivers, and marine reclamation projects, where rapid intervention to retain soil is often necessary.

4.1.1 Machine-Made Gabions

Gabions are baskets made of double-twisted steel woven wire mesh made from machine with higher mechanical characteristics. The hexagonal shape of the mesh is helpful to make a uniform distribution of forces. The double-twisted wire will not unravel, even when cut. These are very easy to assemble and require no specialised labour, and the rocks available at the site can be used for filling. Gabions are available with different levels of protective coating for durability. Gabions can also be provided as toe walls to act as robust bottom drainage walls.

DOI: 10.1201/9781003605621-4

4.1.2 Components of a Gabion Box

i. Double-twisted wire mesh
A non-ravelling mesh made by twisting continuous pairs of wires through three one-half turns (commonly called double-twisted) to form hexagonal shaped openings, which are then interconnected to adjacent wires.

ii. Selvedge wire
A terminal wire used to edge the wire mesh perpendicular to the double twist by mechanically wrapping the mesh wires around it at least 2.5 times.

iii. Edge wire
A terminal wire of the same diameter as the selvedge wire used to edge the wire mesh parallel to the double twist by continuously weaving it mechanically in to the wire mesh.

iv. Lacing wire
A galvanised wire or galvanised wire with PVC coating used to assemble and interconnect empty units, to close and secure stone-filled units, and for internal stiffeners.

v. Diaphragm
An internal partition made of the same wire mesh panel, as used in a gabion, is attached to the bottom, the sides, and the lid of the cage, after the gabion cage is packed with stones.

The tensile strength of the wire used for a double-twisted mesh, lacing wire, and stiffener shall be in accordance with the requirements of IS:280 (1978) for soft wire (350–550 MPa) at a minimum elongation of 10%, performed on a gauge length of test specimen as 200 mm. Different components of a gabion box have been depicted in Figure 4.1 and Figure 4.2.

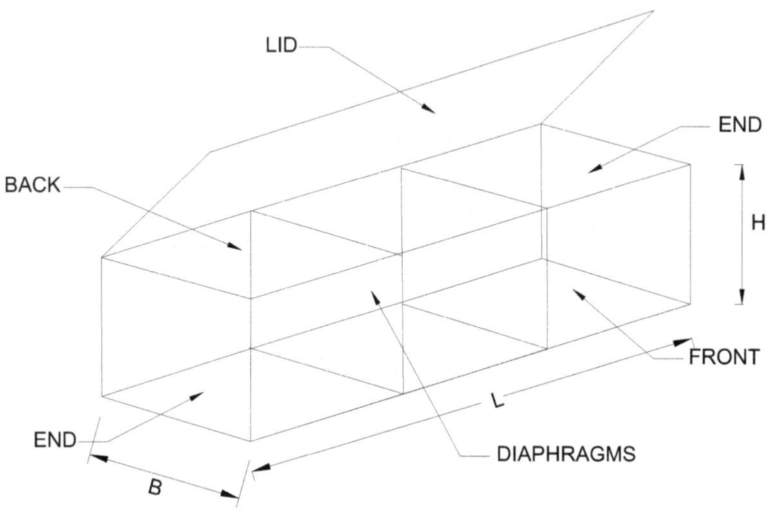

FIGURE 4.1 Typical Gabion Box.

***Source*: IS:16014 (2018)**

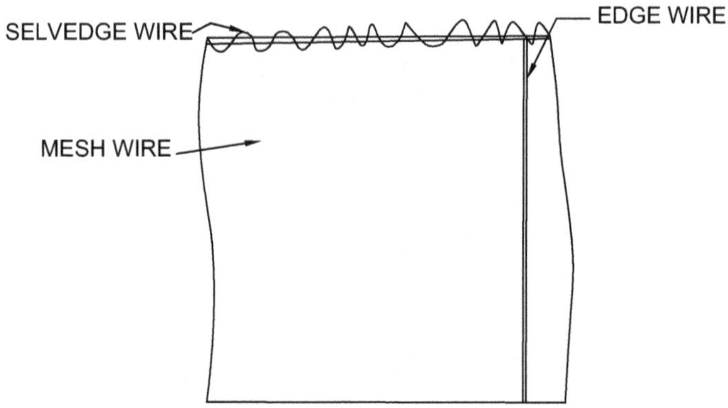

A. Mesh, selvedge, and edge wire

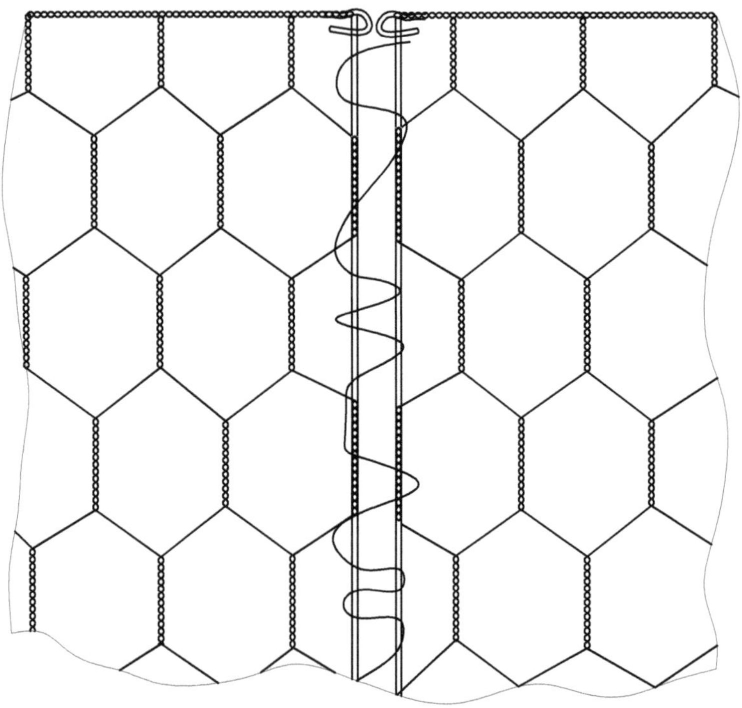

B. Lacing wire for tying adjacent panels

FIGURE 4.2A, 4.2B Components of a Gabion Box.

Source: IS:16014 (2018)

A revet mattress is a wire mesh container uniformly partitioned into internal cells with relatively smaller height in relation to other dimensions, having smaller mesh openings than mesh used for gabions (Figure 4.3). These are generally used for river bank protection, channel lining, and apron.

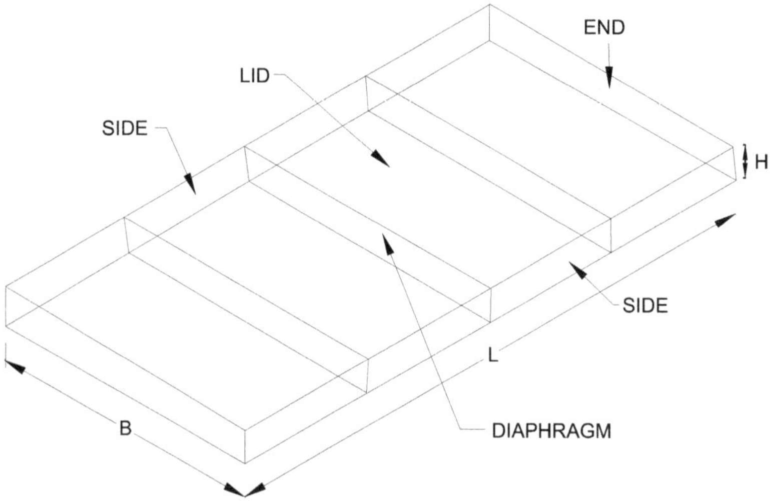

FIGURE 4.3 Typical Revet Mattress.

***Source*: IS:16014 (2018)**

The standard sizes of gabion boxes and revet mattresses, recommended by the Indian Standard Code of Practice for Mechanically Woven, Double-Twisted, Hexagonal Wire Mesh Gabions, Revet Mattresses and Rock Fall Netting (Galvanised Steel Wire or Galvanised Steel Wire with PVC Coating), IS:16014, are given in Table 4.1 and Table 4.2, respectively.

4.1.3 Why Gabion Check Dams?

A gabion check dam structure has following special features:

4.1.3.1 Flexibility

The gabion system, being most flexible soil reinforcing system, helps the structure to accommodate ground settlement without compromising the structural integrity. A gabion structure can be easily bent without breakage and can also be shifted to any point as per site requirement. The gabion structures do not get collapsed like the rigid or semi-rigid structures.

4.1.3.2 Permeability

Empty spaces between the stones filled in gabions allow fluid movements through them which help in soil erosion control. The permeable nature of a gabion structure provides the following main advantages:

i It provides drainage space through the voids left between stones.
ii It facilitates the *bleed-off* of ground water and relieves the hydrostatic pressure, thus preventing the collapse of the structure.
iii A gabion structure does not require a separate drainage system.

TABLE 4.1

Mesh and Box Characteristics for Gabions

Characteristics	Mesh Type							
	10 × 12 D = 100 mm				8 × 10 D = 80 mm			
	Only Zinc/ Zinc Alloy Coated		Zinc/Zinc Alloy + Polymer Coated		Only Zinc/ Zinc Alloy Coated		Zinc/Zinc Alloy + Polymer Coated	
			Inner diameter	Outer diameter			Inner diameter	Outer diameter
i Mesh Wire diameter, mm	2.70	3.00	2.70	3.70	2.70	3.00	2.70	3.70
ii Edge/Selvedge Wire diameter, mm	3.40	3.90	3.40	4.40	3.40	3.90	3.40	4.40
iii Lacing Wire diameter, mm	2.20	2.20	2.20	3.20	2.20	2.20	2.20	3.20
iv PVC Coating thickness, mm								
Nominal			0.50			0.50		
Minimum			0.40			0.40		
v Tolerances in the Size of Gabion Boxes								
Length and Width			+ 5%					
Height = < 300 mm			+ 10%					
Height > 300 mm			+ 5%					

Zinc Alloy shall consist of 90% Zinc + 10% Aluminium or 95% Zinc + 5% Aluminium

vi Typical Standard Sizes of Gabion Box

Length, mm	Width, mm	Height, mm	Number of Diaphragms
4000	1000	1000	3
3000	1000	1000	2
2000	1000	1000	1
1500	1000	1000	0
2000	1000	500	1
3000	1000	500	2
4000	1000	500	3
2000	1000	300	1
3000	1000	300	2
4000	1000	300	3

Source: IS:16014 (2018)

TABLE 4.2
Mesh and Box Characteristics for Revet Mattresses

Characteristics	Mesh Type 6 × 8 D = 60 mm		
	Only Zinc/Zinc Alloy Coated	Zinc/Zinc Alloy + Polymer Coated	
		Inner Diameter	Outer Diameter
i Mesh Wire diameter, mm	2.20	2.20	3.20
ii Edge/Selvedge Wire diameter, mm	2.70	2.70	3.70
iii Lacing Wire diameter, mm	2.20	2.20	3.20
iv PVC Coating thickness, mm			
Nominal		0.50	
Minimum		0.40	
v Tolerances in Size of Gabion Boxes			
Length and Width	+ 5%		
Height = < 300 mm	+ 10%		

Zinc Alloy shall consist of 90% Zinc + 10% Aluminium or 95% Zinc + 5% Aluminium

vi Typical Standard Sizes of Revet Mattresses

Length, mm	Width, mm	Height, mm	Number of Diaphragms
4000	2000	170	3
3000	2000	170	2
2000	2000	170	1
4000	2000	230	3
3000	2000	230	2
2000	2000	230	1
4000	2000	300	3
3000	1000	300	2
2000	2000	300	1

Source: IS:16014 (2018)

4.1.3.3 Stability

A gabion structure is a heavy gravity unit, and the spaces left between stone pieces are also filled with the silt particles in due course of time, which makes the gabion more solid. Further, the soil-filled spaces get covered with vegetation, which creates greater stability. The root systems of vegetation also hold the stone pieces very tightly at their respective positions and thus, further make the structure more stable.

4.1.3.4 Durability

The double twist, in case of a break in any single wire, prevents the unravelling of the mesh and the movement of stones out of the gabion box. The heavy zinc coating on the wires ensures that the eventual deterioration of the netting due to rusting is very slow under normal conditions. Where corrosion is a more severe threat, PVC coating can be provided to extend the wire life.

4.1.3.5 Environmental Impact

A gabion structure has a very positive impact on the environment. Soil pockets can be introduced in between stone fillings, and locally available, healthy species of plants/creepers can be used at intervals to have a lush green effect on gabion facia.

4.1.3.6 Versatility

Gabions can either be manufactured manually or mechanically. If there is a need for rapid integration with local nature and the environment, seeds can be scattered inside gabion facia units.

4.1.4 Economical Alternative

The gabions are more economical as compared to the masonry and concrete structures due to following reasons:

- i Cheaper cost of construction.
- ii The construction of a gabion structure does not require an excavation below the water, driving of piles, etc., and thus reduces the cost of a gabion structure.
- iii It does not involve complicated design and constructional features. An unskilled labour can easily construct it.
- iv Mostly, the boulders and stones are used as fill material in gabion structures, which are often available in the nearby areas at low cost.

4.1.5 Geotextiles

In gabion check dams, the appropriate geotextile should be provided on the upstream face and bed to prevent the seepage/leakage.

4.2 MASONRY CHECK DAMS

The art of masonry construction is very old. From time immemorial, builders in India and elsewhere have been constructing important structures in stone masonry. These structures have withstood the actions of destructive agents of nature, such as rain, wind, sun, etc., for over centuries now. Masonry check dams provide a durable solution that is required for a structure in contact with water and soil and exposed to constant wetting and drying. Despite a very long history of construction, the selection of materials, their sizes and the proportions of mortar in masonry for check dams are still done in an arbitrary manner which is based on the experience of past constructions. Properties and qualities of masonry depend on many variables, and it is desirable to know the effects of all these factors viz. properties of materials, their sizes, proportions, methods of placing and curing, bond strength, etc., on the density, strength, elasticity, and permeability of masonry.

4.2.1 Elements of a Masonry Check Dam

The elements of a masonry check dam typically include the following:

i Foundation
The foundation provides a stable base for the masonry check dam and is prepared by excavating the river bed to the required depth and ensuring the proper compaction of the soil.

ii Wall or barrier
The main element of a masonry check dam is the vertical wall or barrier made of masonry placed in a specific pattern to create a solid wall capable of withstanding water flow and sediment accumulation. Masonry check dams are constructed using boulders, rocks and stones. These materials are stacked and bonded together to form a solid dam structure. Larger boulders are often placed at the base and downstream side to provide strength and stability.

iii Crest of weir
The crest of the masonry weir is usually designed to include a weir or spillway section of sufficient width to pass excess water without causing overflow or damage.

iv Apron
To protect the check dam from downstream erosion, riprap or aprons made of rocks or boulders are placed. The apron acts as an erosion control measure and helps in dissipating the energy of the flowing water.

v Toe or cutoff wall
A toe or cutoff wall is constructed at the downstream end of the apron to prevent water from seeping beneath the structure. It helps maintain the integrity of the check dam by reducing the risk of erosion, scouring, and structural failure.

vi Abutment
The abutment provides support to the check dam structure at its ends, where it meets the natural terrain. The abutment should be properly designed to ensure stability of the structure.

4.2.2 Design of Masonry Check Dams

Design philosophy has already been discussed in Chapter 3. However, it involves following key considerations to ensure the effectiveness and longevity:

i Site assessment
Conduct a thorough site assessment to evaluate the hydrological conditions, bed slope of the stream, soil types, and rate of sediment transport, guiding

the design process to determine the appropriate size, spacing, and number of check dams.

The following considerations be made to select the site:

a. A narrow, straight, well-defined channel confined between banks not submerged in the highest flood should be selected as far as possible, and
b. The availability of construction material within reasonable distance, which affects the cost of the work.

ii Design parameters
Determine the design parameters, including the desired flow velocity reduction, pool depth, and spacing between check dams. These parameters also depend on the site conditions and the specific objectives of the project.

iii Weir construction
Incorporate a weir or spillway into the check dam design to allow excess water to flow through without damaging the check dam.

iv Material selection
Choose suitable materials for the construction of the check dams. Common materials include stones or rocks. It should be ensured that the materials are locally available, durable, and able to withstand the forces exerted by flowing water.

v Key stones
Key stones are larger stones strategically placed at certain intervals along the dam wall. They help distribute the weight of the dam and provide additional stability.

vi Mortar mix proportions
The proportions of cement, sand, and water in the mortar mix should be specified to ensure adequate strength, workability, and durability. The appropriate mortar mix proportions can vary depending on the type of masonry units and local specifications.

vii Foundation preparation
Prepare the foundation of the check dam by excavating and levelling the area to provide a stable base for construction and prevent any movement or shifting of the dam over time.

viii Monitoring, evaluation and maintenance
Monitor the performance of the check dams over time to assess their effectiveness. This can include measuring sediment deposition, observing water velocity and evaluating any changes in the hydrological conditions. Design features should be incorporated to facilitate maintenance, such as access points for sediment removal and regular inspections.

4.3 CONCRETE CHECK DAMS

A concrete check dam is a structure designed and built for the purpose of holding back water across streams to regulate water flow, prevent erosion, and control sedimentation. These dams are constructed using concrete, a durable and strong material that provides stability and longevity. They are used in areas with high water velocity, steep slopes, or where erosion and sedimentation are significant concerns. The elements of a concrete check dam include the following:

i Abutment
The abutment provides support to the check dam structure at its ends, where it meets the natural terrain. The abutment should be properly designed and constructed to ensure stability of the structure.
ii Foundation
The foundation is the base on which the check dam rests. It should be solid, stable, and properly compacted to withstand the forces exerted by the flowing water.
iii Weir
The weir/spillway is an opening or channel within the check dam that allows excess water to pass through during high flow events. It prevents the check dam from being overtopped and reduces the risk of damage. The weir/spillway should be appropriately sized and designed to handle the expected water flow.
iv Apron
To protect the check dam from downstream erosion, measures such as riprap, concrete aprons, or other erosion-resistant materials are placed. The apron acts as an erosion control measure and helps in dissipating the energy of the flowing water.
v Toe or cutoff wall
A toe or cutoff wall is constructed at the downstream end of the apron to prevent water from seeping beneath the structure. It helps maintain the integrity of the check dam by reducing the risk of erosion, scouring, and structural failure.

These elements together contribute to the stability, functioning, and effectiveness of a concrete check dam. The specific design and features of the concrete check dam depend on the site conditions, water flow rates, geological considerations, intended purpose of the structure, etc. Design philosophy has already been discussed in Chapter 3.

4.3.1 Construction

Concrete check dams are typically built by pouring and shaping concrete into a series of steps or tiers across the stream. The design and size of the check dam depend on factors such as the water flow rate, channel width, and local conditions.

4.3.2 Strength and Durability

Concrete check dams are designed to have high strength and durability to withstand the forces exerted by water pressure and other loads. The use of high-strength concrete and advanced construction techniques help in ensuring the stability and longevity.

4.3.3 Maintenance

Regular maintenance is necessary to ensure the effectiveness of concrete check dams. Inspections should be conducted to identify any damage or signs of erosion. Any cracks, surface deterioration or displacement should be repaired promptly to maintain the structural integrity of the dam.

4.3.4 Limitations

Concrete check dams have certain limitations:

- Cost: Concrete construction can be more expensive compared to other materials used for check dams.
- Environmental impact: Construction of concrete check dams may have environmental impacts, such as the alteration of river ecosystems and disturbance during construction.
- Sediment accumulation: Over a period of time, sediment may accumulate upstream of the check dam, requiring periodic maintenance to remove and manage sediment build-up.

The robustness, longevity and versatility of concrete check dams make them a most viable alternative in river and stream management, contributing to sustainable water resource management and protection of downstream environments.

4.4 RUBBER CHECK DAMS

It has been observed that the traditional primary check dams built across small streams quite often get washed away during floods, whereas the permanent check dams constructed get filled up to their crest and need desilting and periodical repairs after the rainy season and also get damaged during floods and earthquakes.

A rubber check dam or flexible check dam is an inflatable structure built across a stream used for water conservation, flood control, and regulating flow of water in the stream. When it is inflated, it serves as a check dam/weir, and when it is deflated, it functions as a flood mitigation device and permits sediment flushing. These rubber dams can also be considered as one of the viable alternatives for constructing check dams. In rubber check dams, the height of the dam can be increased or decreased according to the requirement, which will help in regulating the depth of flow in the channel. The rubber check dam scores over the traditional check dam in several ways:

- While traditional permanent check dams get silted over a period of time, the ease of inflation and deflation allows the rubber check dam to be routinely cleaned from the deposited silt as well as safe against the flood.
- During earthquakes, the traditional permanent check dam could be severely damaged, whereas the rubber head wall of the rubber check dam has little risk of damage.

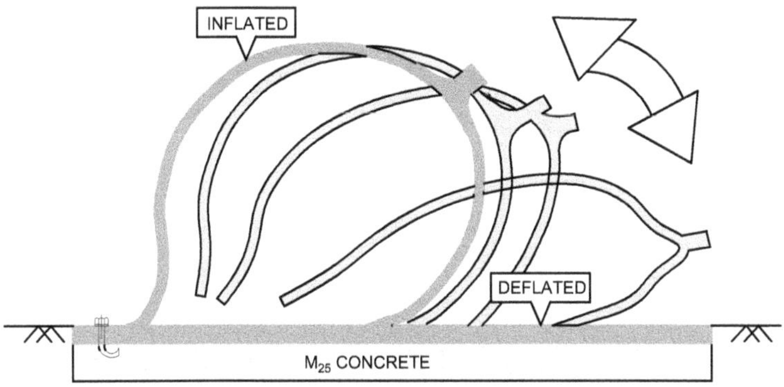

FIGURE 4.4 Inflation and Deflation Phenomenon of Rubber Dam.

Source: Sahoo (2010)

In the recent past, rubber dams have been increasingly accepted and are widely used worldwide. In India, rubber dams have not been used extensively, and till now, only six rubber dams have been installed, whereas more than 20,000 rubber dams have been successfully installed worldwide. The inflation and deflation phenomenon of the rubber check dam is shown in Figure 4.4.

The simplicity and flexibility of the rubber check dam structure and its proven reliability are key considerations in its wide scope of applications. The first inflatable dam, *Fabridam*, was developed in the mid-1950s by an American engineer, Norman Imbertson. It was 1.52 m high and 39.6 m long and was inflated by a combination of water and air. The fabric used was nylon-reinforced neoprene and was manufactured by Firestone Tire and Rubber Co. in Akron, Ohio. India's first rubber check dam was built in 2005 on the Jhanjhavati River, which was completed in six months. The rubber check dam constructed at Bulgwangcheon, Eunpyeong-gu, Seoul, in 2006 is shown in Figure 4.5.

4.4.1 Suitability of Rubber Check Dams over Conventional Check Dams

Rubber check dams have the following advantages over the conventional/traditional permanent check dams:

- *Long span and adaptable to different side slopes*

  Long rubber dams can be installed in broad rivers having spans as long as 100 m without piers. They are adaptable to virtually any side slope angle. Little work is needed to modify river banks.

- *Easy inflation/deflation*

  The rubber check dam body can be fully inflated or deflated automatically with respect to water levels in the upstream with a provision for manual

FIGURE 4.5 Rubber Dam Constructed at Bulgwangcheon, Eunpyeong-gu, Seoul.

Source: Yooil Engineering, South Korea

operation. When the rubber check dam is fully deflated on the base floor, the flood passes without any obstruction. This provides rubber check dams a dominant position over traditional check dams. Besides, the provision of overflow spillway is not needed, as the rubber check dam can itself act as a spillway and pass the discharge above its crest.

- *Short construction period*

  Compared with a conventional check dam, the rubber check dam is lighter and easier to handle. It can be fabricated in one piece at a factory and rolled up for easy transportation to the site. The rubber check dam requires a simple light foundation.

- *Easy maintenance and repair*

  Minimal maintenance is needed for rubber check dams.

- *Useful life of rubber check dams*

Rubber check dams last longer than the traditional permanent check dams. The estimated life span of the rubber check dam is about 30–35 years.

- *Low project life cycle cost*

The life cycle cost of a rubber check dam project is low due to the prefabrication of the dam body, little modification to river banks, light concrete foundation, quick construction and installation, easy operation, and minimal maintenance.

- *Earthquake resistant*

The simple and light upper structure, uniform load on the rubber body and light concrete foundation make a rubber check dam more earthquake resistant than other traditional permanent check dam structures serving in similar situations.

- *Environmentally friendly*

The crest adjustable rubber check dam can form an aesthetically pleasing impoundment in otherwise fluctuating waters. This can create a stable aquatic habitat and provide recreational opportunities for various water activities.

- *Operating cost*

Rubber dams are economical to operate. The only significant running cost is power requirement for the air blower or water pump during bag inflation.

4.4.2 Classification of Rubber Dams

4.4.2.1 Classification Based on Filling Media

Based on filling media, rubber check dams can be classified as follows:

- Air-filled dam
- Water-filled dam
- Hybrid dam (combined air and water filled)

4.4.2.1.1 Air-Filled Rubber Dam

The inflation medium used in these dams is air, and the standard inflating air pressure is kept at about 30 KPa to inflate it fully and gain an elliptical shape.

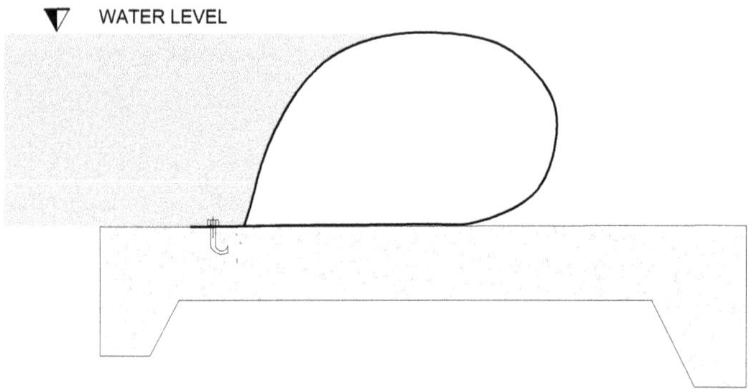

FIGURE 4.6 Air-Filled Rubber Dam.

4.4.2.1.2 Water-Filled Rubber Dam

The inflating medium used in these dams is water. A significant size of water pond/tank is normally provided for water-filled rubber dams in addition to water pump for filling the rubber dams during the inflating operation. The inflation/deflation of water-filled rubber dams is slower than air-filled rubber dams, and the cost of water-filled rubber dams is relatively higher than air-filled rubber dams.

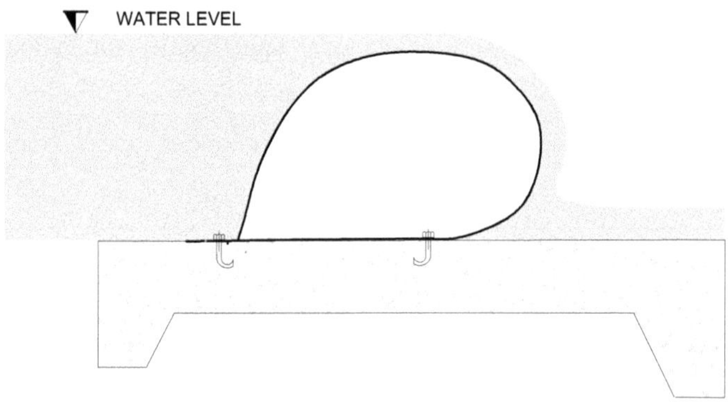

FIGURE 4.7 Water-Filled Rubber Dam.

Air-inflated rubber dams are suitable for low height ranging from 1.0 m to 6.0 m.

4.4.2.2 Classification Based on the Form of Dam Tube

4.4.2.2.1 Pillow-Like Dam

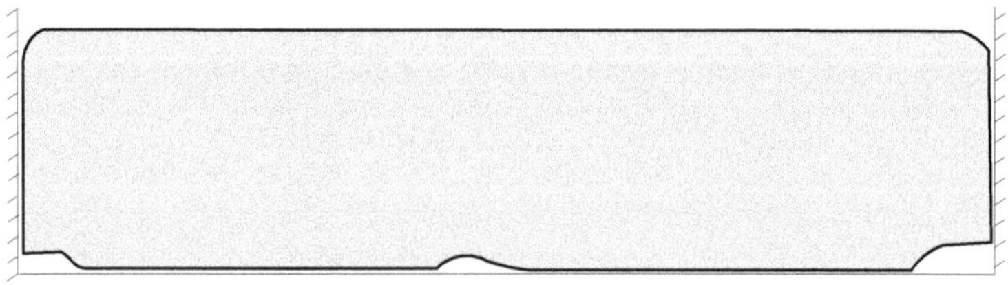

FIGURE 4.8 Pillow Rubber Dam.

4.4.2.2.2 *Inclined Dam*

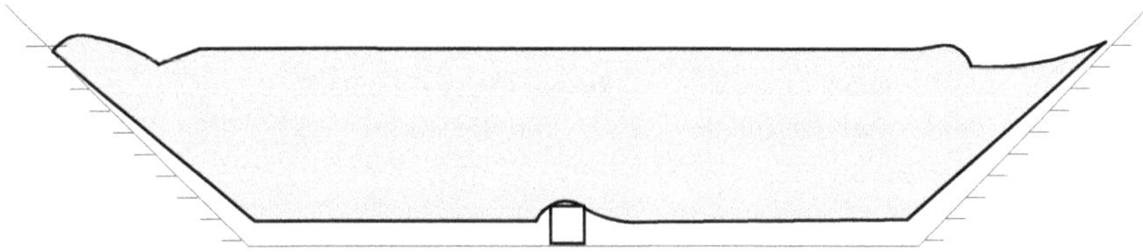

FIGURE 4.9 Inclined Rubber Dam.

4.4.2.3 Classification Based on the Seam of Rubber Bag

Rubber dams are also classified on the seam of rubber bags:

4.4.2.3.1 *Seamless Rubber Dam*

The seamless rubber dam, also known as a non-overlapping rubber dam, is made of framework material (nylon fabric) and inner and outer covering rubber according to the moulding process and one-time vulcanisation process. The thickness of the rubber membrane is even, the strength is consistent, there is no weakness in the lap joint section, and the safety factor is high. The strength of the dam bag in warp and weft direction is consistent, and there is no weakness.

4.4.2.3.2 *Seamed Rubber Dam*

Seamed rubber dam is formed by the adhesion of pieces of adhesive rubber after grinding. Due to human factors and the limitation of technology, the seam joining part usually has defects and is easily damaged, which will affect the life of the dam bag and will cause the tearing of the dam bag under critical condition.

4.4.3 Rubber Check Dam Components

The typical rubber check dam has the following components:

i. Rubber membrane
ii. Foundation and fixing arrangements
iii. Piping arrangement, including pumps, valves
iv. Controls

The rubber sheet is attached to the foundation structure on the bottom of the water course and also to both abutments/piers.

4.4.4 Design Data

The design parameters required for design of a rubber check dam are similar to other check dam structures, which includes design discharge, maximum pond level, average lowest bed level of the river, geotechnical parameters, embankment top level, cross section of the river in U/S and D/S, L-Section of the river, slope of the water surface of the river, catchment area of the stream, etc.

TABLE 4.3
Thickness of Rubber Check Dam Body

Rubber Check Dam Height, m	Rubber Check Dam Body Thickness, mm
0.50 to 1.00	10.0
1.10 to 1.60	11.5
1.70 to 2.40	12.5
2.50 to 2.90	14.5
3.00 to 3.90	15.5
4.00 to 5.00	17.5

i. *Span of a rubber check dam*
There is no limit to the length of a rubber check dam span from the standpoint of stress due to the fact that stress is evenly distributed along the entire length of the anchor line rather than concentrated at selected points.

ii. *Thickness of rubber check dam body*
Table 4.3 shows the usual thickness of the rubber check dam body adopted depending on the height of a dam.

iii. *Overflow height*
Due to the flexible structure of the rubber dam, vibration takes place in the rubber bag when the overflow and downstream water depth increase. Generally, the higher the interior pressure, the more difficult it is for the rubber bag to vibrate. The maximum recommended overflow height from the viewpoint of vibration for the air-inflation type rubber dam is 0.2H (JIID, 1989), where H is the rubber bag height at the time of overflow.

If the expected maximum overflow depth exceeds the aforementioned maximum, vibration is possible. A deflector (fin) should be provided to separate the nappe from the rubber bag to decrease predictable vibration.

iv. *Floor thickness*
Floor thickness at various parts of the structure is selected considering the uplift pressure computed by Khosla's theories and the minimum thickness required for laying the accessories of rubber dams in the foundation. A 1.0 m thick foundation is generally proposed to be laid below low-height rubber check dams.

v. *U/S and D/S protection works*
Protection works for rubber dams are designed and provided on the principles of similar hydraulic structures.

vi. *Control room for operation and maintenance*
One control room is needed to house pump and other accessories for operating rubber check dams. The horsepower of the pump and other things shall be decided at the time of finalising the design of the rubber check dam, and thus, the size of the control rooms may vary as per design requirements

of the rubber check dam system supplier/manufacturer. The application of smart technology like SCADA, PLC (programmable logic controller) and AI (artificial intelligence) in rubber check dams in operation will be very beneficial, and even several rubber check dams can also be controlled from one remote location to have proper control and management of these structures and prevent water losses.

The typical plan and section for rubber check dams are shown in Figure 4.10 and Figure 4.11.

4.4.5 Bridge-cum-Rubber Check Dams for Water Conservation

Looking at the water scarcity all across the world, the proposed sites identified for new bridge structures and existing/abandoned bridge structures can also be

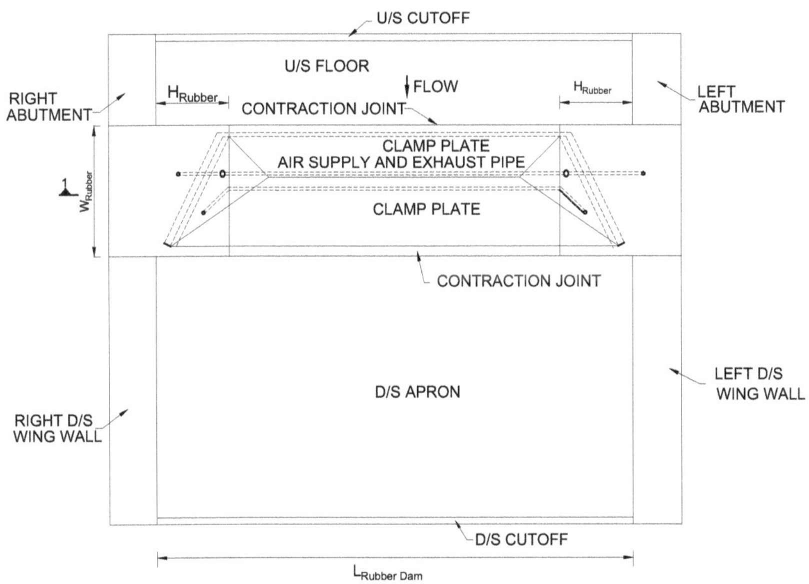

FIGURE 4.10 Plan of Rubber Dam.

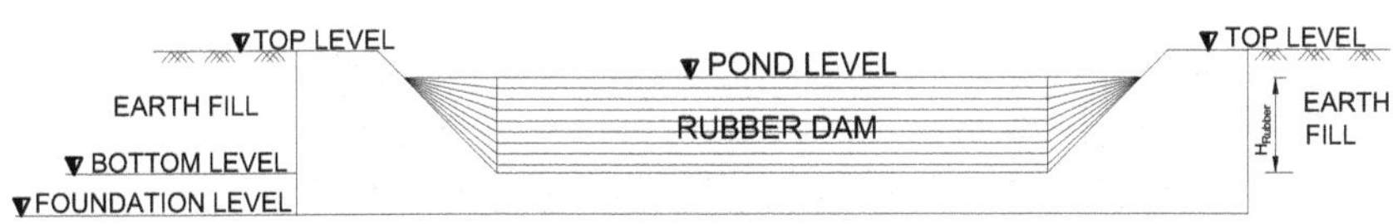

FIGURE 4.11 Cross-Sectional Details of a Rubber Check Dam.

FIGURE 4.12 Bridge-cum-Rubber Check Dam in the Downstream of the Bridge at Daegu Metropolitan City Geumho River Mutaebo, South Korea.

***Source*: Yooil Engineering, South Korea**

used for water conservation for drinking, irrigation, recharging of ground water, etc., by constructing small rubber dams with heights from 1.50 m to 2.00 m in the downstream of bridges to store the water upstream as shown in Figure 4.12. The pond level of these rubber dams is fixed to harness the maximum storage with no submergence and land acquisition. The scheme can be popular, as water stored within the river banks will not require additional land acquisition. Design of rubber dam structures will be more simplified because in case of small spans up to 100 m, single-span rubber check dams can be provided, which will simplify the design of the structure and ease the installation and removal of gates in conventional bridge-cum-check dam structures besides theft, wear and tear of gates and maintenance.

4.4.6 Disadvantages of Rubber Check Dams

Despite several advantages, the rubber check dam has the following disadvantages also:

- The rubber dam needs careful operation and maintenance. Large floating materials like logs, trees, bamboos, etc., may damage the rubber bag, and therefore, at such locations, the rubber bag will have to be designed accordingly.
- Repair to dam bags is done only in dry condition. Underwater repair is not easy.
- The service life of dam bags is about 30–35 years. The replacement of the dam bag may be needed after the designed life.

4.5 SELECTION OF MATERIALS FOR CHECK DAMS

All types of check dams have their advantages and disadvantages, and the suitability of check dams shall have to be done on the basis of the best cost-value ratio at that geographic location, surrounding conditions, availability of materials locally and designed service life. In general, the suitability of type of check dams with advantages and disadvantages is shown in Table 4.4 (Design of a Check Dam, 2017).

TABLE 4.4
Suitability of Materials for Check Dams

Type	Suitability	Advantages	Disadvantages
Brushwood check dam	They are suitable for small streams of 1.00–2.00 m in depth, having a catchment area of ≤ 1 ha.	• Simple in construction. • They are cheap and can be made by farmers with locally available materials. • After the roots and shoots come out, they can form a long-term barrier.	• Takes a long time to develop roots and shoots and gets established.
Log check dam	The main objective of such check dams is to hold fine and coarse material carried by flowing water in a small stream/rivulet. They are suitable for small streams of about 1.50 m in depth, having a catchment area of ≤ 2 ha.	• Simple in construction. • They are cheap and can be made by wooden logs available in abundance.	• White ants are the greatest enemy of these types of structures, and accordingly, the required measures should be taken beforehand.
Loose stone check dam	They are constructed where rocks are available abundantly and suitable for small streams of about 1.50 m in depth, having a catchment area of ≤ 2 ha.	• Simple in construction. • Low cost and can be constructed by local people.	• If not made properly and sizable stones are not used, the stones will be moved by the large water flow and may get damaged quickly.
Boulder check dam	They are constructed where boulders are available abundantly and suitable for streams with depths of about 2.00 m, having a catchment area of ≤ 20 ha. Stability and strength depend on the size of boulders and quality of construction.	• Simple in construction. • Low cost, if boulders are abundantly available. • If properly made, are almost like permanent structure and durable.	• Transportation of big boulders is difficult, especially if not available locally. • Large voids, if not properly filled up, may create water jets, which could be destructive if directed towards banks.

(Continued)

TABLE 4.4 (Continued)

Type	Suitability	Advantages	Disadvantages
Gabion check dam	They are preferable where big boulders are not available and suitable for streams having a catchment area of more than 20.00 ha and if the land mass is unstable.	• Flexible and permeable structures. • Economical compared to other permanent check dams.	• The gabion check dam is to be constructed as a monolithic structure, which is often not followed at the site due to poor workmanship.
Masonry check dam	These check dams are used in streams with high rates of runoff. The construction of these check dams is recommended, where rocks or stones are readily available.	• Permanent, resilient, and solid structure. • Have an aesthetic look.	• Need detailed engineering design and skilled labour for construction.
Concrete check dam	These check dams are used in streams with high rates of runoff. The construction of these check dams is recommended, when inadequate materials are available for constructing the masonry check dams.	• Permanent, resilient and solid structure. • Have an aesthetic look.	• Need detailed engineering design and skilled labour for construction.

4.6 SELECTION OF CHECK DAMS ON THE BASIS OF PARAMETERS

The comparison of gabion, masonry/concrete and rubber check dams and their suitability on the basis of different parameters like flexibility, permeability, etc., is presented in Table 4.5 (MORD:27, 2015).

TABLE 4.5
Comparison of Gabion, Masonry/Concrete and Rubber Check Dams

Parameters	Gabion Check Dam	Masonry/Concrete Check Dam	Rubber Check Dam
Flexibility	It is a flexible structure and can accommodate differential settlement without compromising structural integrity.	It is a rigid structure, and cracks will occur due to differential settlement.	It is a hybrid structure, and cracks will occur only in the concrete base due to differential settlement.

(Continued)

TABLE 4.5 (Continued)

Parameters	Gabion Check Dam	Masonry/Concrete Check Dam	Rubber Check Dam
Permeability	The system is made up of gabion boxes filled with stones, and the permeability of the front face ensures the drainage of the backfill, resulting in less hydrostatic pressure.	It is an impermeable structure, and if sufficient drainage measures are not provided, very high hydrostatic pressure can occur.	It is an impermeable structure.
Simplicity/ Economy	Gabions do not require skilled labour force or special equipment.	Specialised labour, steel formwork, and machinery are required. Material cost is comparatively high.	Specialised labour and machinery are required. Material cost is high for small spans.
Environmental Friendliness	These systems are environmentally friendly. Gabion wall covered with a lush green vegetation topping offer an excellent opportunity to enhance the landscape. In general, the carbon footprint emission for gabion facia structures are relatively less.	Cement concrete construction is non–eco-friendly technology. Growth of vegetation leads to cracks, thereby making the structure more unstable.	These systems are environmentally friendly.
Impact Resistance	Good shock absorption during sudden impact. Flexible facia is well resistant to damages.	Crack develops in the rigid concrete when there is sudden impact. Being a continuous structure, local rectification is not possible.	Good shock absorption during sudden impact and a flexible structure.
Foundation Consideration	Flexible in nature, it can be founded on yielding foundations. Hence, chances of failure due to settlement are minimal.	Systems are rigid in nature and highly sensitive to even the slightest settlement in foundations. They exert huge amounts of pressures in the foundation.	The base of the systems is rigid in nature, and the body is flexible and can easily adjust with pressures. They exert less pressure on the foundation.
Seismic and Dynamic Forces	Flexible in nature, these systems are accommodative to dynamic forces and don't collapse in the event of seismic action.	This system attracts huge inertial forces due to rigidity and can fail instantly in case of seismic action.	Only the base of the system attracts inertial forces due to rigidity, as the body of the rubber check dam is flexible in nature.
Failure Mode	Having a flexible system, failure signs can be detected and rectified.	Having a rigid system, failure is spontaneous.	Having a flexible system, the chances of failure will be less, as the discharge can be passed from the top of rubber check dam.

REFERENCES

IS:16014 (2018) *Indian Standard Mechanically Woven, Double-Twisted, Hexagonal Wire Mesh Gabions, Revet Mattresses, Rock Fall Netting and Other Products for Civil Engineering Purposes* (Galvanized Steel Wire or Galvanized Steel Wire With Polymer Coating) – Specification (First Revision), IS:16014, Bureau of Indian Standards, New Delhi, December.

IS:280 (1978) *Indian Standard Specification for Mild Steel Wire for General Engineering Purposes IS:280*, Bureau of Indian Standards, Water Resources Department, New Delhi.

JIID (1989), *Headworks*, Vol. 2, Japanese Institute of Irrigation and Drainage (JIID), March.

MORD: 27 (2015) *Do It Yourself Gabion Retaining Walls for Rural Roads*, MORD 27, National Rural Roads Development Agency (NRRDA), Ministry of Rural Development, Government of India, New Delhi.

Philippine National Standard (2017) *Design of a Check Dam*, PNS/BAFS/PAES 230:2017, Bureau of Agriculture and Fisheries Standards, Philippines.

5 Easy-to-Use Tables for the Design of Gabion Check Dams

5.1 PARAMETERS CONSIDERED FOR THE DEVELOPMENT OF TABLES

Based on the requirement at the site, the gabion check dams can be designed with several configurations. The typical shapes adopted for developing tables for the design of gabion check dams are shown in Figure 5.1.

In general, the following forces are considered in the design of gabion check dams:

- Self weight of wall;
- Water pressure;
- Uplift pressures; and
- Active earth pressures in the design of abutment and wing walls.

The following parameters were varied for developing the design tables:

Gabion Check Dams

Height of a Check Dam Weir	:	1.50 m, 2.00 m, 2.50 m, and 3.00 m
Catchment Area[1]	:	50 ha, 100 ha, 150 ha, 200 ha, 250 ha, and 300 ha
Width of Stream/Gadhera	:	5.0 m, 10.0 m, 15.0 m, and 20.0 m
Shape of a Gabion Check Dam Section	:	• Vertical D/S Face • Vertical U/S Face • Steps on Both Faces
Unit Weight of Water, γ_{water}	:	1.00 t/m^3 or 9.81 kN/m^3
Unit Weight of Gabion Boxes with Filling Material, γ_{Box}	:	1.60 t/m^3 or 15.70 kN/m^3
Effect of Water for Uplift Pressure	:	100% of Total Height
Coefficient of Friction between the Gabion and Soil/Rock Mass μ	:	0.50

[1]For small width of streams, the design of check dams has been carried out by limiting the head over crest to about 1.30 m.

DOI: 10.1201/9781003605621-5

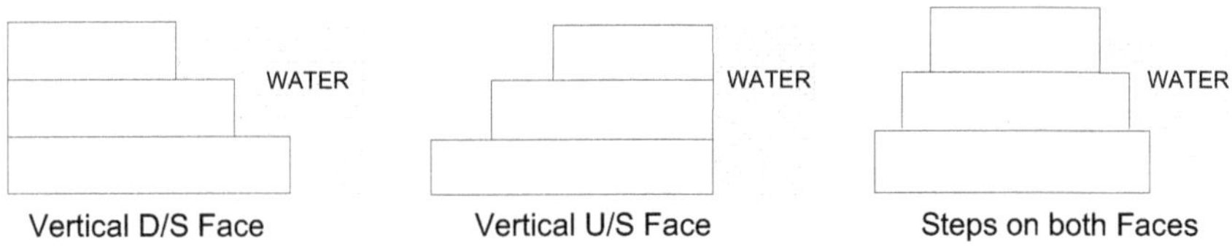

FIGURE 5.1 Typical Shapes of Gabion Check Dams.

Abutment/Wing Walls

Unit Weight of Saturated Earthfill, γ_{sat}:	1.80 t/m³ or 17.65 kN/m³
Angle of Repose of Backfill, ϕ:	30^0
Angle of Friction between the Gabion Wall and Backfill, δ:	$⅔\phi$
Angle of Surcharge of Backfill, ι:	0^0
Coefficient of Friction between the Gabion and Soil/Rock Mass μ:	0.50

5.1.1 Typical Cross Sections Adopted

The generalised cross sections of the gabion check dam weir and non-over flow for heights varying from 1.50 m to 3.00 m for carrying out the analysis for different shapes are depicted in Figure 5.2 to Figure 5.4.

5.1.1.1 Section with Vertical D/S Face

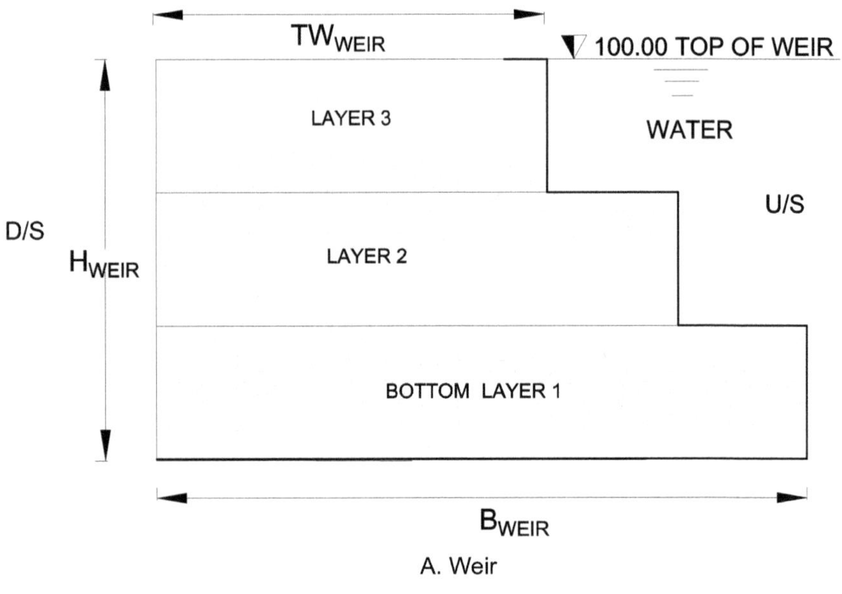

FIGURE 5.2A, 5.2B Typical Cross Sections of Gabion Weir and Non-Over Flow for Vertical D/S Face.

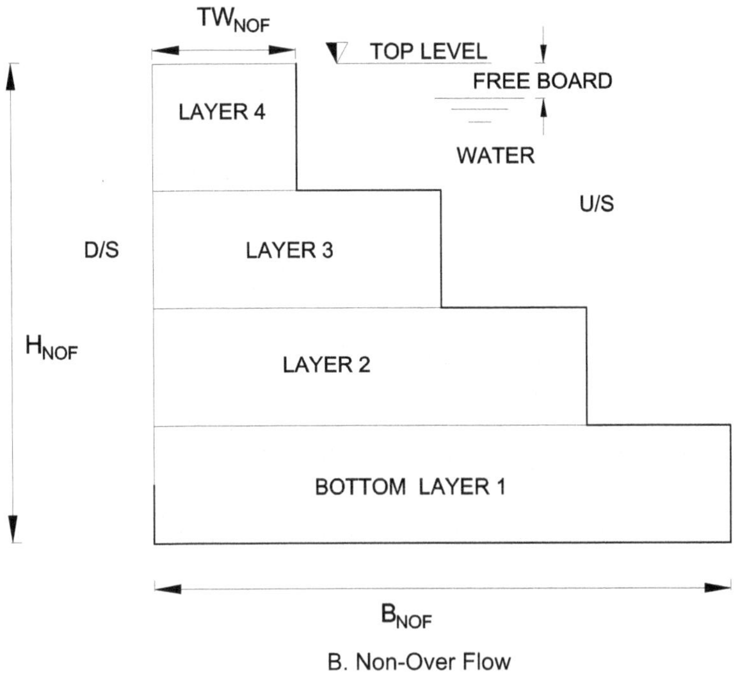

FIGURE 5.2A, 5.2B (Continued)

5.1.1.2 Section with Vertical U/S Face

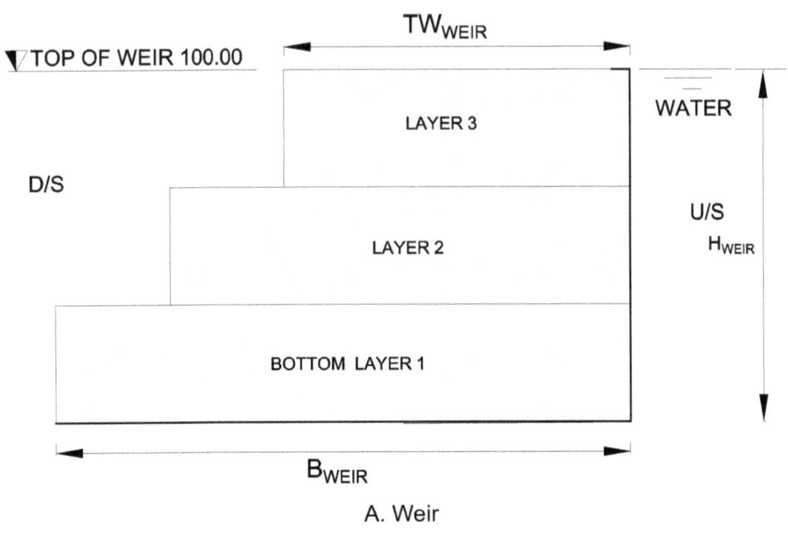

FIGURE 5.3A, 5.3B Typical Cross Sections of Gabion Weir and Non-Over Flow for Vertical U/S Face.

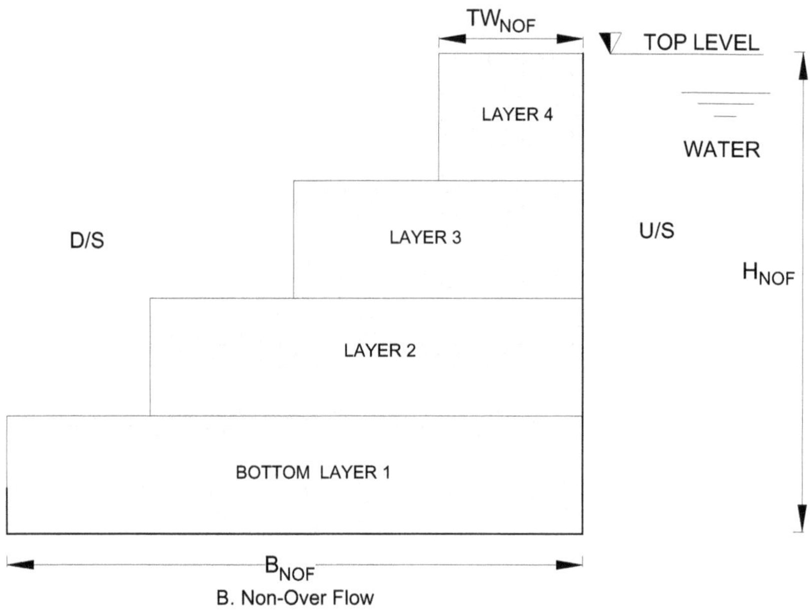

FIGURE 5.3A, 5.3B (Continued)

5.1.1.3 Section with Steps on Both Faces

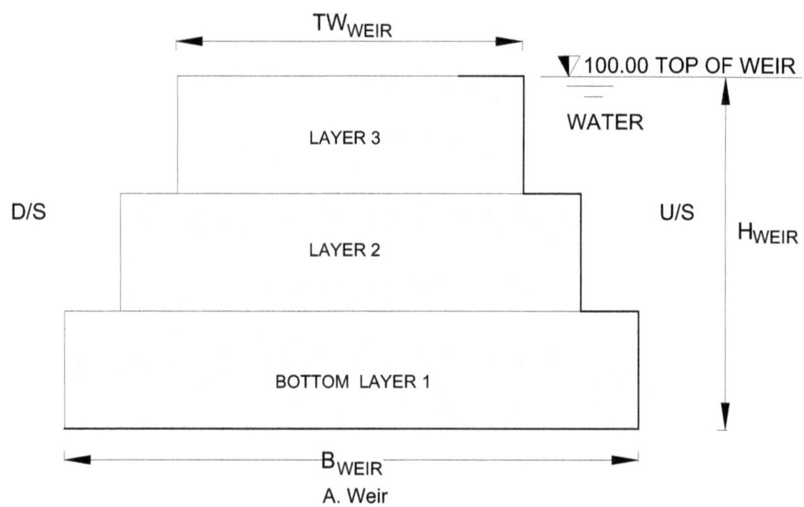

FIGURE 5.4A, 5.4B Typical Cross Sections of Gabion Weir and Non-Over Flow for Steps on Both Faces.

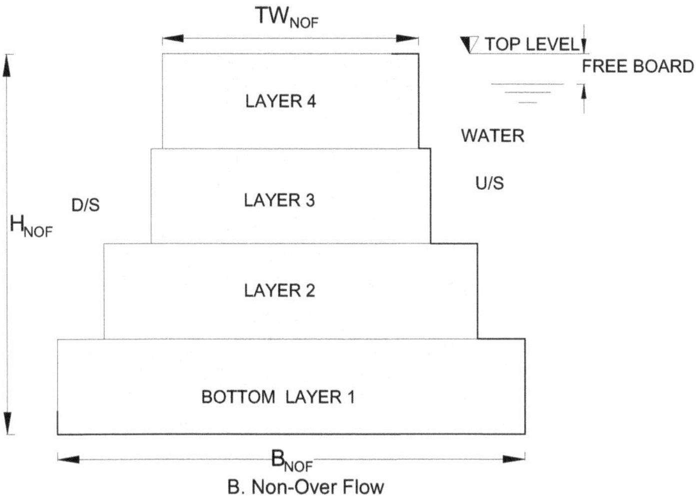

B. Non-Over Flow

FIGURE 5.4A, 5.4B (Continued)

5.1.1.4 Section with Vertical D/S Face for Abutment/Wing Walls

The generalised cross sections of the gabion for wing wall (abutment) and downstream wing wall, for carrying out the analysis of the vertical D/S face, are depicted in Figure 5.5.

5.1.2 Design Approach

For every gabion check dam of specified height, the stability analysis was carried out by conventional approach (IS:6512, 2019), discussed in Chapter 3 on design philosophy, for six sets of catchment areas and four sets of stream widths for each shape to develop the design tables. Wing wall/abutment and d/s wing wall sections were designed with vertical D/S face only. For small width of streams, the design of gabion check dams has been limited to the head over crest to about 1.30 m.

The base width of the gabion check dam (non-over flow and overflow sections), wing wall/abutment, and d/s wing walls has been so proportioned in the analysis that there is nearly zero stress near the heel under static loading; the factor of safety against overturning is equal to or more than 1.50.

5.2 DESIGN OF 1.50 M HIGH GABION CHECK DAMS

5.2.1 Section with Vertical D/S Face

The tables developed after the rigorous design of the different components for 1.50 m high gabion check dams for different combinations are presented from Table 5.1 to Table 5.4.

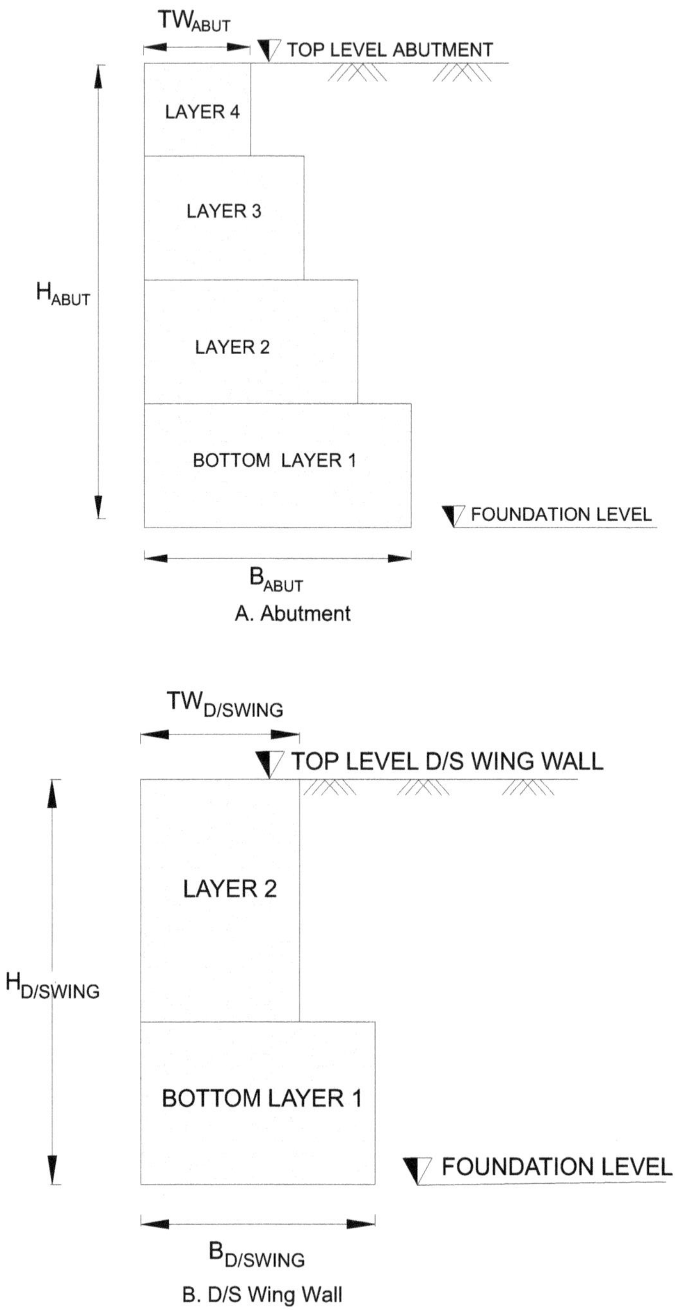

FIGURE 5.5A, 5.5B Typical Cross Sections of Gabion Wing Wall (Abutment) and Downstream Wing Wall for Vertical D/S Face.

TABLE 5.1
Design Results for Gabion Check Dam with Vertical D/S Face Height = 1.50 m and Stream Width L = 5.0 m

Catchment Area, ha	50.00	100.00
A. WEIR		
Span of Weir W, m	4.00	4.00
Flood Discharge Q, m^3/s	6.54	11.00
Head Over Crest, m	0.92	1.31
Top Level, m	100.00	100.00
Foundation Level, m	98.50	98.50
Total Height of Weir H_{WEIR}, m	1.50	1.50
Bottom Width of Weir B_{WEIR}, m	2.50	2.50
Size of Gabion Box (L×B×H), m		
Bottom Layer 1	2.50 × 1.00 × 1.00	2.50 × 1.00 × 1.00
Layer 2	2.50 × 1.00 × 0.50	2.50 × 1.00 × 0.50
Pressure at Toe, t/m^2	4.26	4.79
Pressure at Heel, t/m^2	0.28	0.25
Factor of Safety against Overturning	2.06	2.07
Shear Friction Factor F	5.00	4.16
Total Volume of Gabion Boxes, m^3	15.00	15.00
B. NON-OVER FLOW PORTION		
Length, m	1.00	1.00
Top Level including Free Board, m	101.50	101.80
Foundation Level, m	98.50	98.50
Total Height of Non-Over Flow H_{NOF}, m	3.00	3.30
Bottom Width of Non-Over Flow B_{NOF}, m	2.50	3.00
Size of Gabion Box (L×B×H), m		
Bottom Layer 1	2.50 × 1.00 × 1.00	3.00 × 1.00 × 1.00
Layer 2	2.50 × 1.00 × 1.00	3.00 × 1.00 × 1.00
Layer 3	2.50 × 1.00 × 1.00	2.50 × 1.00 × 1.00
Layer 4	–	2.00 × 1.00 × 0.30
Pressure at Toe, t/m^2	7.38	8.45
Pressure at Heel, t/m^2	0.72	0.07
Factor of Safety against Overturning	2.58	2.47
Shear Friction Factor F	4.49	4.00
Total Volume of Gabion Boxes, m^3	7.50	9.10

TABLE 5.2
Design Results for Gabion Check Dam with Vertical D/S Face Height = 1.50 m and Stream Width L = 10.0 m

Catchment Area, ha	50.00	100.00	150.00	200.00
A. WEIR				
Span of Weir W, m	8.00	8.00	8.00	8.00
Flood Discharge Q, m^3/s	6.54	11.00	14.91	18.50
Head Over Crest, m	0.58	0.82	1.01	1.16
Top Level, m	100.00	100.00	100.00	100.00
Foundation Level, m	98.50	98.50	98.50	98.50
Total Height of Weir H_{WEIR}, m	1.50	1.50	1.50	1.50
Bottom Width of Weir B_{WEIR}, m	2.50	2.50	2.50	2.50
Size of Gabion Box (L×B×H), m				
Bottom Layer 1	2.50 × 1.00 × 1.00	2.50 × 1.00 × 1.00	2.50 × 1.00 × 1.00	2.50 × 1.00 × 1.00
Layer 2	2.50 × 1.00 × 0.50	2.50 × 1.00 × 0.50	2.50 × 1.00 × 0.50	2.50 × 1.00 × 0.50
Pressure at Toe, t/m^2	3.77	4.11	4.37	4.60
Pressure at Heel, t/m^2	0.31	0.28	0.27	0.26
Factor of Safety against Overturning	2.05	2.05	2.06	2.06
Shear Friction Factor F	6.19	5.29	4.79	4.43
Total Volume of Gabion Boxes, m^3	30.00	30.00	30.00	30.00
B. NON-OVER FLOW PORTION				
Length, m	2.00	2.00	2.00	2.00
Top Level including Free Board, m	101.00	101.30	101.50	101.80
Foundation Level, m	98.50	98.50	98.50	98.50
Total Height of Non-Over Flow H_{NOF}, m	2.50	2.80	3.00	3.30
Bottom Width of Non-Over Flow B_{NOF}, m	2.50	2.50	3.00	3.00
Size of Gabion Box (L×B×H), m				
Bottom Layer 1	2.50 × 1.00 × 1.00	2.50 × 1.00 × 1.00	3.00 × 1.00 × 1.00	3.00 × 1.00 × 1.00
Layer 2	2.00 × 1.00 × 1.00	2.50 × 1.00 × 1.00	2.50 × 1.00 × 1.00	2.50 × 1.00 × 1.00
Layer 3	2.00 × 1.00 × 0.50	2.00 × 1.00 × 0.80	2.50 × 1.00 × 1.00	2.50 × 1.00 × 1.00
Layer 4				2.00 × 1.00 × 0.30
Pressure at Toe, t/m^2	5.95	6.96	7.18	8.21
Pressure at Heel, t/m^2	0.05	0.15	0.39	0.06
Factor of Safety against Overturning	2.32	2.39	2.60	2.54
Shear Friction Factor F	5.78	4.76	4.92	4.43
Total Volume of Gabion Boxes, m^3	11.00	13.20	16.00	17.20

TABLE 5.3
Design Results for Gabion Check Dam with Vertical D/S Face Height = 1.50 m and Stream Width L = 15.0 m

Catchment Area, ha	50.00	100.00	150.00	200.00	250.00	300.00
A. WEIR						
Span of Weir W, m	12.00	12.00	12.00	12.00	12.00	12.00
Flood Discharge Q, m^3/s	6.54	11.00	14.91	18.50	21.87	25.07
Head Over Crest, m	0.44	0.63	0.77	0.89	0.99	1.09
Top Level, m	100.00	100.00	100.00	100.00	100.00	100.00
Foundation Level, m	98.50	98.50	98.50	98.50	98.50	98.50
Total Height of Weir H_{WEIR}, m	1.50	1.50	1.50	1.50	1.50	1.50
Bottom Width of Weir B_{WEIR}, m	2.50	2.50	2.50	2.50	2.50	2.50
Size of Gabion Box (L×B×H), m						
Bottom Layer 1	2.50 × 1.00 × 1.00	2.50 × 1.00 × 1.00	2.50 × 1.00 × 1.00	2.50 × 1.00 × 1.00	2.50 × 1.00 × 1.00	2.50 × 1.00× 1.00
Layer 2	2.50 × 1.00 × 0.50	2.50 × 1.00 × 0.50	2.50 × 1.00 × 0.50	2.50 × 1.00 × 0.50	2.50 × 1.00 × 0.50	2.50 × 1.00 × 0.50
Pressure at Toe, t/m^2	3.58	3.83	4.03	4.20	4.35	4.48
Pressure at Heel, t/m^2	0.31	0.30	0.29	0.28	0.27	0.27
Factor of Safety against Overturning	2.04	2.05	2.06	2.06	2.06	2.06
Shear Friction Factor F	6.83	5.99	5.48	5.11	4.82	4.60
Total Volume of Gabion Boxes, m^3	45.00	45.00	45.00	45.00	45.00	45.00
B. NON-OVER FLOW PORTION						
Length, m	3.00	3.00	3.00	3.00	3.00	3.00
Top Level including Free Board, m	100.80	101.00	101.30	101.30	101.50	101.80
Foundation Level, m	98.50	98.50	98.50	98.50	98.50	98.50
Total Height of Non-Over Flow H_{NOF}, m	2.30	2.50	2.80	2.80	3.00	3.30
Bottom Width of Non-Over Flow B_{NOF}, m	2.50	2.50	2.50	2.50	2.50	2.50
Size of Gabion Box (L×B×H), m						
Bottom Layer 1	2.50 × 1.00 × 1.00	2.50 × 1.00 × 1.00	2.50 × 1.00 × 1.00	2.50 × 1.00 × 1.00	2.50 × 1.00 × 1.00	2.50 × 1.00 × 1.00
Layer 2	2 × 1.0 × 1	2.50 × 1.00 × 1.00	2.50 × 1.00 × 1.00	2.50 × 1.00 × 1.00	2.50 × 1.00 × 1.00	2.50 × 1.00 × 1.00
Layer 3	2.00 × 1.00 × 0.30	2.00 × 1.00 × 0.50	2.00 × 1.00 × 0.80	2.50 × 1.00 × 0.80	2.50 × 1.00 × 1.00	2.50 × 1.00 × 1.00
Layer 4						2.00 × 1.00 × 0.30

(Continued)

TABLE 5.3 (Continued)

Catchment Area, ha	50.00	100.00	150.00	200.00	250.00	300.00
Pressure at Toe, t/m^2	5.25	5.87	6.82	6.95	7.64	8.59
Pressure at Heel, t/m^2	0.18	0.39	0.27	0.51	0.46	0.28
Factor of Safety against Overturning	2.31	2.41	2.45	2.46	2.47	2.48
Shear Friction Factor F	6.55	5.55	4.99	4.53	4.22	4.00
Total Volume of Gabion Boxes, m^3	15.30	18.00	19.80	21.00	22.50	24.30

TABLE 5.4
Design Results for Gabion Check Dam with Vertical D/S Face Height = 1.50 m and Stream Width L = 20.0 m

Catchment Area, ha	50.00	100.00	150.00	200.00	250.00	300.00
A. WEIR						
Span of Weir W, m	16.00	16.00	16.00	16.00	16.00	16.00
Flood Discharge Q, m^3/s	6.54	11.00	14.91	18.50	21.87	25.07
Head Over Crest, m	0.37	0.52	0.64	0.73	0.82	0.90
Top Level, m	100.00	100.00	100.00	100.00	100.00	100.00
Foundation Level, m	98.50	98.50	98.50	98.50	98.50	98.50
Total Height of Weir H_{WEIR}, m	1.50	1.50	1.50	1.50	1.50	1.50
Bottom Width of Weir B_{WEIR}, m	2.50	2.50	2.50	2.50	2.50	2.50
Size of Gabion Box (L×B×H), m						
Bottom Layer 1	2.50 × 1.00 × 1.00	2.50 × 1.00 × 1.00	2.50 × 1.00 × 1.00	2.50 × 1.00 × 1.00	2.50 × 1.00 × 1.00	2.50 × 1.00 × 1.00
Layer 2	2.50 × 1.00 × 0.50	2.50 × 1.00 × 0.50	2.50 × 1.00 × 0.50	2.50 × 1.00 × 0.50	2.50 × 1.00 × 0.50	2.50 × 1.00 × 0.50
Pressure at Toe, t/m^2	3.47	3.68	3.84	3.98	4.11	4.22
Pressure at Heel, t/m^2	0.32	0.31	0.30	0.29	0.29	0.28
Factor of Safety against Overturning	2.04	2.05	2.05	2.05	2.06	2.06
Shear Friction Factor F	7.28	6.46	5.97	5.59	5.31	5.07
Total Volume of Gabion Boxes, m^3	60.00	60.00	60.00	60.00	60.00	60.00
B. NON-OVER FLOW PORTION						
Length, m	4.00	4.00	4.00	4.00	4.00	4.00
Top Level including Free Board, m	100.80	101.00	101.30	101.30	101.30	101.30
Foundation Level, m	98.50	98.50	98.50	98.50	98.50	98.50
Total Height of Non-Over Flow H_{NOF}, m	2.30	2.50	2.80	2.80	2.80	2.80

(Continued)

TABLE 5.4 (Continued)

Catchment Area, ha	50.00	100.00	150.00	200.00	250.00	300.00
Bottom Width of Non-Over Flow B_{NOF}, m	2.50	2.50	2.50	2.50	2.50	2.50
Size of Gabion Box (L×B×H), m						
Bottom Layer 1	2.50 × 1.00 × 1.00	2.50 × 1.00 × 1.00	2.50 × 1.00 × 1.00	2.50 × 1.00 × 1.00	2.50 × 1.00 × 1.00	2.50 × 1.00 × 1.00
Layer 2	2.00 × 1.00 × 1.00	2.00 × 1.00 × 1.00	2.00 × 1.00 × 1.00	2.50 × 1.00 × 1.00	2.50 × 1.00 × 1.00	2.50 × 1.00 × 1.00
Layer 3	1.50 × 1.00 × 0.30	2.00 × 1.00 × 0.50	2.00 × 1.00 × 0.80	2.00 × 1.00 × 0.80	2.00 × 1.00 × 0.80	2.50 × 1.00 × 0.80
Pressure at Toe, t/m²	5.15	5.81	6.68	6.74	6.96	6.97
Pressure at Heel, t/m²	0.01	0.16	0.11	0.34	0.15	0.49
Factor of Safety against Overturning	2.25	2.38	2.46	2.49	2.39	2.45
Shear Friction Factor F	7.09	6.15	5.63	5.16	4.75	4.49
Total Volume of Gabion Boxes, m³	19.80	22.00	24.40	26.40	26.40	28.00

5.2.2 Section with Vertical U/S Face

The tables developed after the rigorous design of the different components for 1.50 m high gabion check dams for different combinations are presented from Table 5.5 to Table 5.8.

TABLE 5.5
Design Results for Gabion Check Dam with Vertical U/S Face Height = 1.50 m and Stream Width L = 5.0 m

Catchment Area, ha	50.00	100.00
A. WEIR		
Span of Weir W, m	4.00	4.00
Flood Discharge Q, m³/s	6.54	11.00
Head Over Crest, m	0.92	1.31
Top Level, m	100.00	100.00
Foundation Level, m	98.50	98.50
Total Height of Weir H_{WEIR}, m	1.50	1.50
Bottom Width of Weir B_{WEIR}, m	2.50	2.50
Size of Gabion Box (L×B×H), m		
Bottom Layer 1	2.50 × 1.00 × 1.00	2.50 × 1.00 × 1.00
Layer 2	1.50 × 1.00 × 0.50	1.50 × 1.00 × 0.50
Pressure at Toe, t/m²	2.87	3.20
Pressure at Heel, t/m²	0.53	0.50
Factor of Safety against Overturning	1.86	1.85

(Continued)

TABLE 5.5 (Continued)

Catchment Area, ha	50.00	100.00
Shear Friction Factor F	4.78	3.95
Total Volume of Gabion Boxes, m^3	13.00	13.00
B. NON-OVER FLOW PORTION		
Width, m	1.00	1.00
Top Level including Free Board, m	101.50	101.80
Foundation Level, m	98.50	98.50
Total Height of Non-Over Flow H_{NOF}, m	3.00	3.30
Bottom Width of Non-Over Flow B_{NOF}, m	2.00	2.50
Size of Gabion Box (L×B×H), m		
Bottom Layer 1	2.00 × 1.00 × 1.00	2.50 × 1.00 × 1.00
Layer 2	1.00 × 1.00 × 1.00	1.50 × 1.00 × 1.00
Layer 3	1.00 × 1.00 × 1.00	1.00 × 1.00 × 1.00
Layer 4		1.00 × 1.00 × 0.30
Pressure at Toe, t/m^2	4.82	4.85
Pressure at Heel, t/m^2	0.08	0.43
Factor of Safety against Overturning	1.71	1.80
Shear Friction Factor F	3.20	2.97
Total Volume of Gabion Boxes, m^3	4.00	5.30

TABLE 5.6
Design Results for Gabion Check Dam with Vertical U/S Face Height = 1.50 m and Stream Width L = 10.0 m

Catchment Area, ha	50.00	100.00	150.00	200.00
A. WEIR				
Span of Weir W, m	8.00	8.00	8.00	8.00
Flood Discharge Q, m^3/s	6.54	11.00	14.91	18.50
Head Over Crest, m	0.58	0.82	1.01	1.16
Top Level, m	100.00	100.00	100.00	100.00
Foundation Level, m	98.50	98.50	98.50	98.50
Total Height of Weir H_{WEIR}, m	1.50	1.50	1.50	1.50
Bottom Width of Weir B_{WEIR}, m	2.50	2.50	2.50	2.50
Size of Gabion Box (L×B×H), m				
Bottom Layer 1	2.50 × 1.00 × 1.00	2.50 × 1.00 × 1.00	2.50 × 1.00 × 1.00	2.50 × 1.00 × 1.00
Layer 2	1.50 × 1.00 × 0.50	1.50 × 1.00 × 0.50	1.50 × 1.00 × 0.50	1.50 × 1.00 × 0.50
Pressure at Toe, t/m^2	2.56	2.78	2.94	3.08

(Continued)

TABLE 5.6 (Continued)

Catchment Area, ha	50.00	100.00	150.00	200.00
Pressure at Heel, t/m²	0.56	0.54	0.53	0.51
Factor of Safety against Overturning	1.88	1.87	1.86	1.85
Shear Friction Factor F	5.96	5.07	4.58	4.22
Total Volume of Gabion Boxes, m³	26.00	26.00	26.00	26.00
B. NON-OVER FLOW PORTION				
Width, m	2.00	2.00	2.00	2.00
Top Level including Free Board, m	101.00	101.30	101.50	101.80
Foundation Level, m	98.50	98.50	98.50	98.50
Total Height of Non-Over Flow H_{NOF}, m	2.50	2.80	3.00	3.30
Bottom Width of Non-Over Flow B_{NOF}, m	2.00	2.00	2.50	2.50
Size of Gabion Box (L×B×H), m				
Bottom Layer 1	2.00 × 1.00 × 1.00	2.00 × 1.00 × 1.00	2.50 × 1.00 × 1.00	2.50 × 1.00 × 1.00
Layer 2	1.00 × 1.00 × 1.00	1.00 × 1.00 × 1.00	1.50 × 1.00 × 1.00	1.50 × 1.00 × 1.00
Layer 3	1.00 × 1.00 × 0.50	1.00 × 1.00 × 0.80	1.00 × 1.00 × 1.00	1.00 × 1.00 × 1.00
Layer 4				0.50 × 1.00 × 0.30
Pressure at Toe, t/m²	3.48	4.37	3.77	4.25
Pressure at Heel, t/m²	0.62	0.21	1.13	0.84
Factor of Safety against Overturning	1.86	1.74	2.03	1.92
Shear Friction Factor F	4.31	3.47	3.72	3.30
Total Volume of Gabion Boxes, m³	7.00	7.60	10.00	10.30

TABLE 5.7
Design Results for Gabion Check Dam with Vertical U/S Face Height = 1.50 m and Stream Width L = 15.0 m

Catchment Area, ha	50.00	100.00	150.00	200.00	250.00	300.00
A. WEIR						
Span of Weir W, m	12.00	12.00	12.00	12.00	12.00	12.00
Flood Discharge Q, m³/s	6.54	11.00	14.91	18.50	21.87	25.07
Head Over Crest, m	0.44	0.63	0.77	0.89	0.99	1.09
Top Level, m	100.00	100.00	100.00	100.00	100.00	100.00
Foundation Level, m	98.50	98.50	98.50	98.50	98.50	98.50
Total Height of Weir H_{WEIR}, m	1.50	1.50	1.50	1.50	1.50	1.50

(Continued)

TABLE 5.7 (Continued)

Catchment Area, ha	50.00	100.00	150.00	200.00	250.00	300.00
Bottom Width of Weir B_{WEIR}, m	2.00	2.50	2.50	2.50	2.50	2.50
Size of Gabion Box (L×B×H), m						
Bottom Layer 1	2.00 × 1.00 × 1.00	2.50 × 1.00 × 1.00	2.50 × 1.00 × 1.00	2.50 × 1.00 × 1.00	2.50 × 1.00 × 1.00	2.50 × 1.00 × 1.00
Layer 2	1.50 × 1.00 × 0.50	1.50 × 1.00 × 0.50	1.50 × 1.00 × 0.50	1.50 × 1.00 × 0.50	1.50 × 1.00 × 0.50	1.50 × 1.00 × 0.50
Pressure at Toe, t/m²	3.32	2.60	2.73	2.83	2.93	3.01
Pressure at Heel, t/m²	0.02	0.56	0.55	0.54	0.53	0.52
Factor of Safety against Overturning	1.73	1.88	1.87	1.87	1.86	1.86
Shear Friction Factor F	5.34	5.77	5.26	4.89	4.61	4.39
Total Volume of Gabion Boxes, m³	33.00	39.00	39.00	39.00	39.00	39.00
B. NON-OVER FLOW PORTION						
Width, m	3.00	3.00	3.00	3.00	3.00	3.00
Top Level including Free Board, m	100.80	101.00	101.30	101.30	101.50	101.80
Foundation Level, m	98.50	98.50	98.50	98.50	98.50	98.50
Total Height of Non-Over Flow H_{NOF}, m	2.30	2.50	2.80	2.80	3.00	3.30
Bottom Width of Non-Over Flow B_{NOF}, m	2.00	2.00	2.00	2.00	2.50	2.50
Size of Gabion Box (L×B×H), m						
Bottom Layer 1	2.00 × 1.00 × 1.00	2.00 × 1.00 × 1.00	2.00 × 1.00 × 1.00	2.00 × 1.00 × 1.00	2.50 × 1.00 × 1.00	2.50 × 1.00 × 1.00
Layer 2	1.00 × 1.00 × 1.00	1.00 × 1.00 × 1.00	1.00 × 1.00 × 1.00	1.50 × 1.00 × 1.00	1.50 × 1.00 × 1.00	1.50 × 1.00 × 1.00
Layer 3	1.00 × 1.00 × 0.30	1.00 × 1.00 × 0.50	1.00 × 1.00 × 0.80	1.50 × 1.00 × 0.80	1.00 × 1.00 × 1.00	1.00 × 1.00 × 1.00
Layer 4						1.00 × 1.00 × 0.30
Pressure at Toe, t/m²	3.05	3.65	4.12	5.99	3.73	3.92
Pressure at Heel, t/m²	0.73	0.45	0.46	0.03	1.17	1.37
Factor of Safety against Overturning	1.91	1.81	1.81	1.88	2.04	2.08
Shear Friction Factor F	4.92	4.10	3.64	3.45	3.77	3.53
Total Volume of Gabion Boxes, m³	9.90	10.50	11.40	14.10	15.00	15.90

TABLE 5.8
Design Results for Gabion Check Dam with Vertical U/S Face Height = 1.50 m and Stream Width L = 20.0 m

Catchment Area, ha	50.00	100.00	150.00	200.00	250.00	300.00
A. WEIR						
Span of Weir W, m	16.00	16.00	16.00	16.00	16.00	16.00
Flood Discharge Q, m^3/s	6.54	11.00	14.91	18.50	21.87	25.07
Head Over Crest, m	0.37	0.52	0.64	0.73	0.82	0.90
Top Level, m	100.00	100.00	100.00	100.00	100.00	100.00
Foundation Level, m	98.50	98.50	98.50	98.50	98.50	98.50
Total Height of Weir H_{WEIR}, m	1.50	1.50	1.50	1.50	1.50	1.50
Bottom Width of Weir B_{WEIR}, m	2.00	2.50	2.50	2.50	2.50	2.50
Size of Gabion Box (L×B×H), m						
Bottom Layer 1	2.00 × 1.00 × 1.00	2.50 × 1.00 × 1.00	2.50 × 1.00 × 1.00	2.50 × 1.00 × 1.00	2.50 × 1.00 × 1.00	2.50 × 1.00 × 1.00
Layer 2	1.50 × 1.00 × 0.50	1.50 × 1.00 × 0.50	1.50 × 1.00 × 0.50	1.50 × 1.00 × 0.50	1.50 × 1.00 × 0.50	1.50 × 1.00 × 0.50
Pressure at Toe, t/m^2	3.19	2.51	2.61	2.70	2.77	2.84
Pressure at Heel, t/m^2	0.07	0.57	0.56	0.55	0.54	0.54
Factor of Safety against Overturning	1.75	1.89	1.88	1.88	1.87	1.87
Shear Friction Factor F	5.71	6.23	5.74	5.37	5.09	4.86
Total Volume of Gabion Boxes, m^3	44.00	52.00	52.00	52.00	52.00	52.00
B. NON-OVER FLOW PORTION						
Width, m	4.00	4.00	4.00	4.00	4.00	4.00
Top Level including Free Board, m	100.80	101.00	101.30	101.30	101.30	101.30
Foundation Level, m	98.50	98.50	98.50	98.50	98.50	98.50
Total Height of Non-Over Flow H_{NOF}, m	2.30	2.50	2.80	2.80	2.80	2.80
Bottom Width of Non-Over Flow B_{NOF}, m	2.00	2.00	2.00	2.00	2.00	2.00
Size of Gabion Box (L×B×H), m						
Bottom Layer 1	2.00 × 1.00 × 1.00	2.00 × 1.00 × 1.00	2.00 × 1.00 × 1.00	2.00 × 1.00 × 1.00	2.00 × 1.00 × 1.00	2.00 × 1.00 × 1.00
Layer 2	1.00 × 1.00 × 1.00	1.00 × 1.00 × 1.00	1.00 × 1.00 × 1.00	1.00 × 1.00 × 1.00	1.00 × 1.00 × 1.00	1.50 × 1.00 × 1.00
Layer 3	1.00 × 1.00 × 0.30	1.00 × 1.00 × 0.50	1.00 × 1.00 × 0.80	1.00 × 1.00 × 0.80	1.00 × 1.00 × 0.80	1.50 × 1.00 × 0.80
Pressure at Toe, t/m^2	2.83	3.24	3.58	3.98	4.37	6.02
Pressure at Heel, t/m^2	0.95	0.86	1.00	0.60	0.21	0.00
Factor of Safety against Overturning	2.00	1.95	1.98	1.85	1.74	1.87
Shear Friction Factor F	5.38	4.59	4.15	3.77	3.47	3.41
Total Volume of Gabion Boxes, m^3	13.20	14.00	15.20	15.20	15.20	18.80

5.3 DESIGN OF 2.00 M HIGH GABION CHECK DAMS

5.3.1 Section with Vertical D/S Face

The tables developed after the rigorous design of the different components for 2.00 m high gabion check dams for different combinations are presented from Table 5.9 to Table 5.12.

TABLE 5.9
Design Results for Gabion Check Dam with Vertical D/S Face Height = 2.00 m and Stream Width L = 5.0 m

Catchment Area, ha	50.00	100.00
A. WEIR		
Span of Weir W, m	4.00	4.00
Flood Discharge Q, m^3/s	6.54	11.00
Head Over Crest, m	0.92	1.31
Top Level, m	100.00	100.00
Foundation Level, m	98.00	98.00
Total Height of Weir H_{WEIR}, m	2.00	2.00
Bottom Width of Weir B_{WEIR}, m	3.00	3.50
Size of Gabion Box (L×B×H), m		
Bottom Layer 1	3.00 × 1.00 × 1.00	3.50 × 1.00 × 1.00
Layer 2	3.00 × 1.00 × 1.00	3.50 × 1.00 × 1.00
Pressure at Toe, t/m^2	5.63	5.57
Pressure at Heel, t/m^2	0.00	0.57
Factor of Safety against Overturning	1.92	2.13
Shear Friction Factor F	4.09	4.06
Total Volume of Gabion Boxes, m^3	24.00	28.00
Length of D/S Apron Floor $L_{D/S\ APRON}$, m	5.00	4.50
B. NON-OVER FLOW PORTION		
Length, m	1.00	1.00
Top Level including Free Board, m	101.50	101.80
Foundation Level, m	98.00	98.00
Total Height of Non-Over Flow H_{NOF}, m	3.50	3.80
Bottom Width of Non-Over Flow B_{NOF}, m	3.00	4.00
Size of Gabion Box (L×B×H), m		
Bottom Layer 1	3.00 × 1.00 × 1.00	4.00 × 1.00 × 1.00
Layer 2	3.00 × 1.00 × 1.00	3.50 × 1.00 × 1.00
Layer 3	3.00 × 1.00 × 1.00	3.50 × 1.00 × 1.00
Layer 4	2.50 × 1.00 × 0.50	2.50 × 1.00 × 0.80
Pressure at Toe, t/m^2	8.88	9.17

(Continued)

TABLE 5.9 (Continued)

Catchment Area, ha	**50.00**	**100.00**
Pressure at Heel, t/m^2	0.06	0.07
Factor of Safety against Overturning	2.27	2.42
Shear Friction Factor F	3.86	3.99
Total Volume of Gabion Boxes, m^3	10.25	13.00
C. WING WALLS (ABUTMENTS) ON BOTH BANKS		
Length in U/S $L_{U/S\ ABUT}$, m	3.00	3.50
Length in D/S $L_{D/S\ ABUT}$, m	2.00	2.00
Top Level including Free Board, m	101.50	101.80
Foundation Level, m	98.00	98.00
Total Height H_{ABUT}, m	3.50	3.80
Base Width of Abutment B_{ABUT}, m	2.00	2.50
Size of Gabion Box (L×B×H), m		
Bottom Layer 1	2.00 × 1.00 × 1	2.50 × 1.00 × 1.00
Layer 2	1.00 × 1.00 × 1.00	2.00 × 1.00 × 1.00
Layer 3	1.00× 1.00 × 1.00	1.00 × 1.00 × 1.00
Layer 4	1.00 × 1.00 × 0.50	1.00 × 1.00 × 0.80
Pressure at Toe, t/m^2	10.88	10.41
Pressure at Heel, t/m^2	0.82	2.27
Factor of Safety against Overturning	3.32	4.38
Shear Friction Factor F	4.17	4.55
Total Volume of Gabion Boxes, m^3	45.00	69.30
D. D/S WING WALLS ON BOTH BANKS		
Length in D/S $L_{D/S\ WING}$, m	3.00	2.50
Top Level including Free Board, m	100.00	100.30
Foundation Level, m	98.00	98.00
Total Height $H_{D/S\ WING}$, m	2.00	2.30
Base Width of D/S Wing Wall $B_{D/S\ WING}$, m	1.50	1.50
Size of Gabion Box (L×B×H), m		
Bottom Layer 1	1.50 × 1.00 × 1.00	1.50 × 1.00 × 1.00
Layer 2	1.00 × 1.00 × 1.00	1.00 × 1.00 × 1.00
Layer 3		1.00 × 1.00 × 0.30
Pressure at Toe, t/m^2	4.94	6.32
Pressure at Heel, t/m^2	1.60	1.21
Factor of Safety against Overturning	5.50	4.20
Shear Friction Factor F	8.05	6.33
Total Volume of Gabion Boxes, m^3	15.00	14.00

TABLE 5.10
Design Results for Gabion Check Dam with Vertical D/S Face Height = 2.00 m and Stream Width L = 10.0 m

Catchment Area, ha	50.00	100.00	150.00	200.00
A. WEIR				
Span of Weir W, m	8.00	8.00	8.00	8.00
Flood Discharge Q, m^3/s	6.54	11.00	14.91	18.50
Head Over Crest, m	0.58	0.82	1.01	1.16
Top Level, m	100.00	100.00	100.00	100.00
Foundation Level, m	98.00	98.00	98.00	98.00
Total Height of Weir H_{WEIR}, m	2.00	2.00	2.00	2.00
Bottom Width of Weir B_{WEIR}, m	3.00	3.00	3.50	3.50
Size of Gabion Box (L×B×H), m				
Bottom Layer 1	3.00 × 1.00 × 1.00	3.00 × 1.00 × 1.00	3.50 × 1.00 × 1.00	3.50 × 1.00 × 1.00
Layer 2	3.00 × 1.00 × 1.00	3.00 × 1.00 × 1.00	3.50 × 1.00 × 1.00	3.50 × 1.00 × 1.00
Pressure at Toe, t/m^2	5.06	5.47	5.18	5.39
Pressure at Heel, t/m^2	0.12	0.03	0.56	0.57
Factor of Safety against Overturning	1.93	1.92	2.12	2.12
Shear Friction Factor F	4.90	4.29	4.59	4.29
Total Volume of Gabion Boxes, m^3	48.00	48.00	56.00	56.00
Length of D/S Apron Floor $L_{D/S\ APRON}$, m	5.00	5.00	4.50	4.50
B. NON-OVER FLOW PORTION				
Length, m	2.00	2.00	2.00	2.00
Top Level including Free Board, m	101.00	101.30	101.50	101.80
Foundation Level, m	98.00	98.00	98.00	98.00
Total Height of Non-Over Flow H_{NOF}, m	3.00	3.30	3.50	3.80
Bottom Width of Non-Over Flow B_{NOF}, m	3.00	3.50	3.50	4.00
Size of Gabion Box (L×B×H), m				
Bottom Layer 1	3.00 × 1.00 × 1.00	3.50 × 1.00 × 1.00	3.50 × 1.00 × 1.00	4.00 × 1.00 × 1.00
Layer 2	3.00 × 1.00 × 1.00	3.00 × 1.00 × 1.00	3.50 × 1.00 × 1.00	3.50 × 1.00 × 1.00
Layer 3	2.50 × 1.00 × 1.00	3.00 × 1.00 × 1.00	3.00 × 1.00 × 1.00	3.00 × 1.00 × 1.00
Layer 4		2.50 × 1.00 × 0.30	2.00 × 1.00 × 0.50	2.50 × 1.00 × 0.80
Pressure at Toe, t/m^2	7.09	7.75	8.43	8.95
Pressure at Heel, t/m^2	0.19	0.17	0.02	0.02
Factor of Safety against Overturning	2.23	2.34	2.31	2.44
Shear Friction Factor F	4.77	4.69	4.16	4.35

(Continued)

TABLE 5.10 (Continued)

Catchment Area, ha	50.00	100.00	150.00	200.00
Total Volume of Gabion Boxes, m^3	17.00	20.50	22.00	25.00
C. WING WALLS (ABUTMENTS) ON BOTH BANKS				
Length in U/S $L_{U/S\ ABUT}$, m	3.00	3.00	3.50	3.50
Length in D/S $L_{D/S\ ABUT}$, m	2.00	2.00	2.00	2.00
Top Level including Free Board, m	101.00	101.30	101.50	101.80
Foundation Level, m	98.00	98.00	98.00	98.00
Total Height H_{ABUT}, m	3.00	3.30	3.50	3.80
Base Width of Abutment B_{ABUT}, m	2.00	2.00	2.00	2.50
Size of Gabion Box (L×B×H), m				
Bottom Layer 1	2.00 × 1.00 × 1.00	2.00 × 1.00 × 1.00	2.00 × 1.00 × 1.00	2.50 × 1.00 × 1.00
Layer 2	1.00 × 1.00 × 1.00	1.00 × 1.00 × 1.00	1.00 × 1.00 × 1.00	2.00 × 1.00 × 1.00
Layer 3	1.00 × 1.00 × 1.00	1.00 × 1.00 × 1.00	1.00 × 1.00 × 1.00	1.00 × 1.00 × 1.00
Layer 4		1.00 × 1.00 × 0.30	1.00 × 1.00 × 0.50	1.00 × 1.00 × 0.80
Pressure at Toe, t/m^2	8.09	9.69	10.88	10.41
Pressure at Heel, t/m^2	1.91	1.33	0.82	2.27
Factor of Safety against Overturning	4.51	3.73	3.32	4.38
Shear Friction Factor F	5.39	4.59	4.17	4.55
Total Volume of Gabion Boxes, m^3	40.00	43.00	49.50	69.30
D. D/S WING WALLS ON BOTH BANKS				
Length in D/S $L_{D/S\ WING}$, m	3.00	3.00	2.50	2.50
Top Level including Free Board, m	99.50	99.80	100.00	100.30
Foundation Level, m	98.00	98.00	98.00	98.00
Total Height $H_{D/S\ WING}$, m	1.50	1.80	2.00	2.30
Base Width of D/S Wing Wall $B_{D/S\ WING}$, m	1.00	1.50	1.50	1.50
Size of Gabion Box (L×B×H), m				
Bottom Layer 1	1.00 × 1.00 × 1.00	1.50 × 1.00 × 1.00	1.50 × 1.00 × 1.00	1.50 × 1.00 × 1.00
Layer 2	1.00 × 1.00 × 0.50	1.00 × 1.00 × 0.80	1.00 × 1.00 × 1.00	1.00 × 1.00 × 1.00
Layer 3				1.00 × 1.00 × 0.30
Pressure at Toe, t/m^2	4.11	4.12	4.94	6.32
Pressure at Heel, t/m^2	0.69	1.74	1.60	1.21
Factor of Safety against Overturning	4.21	6.87	5.50	4.20
Shear Friction Factor F	8.93	9.81	8.05	6.33
Total Volume of Gabion Boxes, m^3	9.00	13.80	12.50	14.00

TABLE 5.11
Design Results for Gabion Check Dam with Vertical D/S Face Height = 2.00 m and Stream Width L = 15.0 m

Catchment Area, ha	50.00	100.00	150.00	200.00	250.00	300.00
A. WEIR						
Span of Weir W, m	12.00	12.00	12.00	12.00	12.00	12.00
Flood Discharge Q, m^3/s	6.54	11.00	14.91	18.50	21.87	25.07
Head Over Crest, m	0.44	0.63	0.77	0.89	0.99	1.09
Top Level, m	100.00	100.00	100.00	100.00	100.00	100.00
Foundation Level, m	98.00	98.00	98.00	98.00	98.00	98.00
Total Height of Weir H_{WEIR}, m	2.00	2.00	2.00	2.00	2.00	2.00
Bottom Width of Weir B_{WEIR}, m	3.00	3.00	3.00	3.00	3.50	3.50
Size of Gabion Box (L×B×H), m						
Bottom Layer 1	3.00 × 1.00 × 1.00	3.00 × 1.00 × 1.00	3.00 × 1.00 × 1.00	3.00 × 1.00 × 1.00	3.50 × 1.00 × 1.00	3.50 × 1.00 × 1.00
Layer 2	3.00 × 1.00 × 1.00	3.00 × 1.00 × 1.00	3.00 × 1.00 × 1.00	3.00 × 1.00 × 1.00	3.50 × 1.00 × 1.00	3.50 × 1.00 × 1.00
Pressure at Toe, t/m^2	4.83	5.14	5.38	5.58	5.16	5.29
Pressure at Heel, t/m^2	0.16	0.10	0.05	0.01	0.56	0.56
Factor of Safety against Overturning	1.94	1.93	1.92	1.92	2.12	2.12
Shear Friction Factor F	5.32	4.76	4.42	4.16	4.62	4.43
Total Volume of Gabion Boxes, m^3	72.00	72.00	72.00	72.00	84.00	84.00
Length of D/S Apron Floor $L_{D/S\ APRON}$, m	5.00	5.00	5.00	5.00	4.50	4.50
B. NON-OVER FLOW PORTION						
Length, m	3.00	3.00	3.00	3.00	3.00	3.00
Top Level including Free Board, m	100.80	101.00	101.30	101.30	101.50	101.80
Foundation Level, m	98.00	98.00	98.00	98.00	98.00	98.00

(Continued)

TABLE 5.11 (Continued)

Catchment Area, ha	50.00	100.00	150.00	200.00	250.00	300.00
Total Height of Non-Over Flow H_{NOF}, m	2.80	3.00	3.30	3.30	3.50	3.80
Bottom Width of Non-Over Flow B_{NOF}, m	3.00	3.00	3.00	3.00	3.50	3.50
Size of Gabion Box (L×B×H), m						
Bottom Layer 1	3.00 × 1.00 × 1.00	3.00 × 1.00 × 1.00	3.00 × 1.00 × 1.00	3.00 × 1.00 × 1.00	3.50 × 1.00 × 1.00	3.50 × 1.00 × 1.00
Layer 2	3.00 × 1.00 × 1.00	3.00 × 1.00 × 1.00	3.00 × 1.00 × 1.00	3.00 × 1.00 × 1.00	3.50 × 1.00 × 1.00	3.50 × 1.00 × 1.00
Layer 3	2.50 × 1.00 × 0.80	2.50 × 1.00 × 1.00	3.00 × 1.00 × 1.00	3.00 × 1.00 × 1.00	3.00 × 1.00 × 1.00	3.00 × 1.00 × 1.00
Layer 4			2.00 × 1.00 × 0.30	2.50 × 1.00 × 0.30	2.00 × 1.00 × 0.50	2.50 × 1.00 × 0.80
Pressure at Toe, t/m²	6.40	7.21	8.02	8.37	8.38	9.27
Pressure at Heel, t/m²	0.30	0.09	0.22	0.03	0.05	0.11
Factor of Safety against Overturning	2.22	2.19	2.29	2.21	2.32	2.42
Shear Friction Factor F	5.26	4.59	4.24	3.89	4.20	4.06
Total Volume of Gabion Boxes, m³	24.00	25.50	28.80	29.25	33.00	36.00
C. WING WALLS (ABUTMENTS) ON BOTH BANKS						
Length in U/S $L_{U/S\ ABUT}$, m	3.00	3.00	3.00	3.00	3.50	3.50
Length in D/S $L_{D/S\ ABUT}$, m	2.00	2.00	2.00	2.00	2.00	2.00
Top Level including Free Board, m	100.80	101.00	101.30	101.30	101.50	101.80
Foundation Level, m	98.00	98.00	98.00	98.00	98.00	98.00
Total Height H_{ABUT}, m	2.80	3.00	3.30	3.30	3.50	3.80
Base Width of Abutment B_{ABUT}, m	2.00	2.00	2.00	2.00	2.00	2.50

(Continued)

TABLE 5.11 (Continued)

Catchment Area, ha	**50.00**	**100.00**	**150.00**	**200.00**	**250.00**	**300.00**
Size of Gabion Box (L×B×H), m						
Bottom Layer 1	2.00 × 1.00 × 1.00	2.00 × 1.00 × 1.00	2.00 × 1.00 × 1.00	2.00 × 1.00 × 1.00	2.00 × 1.00 × 1.00	2.50 × 1.00 × 1.00
Layer 2	1.00 × 1.00 × 1.00	1.00 × 1.00 × 1.00	1.00 × 1.00 × 1.00	1.00 × 1.00 × 1.00	1.00 × 1.00 × 1.00	2.00 × 1.00 × 1.00
Layer 3	1.00 × 1.00 × 0.80	1.00 × 1.00 × 1.00	1.00 × 1.00 × 1.00	1.00 × 1.00 × 1.00	1.00 × 1.00 × 1.00	1.00 × 1.00 × 1.00
Layer 4			1.00 × 1.00 × 0.30	1.00 × 1.00 × 0.30	1.00 × 1.00 × 0.50	1.00 × 1.00 × 0.80
Pressure at Toe, t/m²	7.14	8.09	9.69	9.69	10.88	10.41
Pressure at Heel, t/m²	2.18	1.91	1.33	1.33	0.82	2.27
Factor of Safety against Overturning	5.19	4.51	3.73	3.73	3.32	4.38
Shear Friction Factor F	6.05	5.39	4.59	4.59	4.17	4.55
Total Volume of Gabion Boxes, m³	38.00	40.00	43.00	43.00	49.50	69.30
D. D/S WING WALLS ON BOTH BANKS						
Length in D/S $L_{D/S\ WING}$, m	3.00	3.00	3.00	3.00	2.50	2.50
Top Level including Free Board, m	99.30	99.50	99.80	99.80	100.00	100.30
Foundation Level, m	98.00	98.00	98.00	98.00	98.00	98.00
Total Height $H_{D/S\ WING}$, m	1.30	1.50	1.80	1.80	2.00	2.30
Base Width of D/S Wing Wall $B_{D/S\ WING}$, m	1.00	1.00	1.50	1.50	1.50	1.50
Size of Gabion Box (L×B×H), m						
Bottom Layer 1	1.00 × 1.00 × 1.00	1.00 × 1.00 × 1.00	1.50 × 1.00 × 1.00	1.50 × 1.00 × 1.00	1.50 × 1.00 × 1.00	1.50 × 1.00 × 1.00
Layer 2	1.00 × 1.00 ×0.30	1.00 × 1.00 ×0.50	1 × 1.0 × 0.8	1.00 × 1.00 ×0.80	1.00 × 1.00 × 1.00	1.00 × 1.00 × 1.00
Layer 3						1.00 × 1.00 ×0.30
Pressure at Toe, t/m²	3.16	4.11	4.12	4.12	4.94	6.32

(Continued)

TABLE 5.11 (Continued)

Catchment Area, ha	50.00	100.00	150.00	200.00	250.00	300.00
Pressure at Heel, t/m^2	1.00	0.69	1.74	1.74	1.60	1.21
Factor of Safety against Overturning	5.76	4.21	6.87	6.87	5.50	4.20
Shear Friction Factor F	11.83	8.93	9.81	9.81	8.05	6.33
Total Volume of Gabion Boxes, m^3	7.80	9.00	13.80	13.80	12.50	14.00

TABLE 5.12
Design Results for Gabion Check Dam with Vertical D/S Face Height = 2.00 m and Stream Width L = 20.0 m

Catchment Area, ha	50.00	100.00	150.00	200.00	250.00	300.00
A. WEIR						
Span of Weir W, m	16.00	16.00	16.00	16.00	16.00	16.00
Flood Discharge Q, m^3/s	6.54	11.00	14.91	18.50	21.87	25.07
Head Over Crest, m	0.37	0.52	0.64	0.73	0.82	0.90
Top Level, m	100.00	100.00	100.00	100.00	100.00	100.00
Foundation Level, m	98.00	98.00	98.00	98.00	98.00	98.00
Total Height of Weir H_{WEIR}, m	2.00	2.00	2.00	2.00	2.00	2.00
Bottom Width of Weir B_{WEIR}, m	3.00	3.00	3.00	3.00	3.00	3.00
Size of Gabion Box (L×B×H), m						
Bottom Layer 1	3.00 × 1.00 × 1.00	3.00 × 1.00 × 1.00	3.00 × 1.00 × 1.00	3.00 × 1.00 × 1.00	3.00 × 1.00 × 1.00	3.00 × 1.00 × 1.00
Layer 2	3.00 × 1.00 × 1.00	3.00 × 1.00 × 1.00	3.00 × 1.00 × 1.00	3.00 × 1.00 × 1.00	3.00 × 1.00 × 1.00	3.00 × 1.00 × 1.00
Pressure at Toe, t/m^2	4.70	4.96	5.15	5.32	5.46	5.59
Pressure at Heel, t/m^2	0.19	0.13	0.10	0.06	0.03	0.01

(Continued)

TABLE 5.12 (Continued)

Catchment Area, ha	**50.00**	**100.00**	**150.00**	**200.00**	**250.00**	**300.00**
Factor of Safety against Overturning	1.94	1.94	1.93	1.93	1.92	1.92
Shear Friction Factor F	5.61	5.08	4.75	4.50	4.30	4.14
Total Volume of Gabion Boxes, m^3	96.00	96.00	96.00	96.00	96.00	96.00
Length of D/S Apron Floor $L_{D/S\ APRON}$, m	5.00	5.00	5.00	5.00	5.00	5.00
B. NON-OVER FLOW PORTION						
Length, m	4.00	4.00	4.00	4.00	4.00	4.00
Top Level including Free Board, m	100.80	101.00	101.30	101.30	101.30	101.30
Foundation Level, m	98.00	98.00	98.00	98.00	98.00	98.00
Total Height of Non-Over Flow H_{NOF}, m	2.80	3.00	3.30	3.30	3.30	3.30
Bottom Width of Non-Over Flow B_{NOF}, m	3.00	3.00	3.00	3.00	3.00	3.00
Size of Gabion Box (L×B×H), m						
Bottom Layer 1	3.00 × 1.00 × 1.00	3.00 × 1.00 × 1.00	3.00 × 1.00 × 1.00	3.00 × 1.00 × 1.00	3.00 × 1.00 × 1.00	3.00 × 1.00 × 1.00
Layer 2	2.50 × 1.00 × 1.00	3.00 × 1.00 × 1.00	3.00 × 1.00 × 1.00	3.00 × 1.00 × 1.00	3.00 × 1.00 × 1.00	3.00 × 1.00 × 1.00
Layer 3	2.50 × 1.00 × 1.00	2.50 × 1.00 × 1.00	2.50 × 1.00 × 1.00	2.50 × 1.00 × 1.00	3.00 × 1.00 × 1.00	3.00 × 1.00 × 1.00
Layer 4			2.00 × 1.00 × 0.30	2.50 × 1.00 × 0.30	2.00 × 1.00 × 0.30	2.50 × 1.00 × 0.30
Pressure at Toe, t/m^2	6.39	6.96	7.86	8.08	8.18	8.39
Pressure at Heel, t/m^2	0.07	0.30	0.08	0.06	0.06	0.01
Factor of Safety against Overturning	2.18	2.28	2.30	2.26	2.23	2.20

(Continued)

TABLE 5.12 (Continued)

Catchment Area, ha	50.00	100.00	150.00	200.00	250.00	300.00
Shear Friction Factor F	5.60	5.03	4.67	4.34	4.07	3.86
Total Volume of Gabion Boxes, m^3	30.00	34.00	36.40	37.00	38.40	39.00
C. WING WALLS (ABUTMENTS) ON BOTH BANKS						
Length in U/S $L_{U/S\ ABUT}$, m	3.00	3.00	3.00	3.00	3.00	3.00
Length in D/S $L_{D/S\ ABUT}$, m	2.00	2.00	2.00	2.00	2.00	2.00
Top Level including Free Board, m	100.80	101.00	101.30	101.30	101.30	101.30
Foundation Level, m	98.00	98.00	98.00	98.00	98.00	98.00
Total Height H_{ABUT}, m	2.80	3.00	3.30	3.30	3.30	3.30
Base Width of Abutment B_{ABUT}, m	2.00	2.00	2.00	2.00	2.00	2.00
Size of Gabion Box (L×B×H), m						
Bottom Layer 1	2.00 × 1.00 × 1.00	2.00 × 1.00 × 1.00	2.00 × 1.00 × 1.00	2.00 × 1.00 × 1.00	2.00 × 1.00 × 1.00	2.00 × 1.00 × 1.00
Layer 2	1.00 × 1.00 × 1.00	1.00 × 1.00 × 1.00	1.00 × 1.00 × 1.00	1.00 × 1.00 × 1.00	1.00 × 1.00 × 1.00	1.00 × 1.00 × 1.00
Layer 3	1.00 × 1.00 × 0.80	1.00 × 1.00 × 1.00	1.00 × 1.00 × 1.00	1.00 × 1.00 × 1.00	1.00 × 1.00 × 1.00	1.00 × 1.00 × 1.00
Layer 4			1.00 × 1.00 × 0.30	1.00 × 1.00 × 0.30	1.00 × 1.00 × 0.30	1.00 × 1.00 × 0.30
Pressure at Toe, t/m^2	7.14	8.09	9.69	9.69	9.69	9.69
Pressure at Heel, t/m^2	2.18	1.91	1.33	1.33	1.33	1.33
Factor of Safety against Overturning	5.19	4.51	3.73	3.73	3.73	3.73
Shear Friction Factor F	6.05	5.39	4.59	4.59	4.59	4.59
Total Volume of Gabion Boxes, m^3	38.00	40.00	43.00	43.00	43.00	43.00

(Continued)

TABLE 5.12 (Continued)

Catchment Area, ha	50.00	100.00	150.00	200.00	250.00	300.00
D. D/S WING WALLS ON BOTH BANKS						
Length in D/S $L_{D/S\ WING}$, m	3.00	3.00	3.00	3.00	3.00	3.00
Top Level including Free Board, m	99.30	99.50	99.80	99.80	99.80	99.80
Foundation Level, m	98.00	98.00	98.00	98.00	98.00	98.00
Total Height $H_{D/S\ WING}$, m	1.30	1.50	1.80	1.80	1.80	1.80
Base Width of D/S Wing Wall $B_{D/S\ WING}$, m	1.00	1.00	1.50	1.50	1.50	1.50
Size of Gabion Box (L×B×H), m						
Bottom Layer 1	1.00 × 1.00 × 1.00	1.00 × 1.00 × 1.00	1.50 × 1.00 × 1.00	1.50 × 1.00 × 1.00	1.50 × 1.00 × 1.00	1.50 × 1.00 × 1.00
Layer 2	1.00 × 1.00 × 0.30	1.00 × 1.00 × 0.50	1.00 × 1.00 × 0.80	1.00 × 1.00 × 0.80	1.00 × 1.00 × 0.80	1.00 × 1.00 × 0.80
Pressure at Toe, t/m^2	3.16	4.11	4.12	4.12	4.12	4.12
Pressure at Heel, t/m^2	1.00	0.69	1.74	1.74	1.74	1.74
Factor of Safety against Overturning	5.76	4.21	6.87	6.87	6.87	6.87
Shear Friction Factor F	11.83	8.93	9.81	9.81	9.81	9.81
Total Volume of Gabion Boxes, m^3	7.80	9.00	13.80	13.80	13.80	13.80

5.3.2 Section with Vertical U/S Face

The tables developed after the rigorous design of the different components for 2.00 m high gabion check dams for different combinations are presented from Table 5.13 to Table 5.16.

TABLE 5.13
Design Results for Gabion Check Dam with Vertical U/S Face Height = 2.00 m and Stream Width L = 5.0 m

Catchment Area, ha	**50.00**	**100.00**
A. WEIR		
Span of Weir W, m	4.00	4.00
Flood Discharge Q, m^3/s	6.54	11.00
Head Over Crest, m	0.92	1.31
Top Level, m	100.00	100.00
Foundation Level, m	98.00	98.00
Total Height of Weir H_{WEIR}, m	2.00	2.00
Bottom Width of Weir B_{WEIR}, m	3.00	3.00
Size of Gabion Box (L×B×H), m		
Bottom Layer 1	3.00 × 1.00 × 1.00	3.00 × 1.00 × 1.00
Layer 2	2.00 × 1.00 × 1.00	2.00 × 1.00 × 1.00
Pressure at Toe, t/m^2	3.59	4.04
Pressure at Heel, t/m^2	0.57	0.46
Factor of Safety against Overturning	1.77	1.75
Shear Friction Factor F	3.87	3.27
Total Volume of Gabion Boxes, m^3	20.00	20.00
Length of D/S Apron Floor $L_{D/S\ APRON}$, m	4.00	4.00
B. NON-OVER FLOW PORTION		
Width, m	1.00	1.00
Top Level including Free Board, m	101.50	101.80
Foundation Level, m	98.00	98.00
Total Height of Non-Over Flow H_{NOF}, m	3.50	3.80
Bottom Width of Non-Over Flow B_{NOF}, m	3.00	3.00
Size of Gabion Box (L×B×H), m		
Bottom Layer 1	3.00 × 1.00 × 1.00	3.00 × 1.00 × 1.00
Layer 2	2.00 × 1.00 × 1.00	2.00 × 1.00 × 1.00
Layer 3	1.00 × 1.00 × 1.00	1.00 × 1.00 × 1.00
Layer 4	1.00 × 1.00 × 0.50	1.00 × 1.00 × 0.80
Pressure at Toe, t/m^2	3.88	5.21
Pressure at Heel, t/m^2	1.05	0.04
Factor of Safety against Overturning	1.84	1.62
Shear Friction Factor F	3.35	2.61
Total Volume of Gabion Boxes, m^3	6.50	6.80
C. WING WALLS (ABUTMENTS) ON BOTH BANKS		
Length in U/S $L_{U/S\ ABUT}$, m	1.00	1.00
Length in D/S $L_{D/S\ ABUT}$, m	4.00	4.00

(Continued)

TABLE 5.13 (Continued)

Catchment Area, ha	50.00	100.00
Top Level including Free Board, m	101.50	101.80
Foundation Level, m	98.00	98.00
Total Height H_{ABUT}, m	3.50	3.80
Base Width of Abutment B_{ABUT}, m	2.00	2.50
Size of Gabion Box (L×B×H), m		
Bottom Layer 1	2.00 × 1.00 × 1.00	2.50 × 1.00 × 1.00
Layer 2	1.00 × 1.00 × 1.00	2.00 × 1.00 × 1.00
Layer 3	1.00 × 1.00 × 1.00	1.00 × 1.00 × 1.00
Layer 4	1.00 × 1.00 × 0.50	1.00 × 1.00 × 0.80
Pressure at Toe, t/m^2	10.88	10.41
Pressure at Heel, t/m^2	0.82	2.27
Factor of Safety against Overturning	3.32	4.38
Shear Friction Factor F	4.17	4.55
Total Volume of Gabion Boxes, m^3	45.00	63.00
D. D/S WING WALLS ON BOTH BANKS		
Length in D/S $L_{D/S\ WING}$, m	3.00	3.00
Top Level including Free Board, m	100.00	100.30
Foundation Level, m	98.00	98.00
Total Height $H_{D/S\ WING}$, m	2.00	2.30
Base Width of D/S Wing Wall $B_{D/S\ WING}$, m	1.50	1.50
Size of Gabion Box (L×B×H), m		
Bottom Layer 1	1.50 × 1.00 × 1.00	1.50 × 1.00 × 1.00
Layer 2	1.00 × 1.00 × 1.00	1.00 × 1.00 × 1.00
Layer 3		1.00 × 1.00 × 0.30
Pressure at Toe, t/m^2	4.94	6.32
Pressure at Heel, t/m^2	1.60	1.21
Factor of Safety against Overturning	5.50	4.20
Shear Friction Factor F	8.05	6.33
Total Volume of Gabion Boxes, m^3	15.00	16.80

TABLE 5.14
Design Results for Gabion Check Dam with Vertical U/S Face Height = 2.00 m and Stream Width L = 10.0 m

Catchment Area, ha	50.00	100.00	150.00	200.00
A. WEIR				
Span of Weir W, m	8.00	8.00	8.00	8.00
Flood Discharge Q, m^3/s	6.54	11.00	14.91	18.50
Head Over Crest, m	0.58	0.82	1.01	1.16

(Continued)

TABLE 5.14 (Continued)

Catchment Area, ha	50.00	100.00	150.00	200.00
Top Level, m	100.00	100.00	100.00	100.00
Foundation Level, m	98.00	98.00	98.00	98.00
Total Height of Weir H_{WEIR}, m	2.00	2.00	2.00	2.00
Bottom Width of Weir B_{WEIR}, m	3.00	3.00	3.00	3.00
Size of Gabion Box (L×B×H), m				
Bottom Layer 1	3.00 × 1.00 × 1.00	3.00 × 1.00 × 1.00	3.00 × 1.00 × 1.00	3.00 × 1.00 × 1.00
Layer 2	2.00 × 1.00 × 1.00	2.00 × 1.00 × 1.00	2.00 × 1.00 × 1.00	2.00 × 1.00 × 1.00
Pressure at Toe, t/m²	3.18	3.47	3.69	3.87
Pressure at Heel, t/m²	0.67	0.59	0.54	0.49
Factor of Safety against Overturning	1.80	1.78	1.76	1.75
Shear Friction Factor F	4.66	4.07	3.72	3.47
Total Volume of Gabion Boxes, m³	40.00	40.00	40.00	40.00
Length of D/S Apron Floor $L_{D/S\ APRON}$, m	4.00	4.00	4.00	4.00
B. NON-OVER FLOW PORTION				
Width, m	2.00	2.00	2.00	2.00
Top Level including Free Board, m	101.00	101.30	101.50	101.80
Foundation Level, m	98.00	98.00	98.00	98.00
Total Height of Non-Over Flow H_{NOF}, m	3.00	3.30	3.50	3.80
Bottom Width of Non-Over Flow B_{NOF}, m	2.50	2.50	3.00	3.00
Size of Gabion Box (L×B×H), m				
Bottom Layer 1	2.50 × 1.00 × 1.00	2.50 × 1.00 × 1.00	3.00 × 1.00 × 1.00	3.00 × 1.00 × 1.00
Layer 2	1.50 × 1.00 × 1.00	1.50 × 1.00 × 1.00	2.00 × 1.00 × 1.00	2.00 × 1.00 × 1.00
Layer 3	1.00 × 1.00 × 1.00	1.00 × 1.00 × 1.00	1.00 × 1.00 × 1.00	1.00 × 1.00 × 1.00
Layer 4		1.00 × 1.00 × 0.30	1.00 × 1.00 × 0.50	1.00 × 1.00 × 0.80
Pressure at Toe, t/m²	3.86	4.70	4.18	4.61
Pressure at Heel, t/m²	0.54	0.08	0.76	0.64
Factor of Safety against Overturning	1.71	1.61	1.77	1.74
Shear Friction Factor F	3.58	2.99	3.15	2.87
Total Volume of Gabion Boxes, m³	10.00	10.60	13.00	13.60
C. WING WALLS (ABUTMENTS) ON BOTH BANKS				
Length in U/S $L_{U/S\ ABUT}$, m	1.00	1.00	1.00	1.00
Length in D/S $L_{D/S\ ABUT}$, m	3.50	4.00	4.00	4.00

(Continued)

TABLE 5.14 (Continued)

Catchment Area, ha	50.00	100.00	150.00	200.00
Top Level including Free Board, m	101.00	101.30	101.50	101.80
Foundation Level, m	98.00	98.00	98.00	98.00
Total Height H_{ABUT}, m	3.00	3.30	3.50	3.80
Base Width of Abutment B_{ABUT}, m	2.00	2.00	2.00	2.50
Size of Gabion Box (L×B×H), m				
Bottom Layer 1	2.00 × 1.00 × 1.00	2.00 × 1.00 × 1.00	2.00 × 1.00 × 1.00	2.50 × 1.00 × 1.00
Layer 2	1.00 × 1.00 × 1.00	1.00 × 1.00 × 1.00	1.00 × 1.00 × 1.00	2.00 × 1.00 × 1.00
Layer 3	1.00 × 1.00 × 1.00	1.00 × 1.00 × 1.00	1.00 × 1.00 × 1.00	1.00 × 1.00 × 1.00
Layer 4		1.00 × 1.00 × 0.30	1.00 × 1.00 × 0.50	1.00 × 1.00 × 0.80
Pressure at Toe, t/m^2	8.09	9.69	10.88	10.41
Pressure at Heel, t/m^2	1.91	1.33	0.82	2.27
Factor of Safety against Overturning	4.51	3.73	3.32	4.38
Shear Friction Factor F	5.39	4.59	4.17	4.55
Total Volume of Gabion Boxes, m^3	36.00	43.00	45.00	63.00
D. D/S WING WALLS ON BOTH BANKS				
Length in D/S $L_{D/S\ WING}$, m	3.00	3.00	3.00	3.00
Top Level including Free Board, m	99.50	99.80	100.00	100.30
Foundation Level, m	98.00	98.00	98.00	98.00
Total Height $H_{D/S\ WING}$, m	1.50	1.80	2.00	2.30
Base Width of D/S Wing Wall $B_{D/S\ WING}$, m	1.00	1.50	1.50	1.50
Size of Gabion Box (L×B×H), m				
Bottom Layer 1	1.00 × 1.00 × 1.00	1.50 × 1.00 × 1.00	1.50 × 1.00 × 1.00	1.50 × 1.00 × 1.00
Layer 2	1.00 × 1.00 × 0.50	1.00 × 1.00 × 0.80	1.00 × 1.00 × 1.00	1.00 × 1.00 × 1.00
Layer 3				1.00 × 1.00 × 0.30
Pressure at Toe, t/m^2	4.11	4.12	4.94	6.32
Pressure at Heel, t/m^2	0.69	1.74	1.60	1.21
Factor of Safety against Overturning	4.21	6.87	5.50	4.20
Shear Friction Factor F	8.93	9.81	8.05	6.33
Total Volume of Gabion Boxes, m^3	9.00	13.80	15.00	16.80

TABLE 5.15
Design Results for Gabion Check Dam with Vertical U/S Face Height = 2.00 m and Stream Width L = 15.0 m

Catchment Area, ha	50.00	100.00	150.00	200.00	250.00	300.00
A. WEIR						
Span of Weir W, m	12.00	12.00	12.00	12.00	12.00	12.00
Flood Discharge Q, m³/s	6.54	11.00	14.91	18.50	21.87	25.07
Head Over Crest, m	0.44	0.63	0.77	0.89	0.99	1.09
Top Level, m	100.00	100.00	100.00	100.00	100.00	100.00
Foundation Level, m	98.00	98.00	98.00	98.00	98.00	98.00
Total Height of Weir H_{WEIR}, m	2.00	2.00	2.00	2.00	2.00	2.00
Bottom Width of Weir B_{WEIR}, m	3.00	3.00	3.00	3.00	3.00	3.00
Size of Gabion Box (L×B×H), m						
Bottom Layer 1	3.00 × 1.00 × 1.00	3.00 × 1.00 × 1.00	3.00 × 1.00 × 1.00	3.00 × 1.00 × 1.00	3.00 × 1.00 × 1.00	3.00 × 1.00 × 1.00
Layer 2	2.00 × 1.00 × 1.00	2.00 × 1.00 × 1.00	2.00 × 1.00 × 1.00	2.00 × 1.00 × 1.00	2.00 × 1.00 × 1.00	2.00 × 1.00 × 1.00
Pressure at Toe, t/m²	3.02	3.24	3.41	3.55	3.67	3.79
Pressure at Heel, t/m²	0.71	0.65	0.61	0.57	0.54	0.52
Factor of Safety against Overturning	1.81	1.79	1.78	1.77	1.77	1.76
Shear Friction Factor F	5.07	4.52	4.19	3.94	3.74	3.58
Total Volume of Gabion Boxes, m³	60.00	60.00	60.00	60.00	60.00	60.00
Length of D/S Apron Floor $L_{D/S\ APRON}$, m	4.00	4.00	4.00	4.00	4.00	4.00

(Continued)

TABLE 5.15 (Continued)

Catchment Area, ha	50.00	100.00	150.00	200.00	250.00	300.00
B. NON-OVER FLOW PORTION						
Width, m	3.00	3.00	3.00	3.00	3.00	3.00
Top Level including Free Board, m	100.80	101.00	101.30	101.30	101.50	101.80
Foundation Level, m	98.00	98.00	98.00	98.00	98.00	98.00
Total Height of Non-Over Flow H_{NOF}, m	2.80	3.00	3.30	3.30	3.50	3.80
Bottom Width of Non-Over Flow B_{NOF}, m	2.50	2.50	2.50	2.50	3.00	3.00
Size of Gabion Box (L×B×H), m						
Bottom Layer 1	2.50 × 1.00 × 1.00	2.50 × 1.00 × 1.00	2.50 × 1.00 × 1.00	2.50 × 1.00 × 1.00	3.00 × 1.00 × 1.00	3.00 × 1.00 × 1.00
Layer 2	2.00 × 1.00 × 1.00	2.00 × 1.00 × 1.00	2.00 × 1.00 × 1.00	2.00 × 1.00 × 1.00	2.00 × 1.00 × 1.00	2.00 × 1.00 × 1.00
Layer 3	1 × 1.0 × 0.8	1.00 × 1.00 × 1.00	1.00 × 1.00 × 1.00	1.50 × 1.00 × 1.00	1.00 × 1.00 × 1.00	1.00 × 1.00 × 1.00
Layer 4			1.00 × 1.00 × 0.30	1.00 × 1.00 × 0.30	1.00 × 1.00 × 0.50	1.00 × 1.00 × 0.80
Pressure at Toe, t/m²	4.18	4.75	5.16	6.05	4.11	4.31
Pressure at Heel, t/m²	0.60	0.29	0.26	0.01	0.82	0.94
Factor of Safety against Overturning	1.83	1.75	1.74	1.73	1.79	1.81
Shear Friction Factor F	4.09	3.53	3.20	2.99	3.18	3.02
Total Volume of Gabion Boxes, m³	15.90	16.50	17.40	18.90	19.50	20.40
C. WING WALLS (ABUTMENTS) ON BOTH BANKS						
Length in U/S $L_{U/S\ ABUT}$, m	1.00	1.00	1.00	1.00	1.00	1.00

(Continued)

TABLE 5.15 (Continued)

Catchment Area, ha	**50.00**	**100.00**	**150.00**	**200.00**	**250.00**	**300.00**
Length in D/S $L_{D/S\ ABUT}$, m	3.50	3.50	3.50	3.50	4.00	4.00
Top Level including Free Board, m	100.80	101.00	101.30	101.30	101.50	101.80
Foundation Level, m	98.00	98.00	98.00	98.00	98.00	98.00
Total Height H_{ABUT}, m	2.80	3.00	3.30	3.30	3.50	3.80
Base Width of Abutment B_{ABUT}, m	2.00	2.00	2.00	2.00	2.00	2.50
Size of Gabion Box (L×B×H), m						
Bottom Layer 1	2.00 × 1.00 × 1.00	2.00 × 1.00 × 1.00	2.00 × 1.00 × 1.00	2.00 × 1.00 × 1.00	2.00 × 1.00 × 1.00	2.50 × 1.00 × 1.00
Layer 2	1.00 × 1.00 × 1.00	1.00 × 1.00 × 1.00	1.00 × 1.00 × 1.00	1.00 × 1.00 × 1.00	1.00 × 1.00 × 1.00	2.00 × 1.00 × 1.00
Layer 3	1.00 × 1.00 × 0.80	1.00 × 1.00 × 1.00	1.00 × 1.00 × 1.00	1.00 × 1.00 × 1.00	1.00 × 1.00 × 1.00	1.00 × 1.00 × 1.00
Layer 4			1.00 × 1.00 × 0.30	1.00 × 1.00 × 0.30	1.00 × 1.00 × 0.50	1.00 × 1.00 × 0.80
Pressure at Toe, t/m²	7.14	8.09	9.69	9.69	10.88	10.41
Pressure at Heel, t/m²	2.18	1.91	1.33	1.33	0.82	2.27
Factor of Safety against Overturning	5.19	4.51	3.73	3.73	3.32	4.38
Shear Friction Factor F	6.05	5.39	4.59	4.59	4.17	4.55
Total Volume of Gabion Boxes, m³	34.20	36.00	38.70	38.70	45.00	63.00
D. D/S WING WALLS ON BOTH BANKS						
Length in D/S $L_{D/S\ WING}$, m	3.00	3.00	3.00	3.00	3.00	3.00
Top Level including Free Board, m	99.30	99.50	99.80	99.80	100.00	100.30

(Continued)

TABLE 5.15 (Continued)

Catchment Area, ha	50.00	100.00	150.00	200.00	250.00	300.00
Foundation Level, m	98.00	98.00	98.00	98.00	98.00	98.00
Total Height $H_{D/S\ WING}$, m	1.30	1.50	1.80	1.80	2.00	2.30
Base Width of D/S Wing Wall $B_{D/S\ WING}$, m	1.00	1.00	1.50	1.50	1.50	1.50
Size of Gabion Box (L×B×H), m						
Bottom Layer 1	1.00 × 1.00 × 1.00	1.00 × 1.00 × 1.00	1.50 × 1.00 × 1.00	1.50 × 1.00 × 1.00	1.50 × 1.00 × 1.00	1.50 × 1.00 × 1.00
Layer 2	1.00 × 1.00 × 0.30	1.00 × 1.00 × 0.80	1.00 × 1.00 × 0.80	1.00 × 1.00 × 0.80	1.00 × 1.00 × 1.00	1.00 × 1.00 × 1.00
Layer 3						1.00 × 1.00 × 0.30
Pressure at Toe, t/m²	3.16	4.11	4.12	4.12	4.94	6.32
Pressure at Heel, t/m²	1.00	0.69	1.74	1.74	1.60	1.21
Factor of Safety against Overturning	5.76	4.21	6.87	6.87	5.50	4.20
Shear Friction Factor F	11.83	8.93	9.81	9.81	8.05	6.33
Total Volume of Gabion Boxes, m³	7.80	9.00	13.80	13.80	15.00	16.80

TABLE 5.16
Design Results for Gabion Check Dam with Vertical U/S Face Height = 2.00 m and Stream Width L = 20.0 m

Catchment Area, ha	50.00	100.00	150.00	200.00	250.00	300.00
A. WEIR						
Span of Weir W, m	16.00	16.00	16.00	16.00	16.00	16.00
Flood Discharge Q, m³/s	6.54	11.00	14.91	18.50	21.87	25.07
Head Over Crest, m	0.37	0.52	0.64	0.73	0.82	0.90
Top Level, m	100.00	100.00	100.00	100.00	100.00	100.00
Foundation Level, m	98.00	98.00	98.00	98.00	98.00	98.00

(Continued)

TABLE 5.16 (Continued)

Catchment Area, ha	**50.00**	**100.00**	**150.00**	**200.00**	**250.00**	**300.00**
Total Height of Weir H_{WEIR}, m	2.00	2.00	2.00	2.00	2.00	2.00
Bottom Width of Weir B_{WEIR}, m	2.50	3.00	3.00	3.00	3.00	3.00
Size of Gabion Box (L×B×H), m						
Bottom Layer 1	2.50 × 1.00 × 1.00	3.00 × 1.00 × 1.00	3.00 × 1.00 × 1.00	3.00 × 1.00 × 1.00	3.00 × 1.00 × 1.00	3.00 × 1.00 × 1.00
Layer 2	2.00 × 1.00 × 1.00	2.00 × 1.00 × 1.00	2.00 × 1.00 × 1.00	2.00 × 1.00 × 1.00	2.00 × 1.00 × 1.00	2.00 × 1.00 × 1.00
Pressure at Toe, t/m²	4.10	3.11	3.25	3.36	3.47	3.56
Pressure at Heel, t/m²	0.05	0.68	0.65	0.62	0.60	0.57
Factor of Safety against Overturning	1.70	1.80	1.79	1.78	1.78	1.77
Shear Friction Factor F	4.55	4.83	4.51	4.27	4.08	3.92
Total Volume of Gabion Boxes, m³	72.00	80.00	80.00	80.00	80.00	80.00
Length of D/S Apron Floor $L_{D/S\ APRON}$, m	4.40	4.00	4.00	4.00	4.00	4.00
B. NON-OVER FLOW PORTION						
Width, m	4.00	4.00	4.00	4.00	4.00	4.00
Top Level including Free Board, m	100.80	101.00	101.30	101.30	101.30	101.30
Foundation Level, m	98.00	98.00	98.00	98.00	98.00	98.00
Total Height of Non-Over Flow H_{NOF}, m	2.80	3.00	3.30	3.30	3.30	3.30
Bottom Width of Non-Over Flow B_{NOF}, m	2.50	2.50	2.50	2.50	2.50	2.50
Size of Gabion Box (L×B×H), m						
Bottom Layer 1	2.50 × 1.00 × 1.00	2.50 × 1.00 × 1.00	2.50 × 1.00 × 1.00	2.50 × 1.00 × 1.00	2.50 × 1.00 × 1.00	2.50 × 1.00 × 1.00
Layer 2	1.50 × 1.00 × 1.00	1.50 × 1.00 × 1.00	1.50 × 1.00 × 1.00	1.50 × 1.00 × 1.00	1.50 × 1.00 × 1.00	2.00 × 1.00 × 1.00
Layer 3	1.00 × 1.00 × 0.80	1.00 × 1.00 × 1.00	1.00 × 1.00 × 1.00	1.00 × 1.00 × 1.00	1.00 × 1.00 × 1.00	1.50 × 1.00 × 1.00
Layer 4			1.00 × 1.00 × 0.30	1.00 × 1.00 × 0.30	1.00 × 1.00 × 0.30	1.50 × 1.00 × 0.30
Pressure at Toe, t/m²	3.23	3.64	3.91	4.32	4.69	6.18
Pressure at Heel, t/m²	0.92	0.76	0.88	0.46	0.09	0.08
Factor of Safety against Overturning	1.81	1.77	1.79	1.69	1.61	1.76
Shear Friction Factor F	4.28	3.77	3.47	3.21	3.00	2.99
Total Volume of Gabion Boxes, m³	19.20	20.00	21.20	21.20	21.20	25.80

(Continued)

TABLE 5.16 (Continued)

Catchment Area, ha	50.00	100.00	150.00	200.00	250.00	300.00
C. WING WALLS (ABUTMENTS) ON BOTH BANKS						
Length in U/S $L_{U/S\ ABUT}$, m	1.00	1.00	1.00	1.00	1.00	1.00
Length in D/S $L_{D/S\ ABUT}$, m	3.50	3.50	3.50	3.50	3.50	3.50
Top Level including Free Board, m	100.80	101.00	101.30	101.30	101.30	101.30
Foundation Level, m	98.00	98.00	98.00	98.00	98.00	98.00
Total Height H_{ABUT}, m	2.80	3.00	3.30	3.30	3.30	3.30
Base Width of Abutment B_{ABUT}, m	2.00	2.00	2.00	2.00	2.00	2.00
Size of Gabion Box (L×B×H), m						
Bottom Layer 1	2.00 × 1.00 × 1.00	2.00 × 1.00 × 1.00	2.00 × 1.00 × 1.00	2.00 × 1.00 × 1.00	2.00 × 1.00 × 1.00	2.00 × 1.00 × 1.00
Layer 2	1.00 × 1.00 × 1.00	1.00 × 1.00 × 1.00	1.00 × 1.00 × 1.00	1.00 × 1.00 × 1.00	1.00 × 1.00 × 1.00	1.00 × 1.00 × 1.00
Layer 3	1.00 × 1.00 × 0.80	1.00 × 1.00 × 1.00	1.00 × 1.00 × 1.00	1.00 × 1.00 × 1.00	1.00 × 1.00 × 1.00	1.00 × 1.00 × 1.00
Layer 4			1.00 × 1.00 × 0.30	1.00 × 1.00 × 0.30	1.00 × 1.00 × 0.30	1.00 × 1.00 × 0.30
Pressure at Toe, t/m^2	7.14	8.09	9.69	9.69	9.69	9.69
Pressure at Heel, t/m^2	2.18	1.91	1.33	1.33	1.33	1.33
Factor of Safety against Overturning	5.19	4.51	3.73	3.73	3.73	3.73
Shear Friction Factor F	6.05	5.39	4.59	4.59	4.59	4.59
Total Volume of Gabion Boxes, m^3	34.20	36.00	38.70	38.70	38.70	38.70
D. D/S WING WALLS ON BOTH BANKS						
Length in D/S $L_{D/S\ WING}$, m	3.40	3.00	3.00	3.00	3.00	3.00
Top Level including Free Board, m	99.30	99.50	99.80	99.80	99.80	99.80
Foundation Level, m	98.00	98.00	98.00	98.00	98.00	98.00
Total Height $H_{D/S\ WING}$, m	1.30	1.50	1.80	1.80	1.80	1.80
Base Width of D/S Wing Wall $B_{D/S\ WING}$, m	1.00	1.00	1.50	1.50	1.50	1.50
Size of Gabion Box (L×B×H), m						
Bottom Layer 1	1.00 × 1.00 × 1.00	1.00 × 1.00 × 1.00	1.50 × 1.00 × 1.00	1.50 × 1.00 × 1.00	1.50 × 1.00 × 1.00	1.50 × 1.00 × 1.00
Layer 2	1.00 × 1.00 × 0.30	1.00 × 1.00 × 0.50	1.00 × 1.00 × 0.80	1.00 × 1.00 × 0.80	1.00 × 1.00 × 0.80	1.00 × 1.00 × 0.80

(Continued)

TABLE 5.16 (Continued)

Catchment Area, ha	50.00	100.00	150.00	200.00	250.00	300.00
Pressure at Toe, t/m²	3.16	4.11	4.12	4.12	4.12	4.12
Pressure at Heel, t/m²	1.00	0.69	1.74	1.74	1.74	1.74
Factor of Safety against Overturning	5.76	4.21	6.87	6.87	6.87	6.87
Shear Friction Factor F	11.83	8.93	9.81	9.81	9.81	9.81
Total Volume of Gabion Boxes, m³	8.84	9.00	13.80	13.80	13.80	13.80

5.3.3 Section with Steps on Both Faces

The tables developed after the rigorous design of the different components for 2.00 m high gabion check dams for different combinations are presented from Table 5.17 to Table 5.20.

TABLE 5.17
Design Results for Gabion Check Dam with Steps on Both Faces Height = 2.00 m and Stream Width L = 5.0 m

CA, ha	50.00	100.00
A. WEIR		
Span of Weir W, m	4.00	4.00
Flood Discharge Q, m³/s	6.54	11.00
Head Over Crest, m	0.92	1.31
Top Level, m	100.00	100.00
Foundation Level, m	98.00	98.00
Total Height of Weir H_{WEIR}, m	2.00	2.00
Bottom Width of Weir B_{WEIR}, m	3.00	3.50
Size of Gabion Box (L×B×H), m		
Bottom Layer 1	3.00 × 1.00 × 1.00	3.50 × 1.00 × 1.00
Layer 2	3.00 × 1.00 × 1.00	3.00 × 1.00 × 1.00
Pressure at Toe, t/m²	5.63	5.53
Pressure at Heel, t/m²	0.00	0.05
Factor of Safety against Overturning	1.92	1.95
Shear Friction Factor F	4.09	3.98
Total Volume of Gabion Boxes, m³	24.00	26.00
Length of D/S Apron Floor $L_{D/S\ APRON}$, m	4.50	4.10
B. NON-OVER FLOW PORTION		
Length, m	1.00	1.00
Top Level including Free Board, m	101.50	101.80
Foundation Level, m	98.00	98.00
Total Height of Non-Over Flow H_{NOF}, m	3.50	3.80
Bottom Width of Non-Over Flow B_{NOF}, m	3.50	3.50
Size of Gabion Box (L×B×H), m		
Bottom Layer 1	3.50 × 1.00 × 1.00	3.50 × 1.00 × 1.00

(Continued)

TABLE 5.17 (Continued)

CA, ha	50.00	100.00
Layer 2	3.00 × 1.00 × 1.00	3.00 × 1.00 × 1.00
Layer 3	2.00 × 1.00 × 1.00	2.00 × 1.00 × 1.00
Layer 4	1.00 × 1.00 × 0.50	1.00 × 1.00 × 0.80
Pressure at Toe, t/m^2	5.95	7.04
Pressure at Heel, t/m^2	0.86	0.35
Factor of Safety against Overturning	2.23	2.05
Shear Friction Factor F	4.19	3.28
Total Volume of Gabion Boxes, m^3	9.00	9.30
C. WING WALLS (ABUTMENTS) ON BOTH BANKS		
Length in U/S $L_{U/S\ ABUT}$, m	3.50	3.50
Length in D/S $L_{D/S\ ABUT}$, m	2.00	2.00
Top Level including Free Board, m	101.50	101.80
Foundation Level, m	98.00	98.00
Total Height H_{ABUT}, m	3.50	3.80
Base Width of Abutment B_{ABUT}, m	2.00	2.50
Size of Gabion Box (L×B×H), m		
Bottom Layer 1	2.00 × 1.00 × 1.00	2.50 × 1.00 × 1.00
Layer 2	1.00 × 1.00 × 1.00	2.00 × 1.00 × 1.00
Layer 3	1.00 × 1.00 × 1.00	1.00 × 1.00 × 1.00
Layer 4	1.00 × 1.00 × 0.50	1.00 × 1.00 × 0.80
Pressure at Toe, t/m^2	10.88	10.41
Pressure at Heel, t/m^2	0.82	2.27
Factor of Safety against Overturning	3.32	4.38
Shear Friction Factor F	4.17	4.55
Total Volume of Gabion Boxes, m^3	49.50	69.30
D. D/S WING WALLS ON BOTH BANKS		
Length in D/S $L_{D/S\ WING}$, m	2.10	2.10
Top Level including Free Board, m	100.00	100.30
Foundation Level, m	98.00	98.00
Total Height $H_{D/S\ WING}$, m	2.00	2.30
Base Width of D/S Wing Wall $B_{D/S\ WING}$, m	1.50	1.50
Size of Gabion Box (L×B×H), m		
Bottom Layer 1	1.50 × 1.00 × 1.00	1.50 × 1.00 × 1.00
Layer 2	1.00 × 1.00 × 1.00	1.00 × 1.00 × 1.00
Layer 3		1.00 × 1.00 × 0.30
Pressure at Toe, t/m^2	4.94	6.32
Pressure at Heel, t/m^2	1.60	1.21
Factor of Safety against Overturning	5.50	4.20
Shear Friction Factor F	8.05	6.33
Total Volume of Gabion Boxes, m^3	10.50	11.76

TABLE 5.18
Design Results for Gabion Check Dam with Steps on Both Faces Height = 2.00 m and Stream Width L = 10.0 m

Catchment Area, ha	50.00	100.00	150.00	200.00
A. WEIR				
Span of Weir W, m	8.00	8.00	8.00	8.00
Flood Discharge Q, m^3/s	6.54	11.00	14.91	18.50
Head Over Crest, m	0.58	0.82	1.01	1.16
Top Level, m	100.00	100.00	100.00	100.00
Foundation Level, m	98.00	98.00	98.00	98.00
Total Height of Weir H_{WEIR}, m	2.00	2.00	2.00	2.00
Bottom Width of Weir B_{WEIR}, m	3.00	3.00	3.50	3.50
Size of Gabion Box (L×B×H), m				
Bottom Layer 1	3.00 × 1.00 × 1.00	3.00 × 1.00 × 1.00	3.50 × 1.00 × 1.00	3.50 × 1.00 × 1.00
Layer 2	3.00 × 1.00 × 1.00	3.00 × 1.00 × 1.00	3.00 × 1.00 × 1.00	3.00 × 1.00 × 1.00
Pressure at Toe, t/m^2	5.06	5.47	5.07	5.31
Pressure at Heel, t/m^2	0.12	0.03	0.17	0.11
Factor of Safety against Overturning	1.93	1.92	1.96	1.95
Shear Friction Factor F	4.90	4.29	4.50	4.21
Total Volume of Gabion Boxes, m^3	48.00	48.00	52.00	52.00
Length of D/S Apron Floor $L_{D/S\ APRON}$, m	4.10	4.50	4.10	4.10
B. NON-OVER FLOW PORTION				
Length, m	2.00	2.00	2.00	2.00
Top Level including Free Board, m	101.00	101.30	101.50	101.80
Foundation Level, m	98.00	98.00	98.00	98.00
Total Height of Non-Over Flow H_{NOF}, m	3.00	3.30	3.50	3.80
Bottom Width of Non-Over Flow B_{NOF}, m	3.00	3.00	3.50	3.50
Size of Gabion Box (L×B×H), m				
Bottom Layer 1	3.00 × 1.00 × 1.00	3.00 × 1.00 × 1.00	3.50 × 1.00 × 1.00	3.50 × 1.00 × 1.00
Layer 2	2.00 × 1.00 × 1.00	2.00 × 1.00 × 1.00	2.50 × 1.00 × 1.00	2.50 × 1.00 × 1.00
Layer 3	1.00 × 1.00 × 1.00	1.00 × 1.00 × 1.00	1.50 × 1.00 × 1.00	1.50 × 1.00 × 1.00
Layer 4		1.00 × 1.00 × 0.30	1.00 × 1.00 × 0.50	1.00 × 1.00 × 0.80

(Continued)

TABLE 5.18 (Continued)

Catchment Area, ha	50.00	100.00	150.00	200.00
Pressure at Toe, t/m^2	4.79	5.55	5.55	6.07
Pressure at Heel, t/m^2	0.38	0.11	0.71	0.59
Factor of Safety against Overturning	1.92	1.86	2.07	2.05
Shear Friction Factor F	4.42	3.71	3.86	3.52
Total Volume of Gabion Boxes, m^3	12.00	12.60	16.00	16.60
C. WING WALLS (ABUTMENTS) ON BOTH BANKS				
Length in U/S $L_{U/S\ ABUT}$, m	3.00	3.00	3.50	3.50
Length in D/S $L_{D/S\ ABUT}$, m	2.00	2.00	2.00	2.00
Top Level including Free Board, m	101.00	101.30	101.50	101.80
Foundation Level, m	98.00	98.00	98.00	98.00
Total Height H_{ABUT}, m	3.00	3.30	3.50	3.80
Base Width of Abutment B_{ABUT}, m	2.00	2.00	2.00	2.50
Size of Gabion Box (L×B×H), m				
Bottom Layer 1	2.00 × 1.00 × 1.00	2.00 × 1.00 × 1.00	2.00 × 1.00 × 1.00	2.50 × 1.00 × 1.00
Layer 2	1.00 × 1.00 × 1.00	1.00 × 1.00 × 1.00	1.00 × 1.00 × 1.00	2.00 × 1.00 × 1.00
Layer 3	1.00 × 1.00 × 1.00	1.00 × 1.00 × 1.00	1.00 × 1.00 × 1.00	1.00 × 1.00 × 1.00
Layer 4		1.00 × 1.00 × 0.30	1.00 × 1.00 × 0.50	1.00 × 1.00 × 0.80
Pressure at Toe, t/m^2	8.09	9.69	10.88	10.41
Pressure at Heel, t/m^2	1.91	1.33	0.82	2.27
Factor of Safety against Overturning	4.51	3.73	3.32	4.38
Shear Friction Factor F	5.39	4.59	4.17	4.55
Total Volume of Gabion Boxes, m^3	40.00	43.00	49.50	69.30
D. D/S WING WALLS ON BOTH BANKS				
Length in D/S $L_{D/S\ WING}$, m	2.10	2.10	2.10	2.10
Top Level including Free Board, m	99.50	99.80	100.00	100.30
Foundation Level, m	98.00	98.00	98.00	98.00
Total Height $H_{D/S\ WING}$, m	1.50	1.80	2.00	2.30
Base Width of D/S Wing Wall $B_{D/S\ WING}$, m	1.00	1.50	1.50	1.50

(Continued)

TABLE 5.18 (Continued)

Catchment Area, ha	50.00	100.00	150.00	200.00
Size of Gabion Box (L×B×H), m				
Bottom Layer 1	1.00 × 1.00 × 1.00	1.50 × 1.00 × 1.00	1.50 × 1.00 × 1.00	1.50 × 1.00 × 1.00
Layer 2	1.00 × 1.00 × 0.50	1.00 × 1.00 × 0.80	1.00 × 1.00 × 1.00	1.00 × 1.00 × 1.00
Layer 3				1.00 × 1.00 × 0.30
Pressure at Toe, t/m²	4.11	4.12	4.94	6.32
Pressure at Heel, t/m²	0.69	1.74	1.60	1.21
Factor of Safety against Overturning	4.21	6.87	5.50	4.20
Shear Friction Factor F	8.93	9.81	8.05	6.33
Total Volume of Gabion Boxes, m³	6.30	9.66	10.50	11.76

TABLE 5.19
Design Results for Gabion Check Dam with Steps on Both Faces Height = 2.00 m and Stream Width L = 15.0 m

Catchment Area, ha	50.00	100.00	150.00	200.00	250.00	300.00
A. WEIR						
Span of Weir W, m	12.00	12.00	12.00	12.00	12.00	12.00
Flood Discharge Q, m³/s	6.54	11.00	14.91	18.50	21.87	25.07
Head Over Crest, m	0.44	0.63	0.77	0.89	0.99	1.09
Top Level, m	100.00	100.00	100.00	100.00	100.00	100.00
Foundation Level, m	98.00	98.00	98.00	98.00	98.00	98.00
Total Height of Weir H_{WEIR}, m	2.00	2.00	2.00	2.00	2.00	2.00
Bottom Width of Weir B_{WEIR}, m	3.00	3.00	3.00	3.00	3.50	3.50
Size of Gabion Box (L×B×H), m						
Bottom Layer 1	3.00 × 1.00 × 1.00	3.00 × 1.00 × 1.00	3.00 × 1.00 × 1.00	3.00 × 1.00 × 1.00	3.50 × 1.00 × 1.00	3.50 × 1.00 × 1.00
Layer 2	3.00 × 1.00 × 1.00	3.00 × 1.00 × 1.00	3.00 × 1.00 × 1.00	3.00 × 1.00 × 1.00	3.00 × 1.00 × 1.00	3.00 × 1.00 × 1.00
Pressure at Toe, t/m²	4.83	5.14	5.38	5.58	5.04	5.19

(Continued)

TABLE 5.19 (Continued)

Catchment Area, ha	50.00	100.00	150.00	200.00	250.00	300.00
Pressure at Heel, t/m^2	0.16	0.10	0.05	0.01	0.18	0.14
Factor of Safety against Overturning	1.94	1.93	1.92	1.92	1.96	1.95
Shear Friction Factor F	5.32	4.76	4.42	4.16	4.54	4.35
Total Volume of Gabion Boxes, m^3	72.00	72.00	72.00	72.00	78.00	78.00
Length of D/S Apron Floor $L_{D/S\ APRON}$, m	4.50	4.50	4.50	4.50	4.10	4.10
B. NON-OVER FLOW PORTION						
Length, m	3.00	3.00	3.00	3.00	3.00	3.00
Top Level including Free Board, m	100.80	101.00	101.30	101.30	101.50	101.80
Foundation Level, m	98.00	98.00	98.00	98.00	98.00	98.00
Total Height of Non-Over Flow H_{NOF}, m	2.80	3.00	3.30	3.30	3.50	3.80
Bottom Width of Non-Over Flow B_{NOF}, m	3.00	3.00	3.00	3.00	3.00	3.00
Size of Gabion Box (L×B×H), m						
Bottom Layer 1	3.00 × 1.00 × 1.00	3.00 × 1.00 × 1.00	3.00 × 1.00 × 1.00	3.00 × 1.00 × 1.00	3.00 × 1.00 × 1.00	3.00 × 1.00 × 1.00
Layer 2	2.00 × 1.00 × 1.00	2.00 × 1.00 × 1.00	2.00 × 1.00 × 1.00	2.00 × 1.00 × 1.00	2.00 × 1.00 × 1.00	2.00 × 1.00 × 1.00
Layer 3	1.00 × 1.00 × 0.80	1.00 × 1.00 × 1.00	1.00 × 1.00 × 1.00	1.50 × 1.00 × 1.00	2.00 × 1.00 × 1.00	2.00 × 1.00 × 1.00
Layer 4			1.00 × 1.00 × 0.30	1.00 × 1.00 × 0.30	1.50 × 1.00 × 0.50	1.50 × 1.00 × 0.80
Pressure at Toe, t/m^2	4.40	4.90	5.40	6.04	6.91	7.48
Pressure at Heel, t/m^2	0.45	0.30	0.22	0.04	0.07	0.04
Factor of Safety against Overturning	1.93	1.89	1.89	1.88	1.97	1.98
Shear Friction Factor F	4.92	4.26	3.87	3.59	3.43	3.27

(Continued)

TABLE 5.19 (Continued)

Catchment Area, ha	**50.00**	**100.00**	**150.00**	**200.00**	**250.00**	**300.00**
Total Volume of Gabion Boxes, m^3	17.40	18.00	18.90	20.40	23.25	24.60
C. WING WALLS (ABUTMENTS) ON BOTH BANKS						
Length in U/S $L_{U/S\ ABUT}$, m	3.00	3.00	3.00	3.50	3.50	3.50
Length in D/S $L_{D/S\ ABUT}$, m	2.00	2.00	2.00	2.00	2.00	2.00
Top Level including Free Board, m	100.80	101.00	101.30	101.30	101.50	101.80
Foundation Level, m	98.00	98.00	98.00	98.00	98.00	98.00
Total Height H_{ABUT}, m	2.80	3.00	3.30	3.30	3.50	3.80
Base Width of Abutment B_{ABUT}, m	2.00	2.00	2.00	2.00	2.00	2.50
Size of Gabion Box (L×B×H), m						
Bottom Layer 1	2.00 × 1.00 × 1.00	2.00 × 1.00 × 1.00	2.00 × 1.00 × 1.00	2.00 × 1.00 × 1.00	2.00 × 1.00 × 1.00	2.50 × 1.00 × 1.00
Layer 2	1.00 × 1.00 × 1.00	1.00 × 1.00 × 1.00	1.00 × 1.00 × 1.00	1.00 × 1.00 × 1.00	1.00 × 1.00 × 1.00	2.00 × 1.00 × 1.00
Layer 3	1.00 × 1.00 × 0.80	1.00 × 1.00 × 1.00	1.00 × 1.00 × 1.00	1.00 × 1.00 × 1.00	1.00 × 1.00 × 1.00	1.00 × 1.00 × 1.00
Layer 4			1.00 × 1.00 × 0.30	1.00 × 1.00 × 0.30	1.00 × 1.00 × 0.50	1.00 × 1.00 × 0.80
Pressure at Toe, t/m^2	7.14	8.09	9.69	9.69	10.88	10.41
Pressure at Heel, t/m^2	2.18	1.91	1.33	1.33	0.82	2.27
Factor of Safety against Overturning	5.19	4.51	3.73	3.73	3.32	4.38
Shear Friction Factor F	6.05	5.39	4.59	4.59	4.17	4.55
Total Volume of Gabion Boxes, m^3	38.00	40.00	43.00	47.30	49.50	69.30
D. D/S WING WALLS ON BOTH BANKS						
Length in D/S $L_{D/S\ WING}$, m	2.10	2.10	2.10	2.10	2.10	2.10

(Continued)

TABLE 5.19 (Continued)

Catchment Area, ha	50.00	100.00	150.00	200.00	250.00	300.00
Top Level including Free Board, m	99.30	99.50	99.80	99.80	100.00	100.30
Foundation Level, m	98.00	98.00	98.00	98.00	98.00	98.00
Total Height $H_{D/S\ WING}$, m	1.30	1.50	1.80	1.80	2.00	2.30
Base Width of D/S Wing Wall $B_{D/S\ WING}$, m	1.00	1.00	1.50	1.50	1.50	1.50
Size of Gabion Box (L×B×H), m						
Bottom Layer 1	1.00 × 1.00 × 1.00	1.00 × 1.00 × 1.00	1.50 × 1.00 × 1.00	1.50 × 1.00 × 1.00	1.50 × 1.00 × 1.00	1.50 × 1.00 × 1.00
Layer 2	1.00 × 1.00 × 0.30	1.00 × 1.00 × 0.50	1.00 × 1.00 × 0.80	1.00 × 1.00 × 0.80	1.00 × 1.00 × 1.00	1.00 × 1.00 × 1.00
Layer 3						1.00 × 1.00 × 0.30
Pressure at Toe, t/m²	3.16	4.11	4.12	4.12	4.94	6.32
Pressure at Heel, t/m²	1.00	0.69	1.74	1.74	1.60	1.21
Factor of Safety against Overturning	5.76	4.21	6.87	6.87	5.50	4.20
Shear Friction Factor F	11.83	8.93	9.81	9.81	8.05	6.33
Total Volume of Gabion Boxes, m³	5.46	6.30	9.66	9.66	10.50	11.76

TABLE 5.20
Design Results for Gabion Check Dam with Steps on Both Faces Height = 2.00 m and Stream Width L = 20.0 m

Catchment Area, ha	50.00	100.00	150.00	200.00	250.00	300.00
A. WEIR						
Span of Weir W, m	16.00	16.00	16.00	16.00	16.00	16.00
Flood Discharge Q, m³/s	6.54	11.00	14.91	18.50	21.87	25.07

(Continued)

TABLE 5.20 (Continued)

Catchment Area, ha	50.00	100.00	150.00	200.00	250.00	300.00
Head Over Crest, m	0.37	0.52	0.64	0.73	0.82	0.90
Top Level, m	100.00	100.00	100.00	100.00	100.00	100.00
Foundation Level, m	98.00	98.00	98.00	98.00	98.00	98.00
Total Height of Weir H_{WEIR}, m	2.00	2.00	2.00	2.00	2.00	2.00
Bottom Width of Weir B_{WEIR}, m	3.00	3.00	3.00	3.00	3.00	3.00
Size of Gabion Box (L×B×H), m						
Bottom Layer 1	3.00 × 1.00 × 1.00	3.00 × 1.00 × 1.00	3.00 × 1.00 × 1.00	3.00 × 1.00 × 1.00	3.00 × 1.00 × 1.00	3.00 × 1.00 × 1.00
Layer 2	2.50 × 1.00 × 1.00	3.00 × 1.00 × 1.00	3.00 × 1.00 × 1.00	3.00 × 1.00 × 1.00	3.00 × 1.00 × 1.00	3.00 × 1.00 × 1.00
Pressure at Toe, t/m²	4.39	4.96	5.15	5.32	5.46	5.59
Pressure at Heel, t/m²	0.05	0.13	0.10	0.06	0.03	0.01
Factor of Safety against Overturning	1.84	1.94	1.93	1.93	1.92	1.92
Shear Friction Factor F	5.52	5.08	4.75	4.50	4.30	4.14
Total Volume of Gabion Boxes, m³	88.00	96.00	96.00	96.00	96.00	96.00
Length of D/S Apron Floor $L_{D/S\ APRON}$, m	4.50	4.50	4.50	4.50	4.50	4.50
B. NON-OVER FLOW PORTION						
Length, m	4.00	4.00	4.00	4.00	4.00	4.00
Top Level including Free Board, m	100.80	101.00	101.30	101.30	101.30	101.30
Foundation Level, m	98.00	98.00	98.00	98.00	98.00	98.00
Total Height of Non-Over Flow H_{NOF}, m	2.80	3.00	3.30	3.30	3.30	3.30
Bottom Width of Non-Over Flow B_{NOF}, m	2.50	3.00	3.00	3.00	3.00	3.00
Size of Gabion Box (L×B×H), m						

(Continued)

TABLE 5.20 (Continued)

Catchment Area, ha	50.00	100.00	150.00	200.00	250.00	300.00
Bottom Layer 1	2.50 × 1.00 × 1.00	3.00 × 1.00 × 1.00	3.00 × 1.00 × 1.00	3.00 × 1.00 × 1.00	3.00 × 1.00 × 1.00	3.00 × 1.00 × 1.00
Layer 2	2.00 × 1.00 × 1.00	2.00 × 1.00 × 1.00	2.00 × 1.00 × 1.00	2.00 × 1.00 × 1.00	2.00 × 1.00 × 1.00	2.50 × 1.00 × 1.00
Layer 3	1.50 × 1.00 × 0.80	1.00 × 1.00 × 1.00	1.00 × 1.00 × 1.00	1.00 × 1.00 × 1.00	1.00 × 1.00 × 1.00	2.00 × 1.00 × 1.00
Layer 4			1.00 × 1.00 × 0.30	1.00 × 1.00 × 0.30	1.00 × 1.00 × 0.30	1.00 × 1.00 × 0.30
Pressure at Toe, t/m^2	5.59	4.66	5.07	5.32	5.55	6.76
Pressure at Heel, t/m^2	0.06	0.46	0.45	0.28	0.11	0.06
Factor of Safety against Overturning	1.92	1.94	1.96	1.91	1.86	1.98
Shear Friction Factor F	4.53	4.65	4.29	3.97	3.72	3.66
Total Volume of Gabion Boxes, m^3	22.80	24.00	25.20	25.20	25.20	31.20
C. WING WALLS (ABUTMENTS) ON BOTH BANKS						
Length in U/S $L_{U/S\ ABUT}$, m	3.00	3.00	3.00	3.00	3.50	3.50
Length in D/S $L_{D/S\ ABUT}$, m	2.00	2.00	2.00	2.00	2.00	2.00
Top Level including Free Board, m	100.80	101.00	101.30	101.30	101.30	101.30
Foundation Level, m	98.00	98.00	98.00	98.00	98.00	98.00
Total Height H_{ABUT}, m	2.80	3.00	3.30	3.30	3.30	3.30
Base Width of Abutment B_{ABUT}, m	2.00	2.00	2.00	2.00	2.00	2.00
Size of Gabion Box (L×B×H), m						
Bottom Layer 1	2.00 × 1.00 × 1.00	2.00 × 1.00 × 1.00	2.00 × 1.00 × 1.00	2.00 × 1.00 × 1.00	2.00 × 1.00 × 1.00	2.00 × 1.00 × 1.00
Layer 2	1.00 × 1.00 × 1.00	1.00 × 1.00 × 1.00	1.00 × 1.00 × 1.00	1.00 × 1.00 × 1.00	1.00 × 1.00 × 1.00	1.00 × 1.00 × 1.00
Layer 3	1.00 × 1.00 × 0.80	1.00 × 1.00 × 1.00	1.00 × 1.00 × 1.00	1.00 × 1.00 × 1.00	1.00 × 1.00 × 1.00	1.00 × 1.00 × 1.00
Layer 4			1.00 × 1.00 × 0.30	1.00 × 1.00 × 0.30	1.00 × 1.00 × 0.30	1.00 × 1.00 × 0.30

(Continued)

TABLE 5.20 (Continued)

Catchment Area, ha	50.00	100.00	150.00	200.00	250.00	300.00
Pressure at Toe, t/m^2	7.14	8.09	9.69	9.69	9.69	9.69
Pressure at Heel, t/m^2	2.18	1.91	1.33	1.33	1.33	1.33
Factor of Safety against Overturning	5.19	4.51	3.73	3.73	3.73	3.73
Shear Friction Factor F	6.05	5.39	4.59	4.59	4.59	4.59
Total Volume of Gabion Boxes, m^3	38.00	40.00	43.00	43.00	47.30	47.30
D. D/S WING WALLS ON BOTH BANKS						
Length in D/S $L_{D/S\ WING}$, m	2.10	2.10	2.10	2.10	2.10	2.10
Top Level including Free Board, m	99.30	99.50	99.80	99.80	99.80	99.80
Foundation Level, m	98.00	98.00	98.00	98.00	98.00	98.00
Total Height $H_{D/S\ WING}$, m	1.30	1.50	1.80	1.80	1.80	1.80
Base Width of D/S Wing Wall $B_{D/S\ WING}$, m	1.00	1.00	1.50	1.50	1.50	1.50
Size of Gabion Box (L×B×H), m						
Bottom Layer 1	1.00 × 1.00 × 1.00	1.00 × 1.00 × 1.00	1.50 × 1.00 × 1.00	1.50 × 1.00 × 1.00	1.50 × 1.00 × 1.00	1.50 × 1.00 × 1.00
Layer 2	1.00 × 1.00 × 0.30	1.00 × 1.00 × 0.50	1.00 × 1.00 × 0.80	1.00 × 1.00 × 0.80	1.00 × 1.00 × 0.80	1.00 × 1.00 × 0.80
Pressure at Toe, t/m^2	3.16	4.11	4.12	4.12	4.12	4.12
Pressure at Heel, t/m^2	1.00	0.69	1.74	1.74	1.74	1.74
Factor of Safety against Overturning	5.76	4.21	6.87	6.87	6.87	6.87
Shear Friction Factor F	11.83	8.93	9.81	9.81	9.81	9.81
Total Volume of Gabion Boxes, m^3	5.46	6.30	9.66	9.66	9.66	9.66

5.4 DESIGN OF 2.50 M HIGH GABION CHECK DAMS

5.4.1 Section with Vertical D/S Face

The tables developed after the rigorous design of the different components for 2.50 m high gabion check dams for different combinations are presented from Table 5.21 to Table 5.24.

TABLE 5.21
Design Results for Gabion Check Dam with Vertical D/S Face Height = 2.50 m and Stream Width L = 5.0 m

Catchment Area, ha	50.00	100.00
A. WEIR		
Span of Weir W, m	4.00	4.00
Flood Discharge Q, m^3/s	6.54	11.00
Head Over Crest, m	0.92	1.31
Top Level, m	100.00	100.00
Foundation Level, m	97.50	97.50
Total Height of Weir H_{WEIR}, m	2.50	2.50
Bottom Width of Weir B_{WEIR}, m	4.00	4.00
Size of Gabion Box (L×B×H), m		
Bottom Layer 1	4.00 × 1.00 × 1.00	4.00 × 1.00 × 1.00
Layer 2	4.00 × 1.00 × 1.00	4.00 × 1.00 × 1.00
Layer 3	4.00 × 1.00 × 0.50	4.00 × 1.00 × 0.50
Pressure at Toe, t/m^2	6.36	6.94
Pressure at Heel, t/m^2	0.37	0.30
Factor of Safety against Overturning	2.01	2.01
Shear Friction Factor F	4.02	3.47
Total Volume of Gabion Boxes, m^3	40.00	40.00
Length of D/S Apron Floor $L_{D/S\ APRON}$, m	6.00	6.00
B. NON-OVER FLOW PORTION		
Length, m	1.00	1.00
Top Level including Free Board, m	101.50	101.80
Foundation Level, m	97.50	97.50
Total Height of Non-Over Flow H_{NOF}, m	4.00	4.30
Bottom Width of Non-Over Flow B_{NOF}, m	4.00	4.00
Size of Gabion Box (L×B×H), m		
Bottom Layer 1	4.00 × 1.00 × 1.00	4.00 × 1.00 × 1.00
Layer 2	4.00 × 1.00 × 1.00	4.00 × 1.00 × 1.00
Layer 3	3.50 × 1.00 × 1.00	4.00 × 1.00 × 1.00
Layer 4	3.00 × 1.00 × 1.00	3.50 × 1.00 × 1.30

(Continued)

TABLE 5.21 (Continued)

Catchment Area, ha	**50.00**	**100.00**
Pressure at Toe, t/m^2	9.61	10.97
Pressure at Heel, t/m^2	0.00	0.00
Factor of Safety against Overturning	2.24	2.24
Shear Friction Factor F	3.87	3.22
Total Volume of Gabion Boxes, m^3	14.50	16.55
C. WING WALLS (ABUTMENTS) ON BOTH BANKS		
Length in U/S $L_{U/S\ ABUT}$, m	4.00	4.00
Length in D/S $L_{D/S\ ABUT}$, m	2.00	2.00
Top Level including Free Board, m	101.50	101.80
Foundation Level, m	97.50	97.50
Total Height H_{ABUT}, m	4.00	4.30
Base Width of Abutment B_{ABUT}, m	2.50	2.50
Size of Gabion Box (L×B×H), m		
Bottom Layer 1	2.50 × 1.00 × 1.00	2.50 × 1.00 × 1.00
Layer 2	2.00 × 1.00 × 1.00	2.00 × 1.00 × 1.00
Layer 3	1.00 × 1.00 × 1.00	1.00 × 1.00 × 1.00
Layer 4	1.00 × 1.00 × 1.00	1.00 × 1.00 × 1.30
Pressure at Toe, t/m^2	11.43	13.15
Pressure at Heel, t/m^2	1.93	1.24
Factor of Safety against Overturning	3.98	3.45
Shear Friction Factor F	4.19	3.73
Total Volume of Gabion Boxes, m^3	78.00	81.60
D. D/S WING WALLS ON BOTH BANKS		
Length in D/S $L_{D/S\ WING}$, m	4.00	4.00
Top Level including Free Board, m	99.50	99.80
Foundation Level, m	97.50	97.50
Total Height $H_{D/S\ WING}$, m	2.00	2.30
Base Width of D/S Wing Wall $B_{D/S\ WING}$, m	1.50	1.50
Size of Gabion Box (L×B×H), m		
Bottom Layer 1	1.50 × 1.00 × 1.00	1.50 × 1.00 × 1.00
Layer 2	1.00 × 1.00 × 1.00	1.00 × 1.00 × 1.00
Layer 3		1.00 × 1.00 × 0.30
Pressure at Toe, t/m^2	4.94	6.32
Pressure at Heel, t/m^2	1.60	1.21
Factor of Safety against Overturning	5.50	4.20
Shear Friction Factor F	8.05	6.33
Total Volume of Gabion Boxes, m^3	20.00	22.40

(Continued)

TABLE 5.22
Design Results for Gabion Check Dam with Vertical D/S Face Height = 2.50 m and Stream Width L = 10.0 m

Catchment Area, ha	50.00	100.00	150.00	200.00
A. WEIR				
Span of Weir W, m	8.00	8.00	8.00	8.00
Flood Discharge Q, m^3/s	6.54	11.00	14.91	18.50
Head Over Crest, m	0.58	0.82	1.01	1.16
Top Level, m	100.00	100.00	100.00	100.00
Foundation Level, m	97.50	97.50	97.50	97.50
Total Height of Weir H_{WEIR}, m	2.50	2.50	2.50	2.50
Bottom Width of Weir B_{WEIR}, m	4.00	4.00	4.00	4.00
Size of Gabion Box (L×B×H), m				
Bottom Layer 1	4.00 × 1.00 × 1.00	4.00 × 1.00 × 1.00	4.00 × 1.00 × 1.00	4.00 × 1.00 × 1.00
Layer 2	4.00 × 1.00 × 1.00	4.00 × 1.00 × 1.00	4.00 × 1.00 × 1.00	4.00 × 1.00 × 1.00
Layer 3	3.50 × 1.00 × 0.50	4.00 × 1.00 × 0.50	4.00 × 1.00 × 0.50	4.00 × 1.00 × 0.50
Pressure at Toe, t/m^2	6.04	6.21	6.49	6.73
Pressure at Heel, t/m^2	0.07	0.38	0.35	0.33
Factor of Safety against Overturning	1.93	2.01	2.01	2.01
Shear Friction Factor F	4.65	4.19	3.88	3.66
Total Volume of Gabion Boxes, m^3	78.00	80.00	80.00	80.00
Length of D/S Apron Floor $L_{D/S\ APRON}$, m	6.00	6.00	6.00	6.00
B. NON-OVER FLOW PORTION				
Length, m	2.00	2.00	2.00	2.00
Top Level including Free Board, m	101.00	101.30	101.50	101.80
Foundation Level, m	97.50	97.50	97.50	97.50
Total Height of Non-Over Flow H_{NOF}, m	3.50	3.80	4.00	4.30
Bottom Width of Non-Over Flow B_{NOF}, m	4.00	4.00	4.00	4.00
Size of Gabion Box (L×B×H), m				
Bottom Layer 1	4.00 × 1.00 × 1.00	4.00 × 1.00 × 1.00	4.00 × 1.00 × 1.00	4.00 × 1.00 × 1.00
Layer 2	4.00 × 1.00 × 1.00	4.00 × 1.00 × 1.00	4.00 × 1.00 × 1.00	4.00 × 1.00 × 1.00
Layer 3	3.00 × 1.00 × 1.00	3.50 × 1.00 × 1.00	3.50 × 1.00 × 1.00	3.50 × 1.00 × 1.00
Layer 4	3.00 × 1.00 × 0.50	3.00 × 1.00 × 0.80	3.50 × 1.00 × 1.00	3.50 × 1.00 × 1.30

(Continued)

TABLE 5.22 (Continued)

Catchment Area, ha	50.00	100.00	150.00	200.00
Pressure at Toe, t/m^2	7.94	8.96	9.69	10.66
Pressure at Heel, t/m^2	0.13	0.11	0.21	0.13
Factor of Safety against Overturning	2.18	2.23	2.27	2.30
Shear Friction Factor F	4.62	4.06	3.71	3.47
Total Volume of Gabion Boxes, m^3	25.00	27.80	30.00	32.10
C. WING WALLS (ABUTMENTS) ON BOTH BANKS				
Length in U/S $L_{U/S\ ABUT}$, m	4.00	4.00	4.00	4.00
Length in D/S $L_{D/S\ ABUT}$, m	2.00	2.00	2.00	2.00
Top Level including Free Board, m	101.00	101.30	101.50	101.80
Foundation Level, m	97.50	97.50	97.50	97.50
Total Height H_{ABUT}, m	3.50	3.80	4.00	4.30
Base Width of Abutment B_{ABUT}, m	2.00	2.50	2.50	2.50
Size of Gabion Box (L×B×H), m				
Bottom Layer 1	2.00 × 1.00 × 1.00	2.50 × 1.00 × 1.00	2.50 × 1.00 × 1.00	2.50 × 1.00 × 1.00
Layer 2	1.00 × 1.00 × 1.00	2.00 × 1.00 × 1.00	2.00 × 1.00 × 1.00	2.00 × 1.00 × 1.00
Layer 3	1.00 × 1.00 × 1.00	1.00 × 1.00 × 1.00	1.00 × 1.00 × 1.00	1.00 × 1.00 × 1.00
Layer 4	1.00 × 1.00 × 0.50	1.00 × 1.00 × 0.80	1.00 × 1.00 × 1.00	1.00 × 1.00 × 1.30
Pressure at Toe, t/m^2	10.88	10.41	11.43	13.15
Pressure at Heel, t/m^2	0.82	2.27	1.93	1.24
Factor of Safety against Overturning	3.32	4.38	3.98	3.45
Shear Friction Factor F	4.17	4.55	4.19	3.73
Total Volume of Gabion Boxes, m^3	54.00	75.60	78.00	81.60
D. D/S WING WALLS ON BOTH BANKS				
Length in D/S $L_{D/S\ WING}$, m	4.00	4.00	4.00	4.00
Top Level including Free Board, m	99.00	99.30	99.50	99.80
Foundation Level, m	97.50	97.50	97.50	97.50
Total Height $H_{D/S\ WING}$, m	1.50	1.80	2.00	2.30

(Continued)

TABLE 5.22 (Continued)

Catchment Area, ha	50.00	100.00	150.00	200.00
Base Width of D/S Wing Wall $B_{D/S\ WING}$, m	1.00	1.50	1.50	1.50
Size of Gabion Box (L×B×H), m				
Bottom Layer 1	1.00 × 1.00 × 1.00	1.50 × 1.00 × 1.00	1.50 × 1.00 × 1.00	1.50 × 1.00 × 1.00
Layer 2	1.00 × 1.00 × 0.50	1.00 × 1.00 × 0.80	1.00 × 1.00 × 1.00	1.00 × 1.00 × 1.00
Layer 3				1.00 × 1.00 × 0.30
Pressure at Toe, t/m²	4.11	4.12	4.94	6.32
Pressure at Heel, t/m²	0.69	1.74	1.60	1.21
Factor of Safety against Overturning	4.21	6.87	5.50	4.20
Shear Friction Factor F	8.93	9.81	8.05	6.33
Total Volume of Gabion Boxes, m³	12.00	18.40	20.00	22.40

TABLE 5.23
Design Results for Gabion Check Dam with Vertical D/S Face Height = 2.50 m and Stream Width L = 15.0 m

Catchment Area, ha	50.00	100.00	150.00	200.00	250.00	300.00
A. WEIR						
Span of Weir W, m	12.00	12.00	12.00	12.00	12.00	12.00
Flood Discharge Q, m³/s	6.54	11.00	14.91	18.50	21.87	25.07
Head Over Crest, m	0.44	0.63	0.77	0.89	0.99	1.09
Top Level, m	100.00	100.00	100.00	100.00	100.00	100.00
Foundation Level, m	97.50	97.50	97.50	97.50	97.50	97.50
Total Height of Weir H_{WEIR}, m	2.50	2.50	2.50	2.50	2.50	2.50
Bottom Width of Weir B_{WEIR}, m	4.00	4.00	4.00	4.00	4.00	4.00
Size of Gabion Box (L×B×H), m						
Bottom Layer 1	4.00 × 1.00 × 1.00	4.00 × 1.00 × 1.00	4.00 × 1.00 × 1.00	4.00 × 1.00 × 1.00	4.00 × 1.00 × 1.00	4.00 × 1.00 × 1.00
Layer 2	4.00 × 1.00 × 1.00	4.00 × 1.00 × 1.00	4.00 × 1.00 × 1.00	4.00 × 1.00 × 1.00	4.00 × 1.00 × 1.00	4.00 × 1.00 × 1.00

(Continued)

TABLE 5.23 (Continued)

Catchment Area, ha	50.00	100.00	150.00	200.00	250.00	300.00
Layer 3	3.50 × 1.00 × 0.50	3.50 × 1.00 × 0.50	4.00 × 1.00 × 0.50	4.00 × 1.00 × 0.50	4.00 × 1.00 × 0.50	4.00 × 1.00 × 0.50
Pressure at Toe, t/m^2	5.80	6.11	6.13	6.31	6.47	6.61
Pressure at Heel, t/m^2	0.15	0.04	0.40	0.38	0.36	0.34
Factor of Safety against Overturning	1.94	1.92	2.01	2.01	2.01	2.01
Shear Friction Factor F	5.01	4.55	4.29	4.08	3.91	3.76
Total Volume of Gabion Boxes, m^3	117.00	117.00	120.00	120.00	120.00	120.00
Length of D/S Apron Floor $L_{D/S\ APRON}$, m	6.00	6.00	6.00	6.00	6.00	6.00
B. NON-OVER FLOW PORTION						
Length, m	3.00	3.00	3.00	3.00	3.00	3.00
Top Level including Free Board, m	100.80	101.00	101.30	101.30	101.50	101.80
Foundation Level, m	97.50	97.50	97.50	97.50	97.50	97.50
Total Height of Non-Over Flow H_{NOF}, m	3.30	3.50	3.80	3.80	4.00	4.30
Bottom Width of Non-Over Flow B_{NOF}, m	4.00	4.00	4.00	4.00	4.00	4.00
Size of Gabion Box (L×B×H), m						
Bottom Layer 1	4.00 × 1.00 × 1.00	4.00 × 1.00 × 1.00	4.00 × 1.00 × 1.00	4.00 × 1.00 × 1.00	4.00 × 1.00 × 1.00	4.00 × 1.00 × 1.00
Layer 2	4.00 × 1.00 × 1.00	4.00 × 1.00 × 1.00	4.00 × 1.00 × 1.00	4.00 × 1.00 × 1.00	4.00 × 1.00 × 1.00	4.00 × 1.00 × 1.00
Layer 3	3.00 × 1.00 × 1.00	3.00 × 1.00 × 1.00	3.50 × 1.00 × 1.00	3.50 × 1.00 × 1.00	3.50 × 1.00 × 1.00	3.50 × 1.00 × 1.00
Layer 4	2.50 × 1.00 × 0.30	3.00 × 1.00 × 0.50	3.00 × 1.00 × 0.80	3.00 × 1.00 × 0.80	3.50 × 1.00 × 1.00	3.50 × 1.00 × 1.30
Pressure at Toe, t/m^2	7.31	8.02	8.85	9.09	9.64	10.48
Pressure at Heel, t/m^2	0.09	0.07	0.19	0.02	0.26	0.29

(Continued)

TABLE 5.23 (Continued)

Catchment Area, ha	50.00	100.00	150.00	200.00	250.00	300.00
Factor of Safety against Overturning	2.12	2.16	2.25	2.19	2.28	2.35
Shear Friction Factor F	4.99	4.48	4.20	3.89	3.74	3.63
Total Volume of Gabion Boxes, m^3	35.25	37.50	41.70	41.70	45.00	48.15
C. WING WALLS (ABUTMENTS) ON BOTH BANKS						
Length in U/S $L_{U/S\ ABUT}$, m	4.00	4.00	4.00	4.00	4.00	4.00
Length in D/S $L_{D/S\ ABUT}$, m	2.00	2.00	2.00	2.00	2.00	2.00
Top Level including Free Board, m	100.80	101.00	101.30	101.30	101.50	101.80
Foundation Level, m	97.50	97.50	97.50	97.50	97.50	97.50
Total Height H_{ABUT}, m	3.30	3.50	3.80	3.80	4.00	4.30
Base Width of Abutment B_{ABUT}, m	2.00	2.00	2.50	2.50	2.50	2.50
Size of Gabion Box (L×B×H), m						
Bottom Layer 1	2.00 × 1.00 × 1.00	2.00 × 1.00 × 1.00	2.50 × 1.00 × 1.00	2.50 × 1.00 × 1.00	2.50 × 1.00 × 1.00	2.50 × 1.00 × 1.00
Layer 2	1.00 × 1.00 × 1.00	1.00 × 1.00 × 1.00	1.50 × 1.00 × 1.00	1.50 × 1.00 × 1.00	1.50 × 1.00 × 1.00	2.00 × 1.00 × 1.00
Layer 3	1.00 × 1.00 × 1.00	1.00 × 1.00 × 1.00	1.00 × 1.00 × 1.00	1.00 × 1.00 × 1.00	1.00 × 1.00 × 1.00	1.00 × 1.00 × 1.00
Layer 4	1.00 × 1.00 × 0.30	1.00 × 1.00 × 0.50	1.00 × 1.00 × 0.80	1.00 × 1.00 × 0.80	1.00 × 1.00 × 1.00	1.00 × 1.00 × 1.30
Pressure at Toe, t/m^2	9.69	10.88	10.40	10.40	11.42	13.15
Pressure at Heel, t/m^2	1.33	0.82	2.35	2.35	2.02	1.24
Factor of Safety against Overturning	3.73	3.32	4.41	4.41	4.01	3.45
Shear Friction Factor F	4.59	4.17	4.56	4.56	4.20	3.73
Total Volume of Gabion Boxes, m^3	51.60	54.00	69.60	69.60	72.00	81.60

(Continued)

TABLE 5.23 (Continued)

Catchment Area, ha	50.00	100.00	150.00	200.00	250.00	300.00
D. D/S WING WALLS ON BOTH BANKS						
Length in D/S $L_{D/S\ WING}$, m	4.00	4.00	4.00	4.00	4.00	4.00
Top Level including Free Board, m	98.80	99.00	99.30	99.30	99.50	99.80
Foundation Level, m	97.50	97.50	97.50	97.50	97.50	97.50
Total Height $H_{D/S\ WING}$, m	1.30	1.50	1.80	1.80	2.00	2.30
Base Width of D/S Wing Wall $B_{D/S\ WING}$, m	1.00	1.00	1.50	1.50	1.50	1.50
Size of Gabion Box (L×B×H), m						
Bottom Layer 1	1.00 × 1.00 × 1.00	1.00 × 1.00 × 1.00	1.50 × 1.00 × 1.00	1.50 × 1.00 × 1.00	1.50 × 1.00 × 1.00	1.50 × 1.00 × 1.00
Layer 2	1.00 × 1.00 × 0.30	1.00 × 1.00 × 0.50	1.00 × 1.00 × 0.80	1.00 × 1.00 × 0.80	1.00 × 1.00 × 1.00	1.00 × 1.00 × 1.00
Layer 3						1.00 × 1.00 × 0.30
Pressure at Toe, t/m²	3.16	4.11	4.12	4.12	4.94	6.32
Pressure at Heel, t/m²	1.00	0.69	1.74	1.74	1.60	1.21
Factor of Safety against Overturning	5.76	4.21	6.87	6.87	5.50	4.20
Shear Friction Factor F	11.83	8.93	9.81	9.81	8.05	6.33
Total Volume of Gabion Boxes, m³	10.40	12.00	18.40	18.40	20.00	22.40

TABLE 5.24
Design Results for Gabion Check Dam with Vertical D/S Face Height = 2.50 m and Stream Width L = 20.0 m

Catchment Area, ha	50.00	100.00	150.00	200.00	250.00	300.00
A. WEIR						
Span of Weir W, m	16.00	16.00	16.00	16.00	16.00	16.00

(Continued)

TABLE 5.24 (Continued)

Catchment Area, ha	50.00	100.00	150.00	200.00	250.00	300.00
Flood Discharge Q, m^3/s	6.54	11.00	14.91	18.50	21.87	25.07
Head Over Crest, m	0.37	0.52	0.64	0.73	0.82	0.90
Top Level, m	100.00	100.00	100.00	100.00	100.00	100.00
Foundation Level, m	97.50	97.50	97.50	97.50	97.50	97.50
Total Height of Weir H_{WEIR}, m	2.50	2.50	2.50	2.50	2.50	2.50
Bottom Width of Weir B_{WEIR}, m	4.00	4.00	4.00	4.00	4.00	4.00
Size of Gabion Box (L×B×H), m						
Bottom Layer 1	4.00 × 1.00 × 1.00	4.00 × 1.00 × 1.00	4.00 × 1.00 × 1.00	4.00 × 1.00 × 1.00	4.00 × 1.00 × 1.00	4.00 × 1.00 × 1.00
Layer 2	4.00 × 1.00 × 1.00	4.00 × 1.00 × 1.00	4.00 × 1.00 × 1.00	4.00 × 1.00 × 1.00	4.00 × 1.00 × 1.00	4.00 × 1.00 × 1.00
Layer 3	3.50 × 1.00 × 0.50	3.50 × 1.00 × 0.50	3.50 × 1.00 × 0.50	4.00 × 1.00 × 0.50	4.00 × 1.00 × 0.50	4.00 × 1.00 × 0.50
Pressure at Toe, t/m^2	5.67	5.93	6.13	6.08	6.21	6.33
Pressure at Heel, t/m^2	0.19	0.10	0.04	0.40	0.39	0.37
Factor of Safety against Overturning	1.94	1.93	1.92	2.01	2.01	2.01
Shear Friction Factor F	5.23	4.81	4.54	4.37	4.20	4.06
Total Volume of Gabion Boxes, m^3	156.00	156.00	156.00	160.00	160.00	160.00
Length of D/S Apron Floor $L_{D/S\ APRON}$, m	6.00	6.00	6.00	6.00	6.00	6.00
B. NON-OVER FLOW PORTION						
Length, m	4.00	4.00	4.00	4.00	4.00	4.00
Top Level including Free Board, m	100.80	101.00	101.30	101.30	101.30	101.30
Foundation Level, m	97.50	97.50	97.50	97.50	97.50	97.50
Total Height of Non-Over Flow H_{NOF}, m	3.30	3.50	3.80	3.80	3.80	3.80
Bottom Width of Non-Over Flow B_{NOF}, m	4.00	4.00	4.00	4.00	4.00	4.00

(Continued)

TABLE 5.24 (Continued)

Catchment Area, ha	**50.00**	**100.00**	**150.00**	**200.00**	**250.00**	**300.00**
Size of Gabion Box (L×B×H), m						
Bottom Layer 1	4.00 × 1.00 × 1.00	4.00 × 1.00 × 1.00	4.00 × 1.00 × 1.00	4.00 × 1.00 × 1.00	4.00 × 1.00 × 1.00	4.00 × 1.00 × 1.00
Layer 2	3.50 × 1.00 × 1.00	3.50 × 1.00 × 1.00	3.50 × 1.00 × 1.00	3.50 × 1.00 × 1.00	3.50 × 1.00 × 1.00	3.50 × 1.00 × 1.00
Layer 3	3.00 × 1.00 × 1.00	3.50 × 1.00 × 1.00	3.50 × 1.00 × 1.00	3.50 × 1.00 × 1.00	3.50 × 1.00 × 1.00	3.50 × 1.00 × 1.00
Layer 4	3.00 × 1.00 × 0.30	3.00 × 1.00 × 0.50	3.00 × 1.00 × 0.80	3.50 × 1.00 × 0.80	3.50 × 1.00 × 0.80	3.50 × 1.00 × 0.80
Pressure at Toe, t/m^2	7.31	7.90	8.72	8.78	8.99	9.17
Pressure at Heel, t/m^2	0.02	0.14	0.10	0.34	0.16	0.00
Factor of Safety against Overturning	2.12	2.20	2.26	2.30	2.24	2.19
Shear Friction Factor F	5.27	4.83	4.56	4.31	4.07	3.88
Total Volume of Gabion Boxes, m^3	45.60	50.00	53.60	55.20	55.20	55.20
C. WING WALLS (ABUTMENTS) ON BOTH BANKS						
Length in U/SL$_{U/S\ ABUT}$, m	4.00	4.00	4.00	4.00	4.00	4.00
Length in D/S L$_{D/S\ ABUT}$, m	2.00	2.00	2.00	2.00	2.00	2.00
Top Level including Free Board, m	100.80	101.00	101.30	101.30	101.30	101.30
Foundation Level, m	97.50	97.50	97.50	97.50	97.50	97.50
Total Height H$_{ABUT}$, m	3.30	3.50	3.80	3.80	3.80	3.80
Base Width of Abutment B$_{ABUT}$, m	2.00	2.00	2.50	2.50	2.50	2.50
Size of Gabion Box (L×B×H), m						
Bottom Layer 1	2.00 × 1.00 × 1.00	2.00 × 1.00 × 1.00	2.50 × 1.00 × 1.00	2.50 × 1.00 × 1.00	2.50 × 1.00 × 1.00	2.50 × 1.00 × 1.00
Layer 2	1.00 × 1.00 × 1.00	1.00 × 1.00 × 1.00	1.50 × 1.00 × 1.00	1.50 × 1.00 × 1.00	1.50 × 1.00 × 1.00	1.50 × 1.00 × 1.00
Layer 3	1.00 × 1.00 × 1.00	1.00 × 1.00 × 1.00	1.00 × 1.00 × 1.00	1.00 × 1.00 × 1.00	1.00 × 1.00 × 1.00	1.00 × 1.00 × 1.00
Layer 4	1.00 × 1.00 × 0.30	1.00 × 1.00 × 0.50	1.00 × 1.00 × 0.80	1.00 × 1.00 × 0.80	1.00 × 1.00 × 0.80	1.00 × 1.00 × 0.80

(Continued)

TABLE 5.24 (Continued)

Catchment Area, ha	50.00	100.00	150.00	200.00	250.00	300.00
Pressure at Toe, t/m²	9.69	10.88	10.40	10.40	10.40	10.40
Pressure at Heel, t/m²	1.33	0.82	2.35	2.35	2.35	2.35
Factor of Safety against Overturning	3.73	3.32	4.41	4.41	4.41	4.41
Shear Friction Factor F	4.59	4.17	4.56	4.56	4.56	4.56
Total Volume of Gabion Boxes, m³	51.60	54.00	69.60	69.60	69.60	69.60
D. D/S WING WALLS ON BOTH BANKS						
Length in D/S $L_{D/S\ WING}$, m	4.00	4.00	4.00	4.00	4.00	4.00
Top Level including Free Board, m	98.80	99.00	99.30	99.30	99.30	99.30
Foundation Level, m	97.50	97.50	97.50	97.50	97.50	97.50
Total Height $H_{D/S\ WING}$, m	1.30	1.50	1.80	1.80	1.80	1.80
Base Width of D/S Wing Wall $B_{D/S\ WING}$, m	1.00	1.00	1.50	1.50	1.50	1.50
Size of Gabion Box (L×B×H), m						
Bottom Layer 1	1.00 × 1.00 ×1.00	1.00 × 1.00 ×1.00	1.50 × 1.00 ×1.00	1.50 × 1.00 ×1.00	1.50 × 1.00 ×1.00	1.50 × 1.00 ×1.00
Layer 2	1.00 × 1.00 × 0.30	1.00 × 1.00 × 0.50	1.00 × 1.00 × 0.80	1.00 × 1.00 × 0.80	1.00 × 1.00 × 0.80	1.00 × 1.00 × 0.80
Pressure at Toe, t/m²	3.16	4.11	4.12	4.12	4.12	4.12
Pressure at Heel, t/m²	1.00	0.69	1.74	1.74	1.74	1.74
Factor of Safety against Overturning	5.76	4.21	6.87	6.87	6.87	6.87
Shear Friction Factor F	11.83	8.93	9.81	9.81	9.81	9.81
Total Volume of Gabion Boxes, m³	10.40	12.00	18.40	18.40	18.40	18.40

5.4.2 Section with Vertical U/S Face

The tables developed after the rigorous design of the different components for 2.50 m high gabion check dams for different combinations are presented from Table 5.25 to Table 5.28.

TABLE 5.25
Design Results for Gabion Check Dam with Vertical U/S Face Height = 2.50 m and Stream Width L = 5.0 m

Catchment Area, ha	50.00	100.00
A. WEIR		
Span of Weir W, m	4.00	4.00
Flood Discharge Q, m^3/s	6.54	11.00
Head Over Crest, m	0.92	1.31
Top Level, m	100.00	100.00
Foundation Level, m	97.50	97.50
Total Height of Weir H_{WEIR}, m	2.50	2.50
Bottom Width of Weir B_{WEIR}, m	3.50	3.50
Size of Gabion Box (L×B×H), m		
Bottom Layer 1	3.50 × 1.00 × 1.00	3.50 × 1.00 × 1.00
Layer 2	3.00 × 1.00 × 1.00	3.00 × 1.00 × 1.00
Layer 3	2.00 × 1.00 × 0.50	2.00 × 1.00 × 0.50
Pressure at Toe, t/m^2	4.74	5.24
Pressure at Heel, t/m^2	0.32	0.11
Factor of Safety against Overturning	1.70	1.66
Shear Friction Factor F	3.31	2.84
Total Volume of Gabion Boxes, m^3	30.00	30.00
Length of D/S Apron Floor $L_{D/S\ APRON}$, m	5.20	5.20
B. NON-OVER FLOW PORTION		
Width, m	1.00	1.00
Top Level including Free Board, m	101.50	101.80
Foundation Level, m	97.50	97.50
Total Height of Non-Over Flow H_{NOF}, m	4.00	4.30
Bottom Width of Non-Over Flow B_{NOF}, m	3.00	3.50
Size of Gabion Box (L×B×H), m		
Bottom Layer 1	3.00 × 1.00 × 1.00	3.50 × 1.00 × 1.00
Layer 2	2.50 × 1.00 × 1.00	3.00 × 1.00 × 1.00
Layer 3	2.00 × 1.00 × 1.00	2.00 × 1.00 × 1.00
Layer 4	1.50 × 1.00 × 1.00	1.50 × 1.00 × 1.30
Pressure at Toe, t/m^2	6.76	6.54
Pressure at Heel, t/m^2	0.34	0.52
Factor of Safety against Overturning	1.75	1.75
Shear Friction Factor F	2.67	2.48
Total Volume of Gabion Boxes, m^3	9.00	10.45
C. WING WALLS (ABUTMENTS) ON BOTH BANKS		
Length in U/S $L_{U/S\ ABUT}$, m	1.00	1.00
Length in D/S $L_{D/S\ ABUT}$, m	4.00	4.50
Top Level including Free Board, m	101.50	101.80

(Continued)

TABLE 5.25 (Continued)

Catchment Area, ha	50.00	100.00
Foundation Level, m	97.50	97.50
Total Height H_{ABUT}, m	4.00	4.30
Base Width of Abutment B_{ABUT}, m	2.50	2.50
Size of Gabion Box (L×B×H), m		
Bottom Layer 1	2.50 × 1.00 × 1.00	2.50 × 1.00 × 1.00
Layer 2	2.00 × 1.00 × 1.00	2.00 × 1.00 × 1.00
Layer 3	1.00 × 1.00 × 1.00	1.00 × 1.00 × 1.00
Layer 4	1.00 × 1.00 × 1.00	1.00 × 1.00 × 1.30
Pressure at Toe, t/m²	11.43	13.15
Pressure at Heel, t/m²	1.93	1.24
Factor of Safety against Overturning	3.98	3.45
Shear Friction Factor F	4.19	3.73
Total Volume of Gabion Boxes, m³	65.00	74.80
D. D/S WING WALLS ON BOTH BANKS		
Length in D/S $L_{D/S\ WING}$, m	4.20	4.20
Top Level including Free Board, m	99.50	99.80
Foundation Level, m	97.50	97.50
Total Height $H_{D/S\ WING}$, m	2.00	2.30
Base Width of D/S Wing Wall $B_{D/S\ WING}$, m	1.50	1.50
Size of Gabion Box (L×B×H), m		
Bottom Layer 1	1.50 × 1.00 × 1.00	1.50 × 1.00 × 1.00
Layer 2	1.00 × 1.00 × 1.00	1.00 × 1.00 × 1.00
Layer 3		1.00 × 1.00 × 0.30
Pressure at Toe, t/m²	4.94	6.32
Pressure at Heel, t/m²	1.60	1.21
Factor of Safety against Overturning	5.50	4.20
Shear Friction Factor F	8.05	6.33
Total Volume of Gabion Boxes, m³	21.00	23.52

TABLE 5.26

Design Results for Gabion Check Dam with Vertical U/S Face Height = 2.50 m and Stream Width L = 10.0 m

Catchment Area, ha	50.00	100.00	150.00	200.00
A. WEIR				
Span of Weir W, m	8.00	8.00	8.00	8.00
Flood Discharge Q, m³/s	6.54	11.00	14.91	18.50
Head Over Crest, m	0.58	0.82	1.01	1.16
Top Level, m	100.00	100.00	100.00	100.00

(Continued)

TABLE 5.26 (Continued)

Catchment Area, ha	50.00	100.00	150.00	200.00
Foundation Level, m	97.50	97.50	97.50	97.50
Total Height of Weir H_{WEIR}, m	2.50	2.50	2.50	2.50
Bottom Width of Weir B_{WEIR}, m	3.50	3.50	3.50	3.50
Size of Gabion Box (L×B×H), m				
Bottom Layer 1	3.50 × 1.00 × 1.00	3.50 × 1.00 × 1.00	3.50 × 1.00 × 1.00	3.50 × 1.00 × 1.00
Layer 2	2.50 × 1.00 × 1.00	2.50 × 1.00 × 1.00	2.50 × 1.00 × 1.00	2.50 × 1.00 × 1.00
Layer 3	1.50 × 1.00 × 0.50	1.50 × 1.00 × 0.50	1.50 × 1.00 × 0.50	1.50 × 1.00 × 0.50
Pressure at Toe, t/m²	3.53	3.83	4.07	4.26
Pressure at Heel, t/m²	0.47	0.31	0.18	0.07
Factor of Safety against Overturning	1.62	1.59	1.57	1.55
Shear Friction Factor F	3.76	3.35	3.09	2.90
Total Volume of Gabion Boxes, m³	54.00	54.00	54.00	54.00
Length of D/S Apron Floor $L_{D/S\ APRON}$, m	5.20	5.20	5.20	5.20
B. NON-OVER FLOW PORTION				
Width, m	2.00	2.00	2.00	2.00
Top Level including Free Board, m	101.00	101.30	101.50	101.80
Foundation Level, m	97.50	97.50	97.50	97.50
Total Height of Non-Over Flow H_{NOF}, m	3.50	3.80	4.00	4.30
Bottom Width of Non-Over Flow B_{NOF}, m	3.00	3.00	3.50	3.50
Size of Gabion Box (L×B×H), m				
Bottom Layer 1	3.00 × 1.00 × 1.00	3.00 × 1.00 × 1.00	3.50 × 1.00 × 1.00	3.50 × 1.00 × 1.00
Layer 2	2.00 × 1.00 × 1.00	2.50 × 1.00 × 1.00	3.00 × 1.00 × 1.00	3.00 × 1.00 × 1.00
Layer 3	1.00 × 1.00 × 1.00	2.00 × 1.00 × 1.00	2.00 × 1.00 × 1.00	2.00 × 1.00 × 1.00
Layer 4	1.00 × 1.00 × 0.50	1.00 × 1.00 × 0.80	1.00 × 1.00 × 1.00	1.00 × 1.00 × 1.30
Pressure at Toe, t/m²	4.27	6.27	5.48	5.88
Pressure at Heel, t/m²	0.17	0.08	0.70	0.58
Factor of Safety against Overturning	1.54	1.68	1.78	1.75
Shear Friction Factor F	3.04	2.77	2.86	2.63

(Continued)

TABLE 5.26 (Continued)

Catchment Area, ha	**50.00**	**100.00**	**150.00**	**200.00**
Total Volume of Gabion Boxes, m^3	13.00	16.60	19.00	19.60
C. WING WALLS (ABUTMENTS) ON BOTH BANKS				
Length in U/S $L_{U/S\ ABUT}$, m	1.00	1.00	1.00	1.00
Length in D/S $L_{D/S\ ABUT}$, m	4.00	4.00	4.50	4.50
Top Level including Free Board, m	101.00	101.30	101.50	101.80
Foundation Level, m	97.50	97.50	97.50	97.50
Total Height H_{ABUT}, m	3.50	3.80	4.00	4.30
Base Width of Abutment B_{ABUT}, m	2.00	2.50	2.50	2.50
Size of Gabion Box (L×B×H), m				
Bottom Layer 1	2.00 × 1.00 × 1.00	2.50 × 1.00 × 1.00	2.50 × 1.00 × 1.00	2.50 × 1.00 × 1.00
Layer 2	1.00 × 1.00 × 1.00	2.00 × 1.00 × 1.00	2.00 × 1.00 × 1.00	2.00 × 1.00 × 1.00
Layer 3	1.00 × 1.00 × 1.00	1.00 × 1.00 × 1.00	1.00 × 1.00 × 1.00	1.00 × 1.00 × 1.00
Layer 4	1.00 × 1.00 × 0.50	1.00 × 1.00 × 0.80	1.00 × 1.00 × 1.00	1.00 × 1.00 × 1.30
Pressure at Toe, t/m^2	10.88	10.41	11.43	13.15
Pressure at Heel, t/m^2	0.82	2.27	1.93	1.24
Factor of Safety against Overturning	3.32	4.38	3.98	3.45
Shear Friction Factor F	4.17	4.55	4.19	3.73
Total Volume of Gabion Boxes, m^3	45.00	63.00	71.50	74.80
D. D/S WING WALLS ON BOTH BANKS				
Length in D/S $L_{D/S\ WING}$, m	4.20	4.20	4.20	4.20
Top Level including Free Board, m	99.00	99.30	99.50	99.80
Foundation Level, m	97.50	97.50	97.50	97.50
Total Height $H_{D/S\ WING}$, m	1.50	1.80	2.00	2.30
Base Width of D/S Wing Wall $B_{D/S\ WING}$, m	1.00	1.50	1.50	1.50
Size of Gabion Box (L×B×H), m				
Bottom Layer 1	1.00 × 1.00 × 1.00	1.50 × 1.00 × 1.00	1.50 × 1.00 × 1.00	1.50 × 1.00 × 1.00
Layer 2	1.00 × 1.00 × 0.50	1.00 × 1.00 × 0.80	1.00 × 1.00 × 1.00	1.00 × 1.00 × 1.00
Layer 3				1.00 × 1.00 × 0.30

(Continued)

TABLE 5.26 (Continued)

Catchment Area, ha	50.00	100.00	150.00	200.00
Pressure at Toe, t/m^2	4.11	4.12	4.94	6.32
Pressure at Heel, t/m^2	0.69	1.74	1.60	1.21
Factor of Safety against Overturning	4.21	6.87	5.50	4.20
Shear Friction Factor F	8.93	9.81	8.05	6.33
Total Volume of Gabion Boxes, m^3	12.60	19.32	21.00	23.52

TABLE 5.27
Design Results for Gabion Check Dam with Vertical U/S Face Height = 2.50 m and Stream Width L = 15.0 m

Catchment Area, ha	50.00	100.00	150.00	200.00	250.00	300.00
A. WEIR						
Span of Weir W, m	12.00	12.00	12.00	12.00	12.00	12.00
Flood Discharge Q, m^3/s	6.54	11.00	14.91	18.50	21.87	25.07
Head Over Crest, m	0.44	0.63	0.77	0.89	0.99	1.09
Top Level, m	100.00	100.00	100.00	100.00	100.00	100.00
Foundation Level, m	97.50	97.50	97.50	97.50	97.50	97.50
Total Height of Weir H_{WEIR}, m	2.50	2.50	2.50	2.50	2.50	2.50
Bottom Width of Weir B_{WEIR}, m	3.50	3.50	3.50	3.50	3.50	3.50
Size of Gabion Box (L×B×H), m						
Bottom Layer 1	3.50 × 1.00 × 1.00	3.50 × 1.00 × 1.00	3.50 × 1.00 × 1.00	3.50 × 1.00 × 1.00	3.50 × 1.00 × 1.00	3.50 × 1.00 × 1.00
Layer 2	2.50 × 1.00 × 1.00	2.50 × 1.00 × 1.00	2.50 × 1.00 × 1.00	2.50 × 1.00 × 1.00	2.50 × 1.00 × 1.00	2.50 × 1.00 × 1.00
Layer 3	1.50 × 1.00 × 0.50	1.50 × 1.00 × 0.50	1.50 × 1.00 × 0.50	1.50 × 1.00 × 0.50	1.50 × 1.00 × 0.50	1.50 × 1.00 × 0.50
Pressure at Toe, t/m^2	3.35	3.59	3.76	3.92	4.05	4.17

(Continued)

TABLE 5.27 (Continued)

Catchment Area, ha	50.00	100.00	150.00	200.00	250.00	300.00
Pressure at Heel, t/m^2	0.57	0.44	0.35	0.26	0.19	0.13
Factor of Safety against Overturning	1.65	1.62	1.60	1.58	1.57	1.56
Shear Friction Factor F	4.06	3.68	3.44	3.25	3.11	2.99
Total Volume of Gabion Boxes, m^3	81.00	81.00	81.00	81.00	81.00	81.00
Length of D/S Apron Floor $L_{D/S\ APRON}$, m	5.20	5.20	5.20	5.20	5.20	5.20
B. NON-OVER FLOW PORTION						
Width, m	3.00	3.00	3.00	3.00	3.00	3.00
Top Level including Free Board, m	100.80	101.00	101.30	101.30	101.50	101.80
Foundation Level, m	97.50	97.50	97.50	97.50	97.50	97.50
Total Height of Non-Over Flow H_{NOF}, m	3.30	3.50	3.80	3.80	4.00	4.30
Bottom Width of Non-Over Flow B_{NOF}, m	3.00	3.00	3.00	3.50	3.50	3.50
Size of Gabion Box (L×B×H), m						
Bottom Layer 1	3.00 × 1.00 × 1.00	3.00 × 1.00 × 1.00	3.00 × 1.00 × 1.00	3.50 × 1.00 × 1.00	3.50 × 1.00 × 1.00	3.50 × 1.00 × 1.00
Layer 2	2.00 × 1.00 × 1.00	2.00 × 1.00 × 1.00	2.00 × 1.00 × 1.00	2.50 × 1.00 × 1.00	2.50 × 1.00 × 1.00	2.50 × 1.00 × 1.00
Layer 3	1.00 × 1.00 × 1.00	1.50 × 1.00 × 1.00	1.50 × 1.00 × 1.00	1.50 × 1.00 × 1.00	1.50 × 1.00 × 1.00	1.50 × 1.00 × 1.00
Layer 4	1.00 × 1.00 × 0.30	1.00 × 1.00 × 0.50	1.00 × 1.00 × 0.80	1.00 × 1.00 × 0.80	1.00 × 1.00 × 1.00	1.00 × 1.00 × 1.30
Pressure at Toe, t/m^2	3.89	4.58	4.98	4.31	4.57	4.77
Pressure at Heel, t/m^2	0.33	0.39	0.31	0.78	0.71	0.77
Factor of Safety against Overturning	1.57	1.62	1.61	1.69	1.68	1.69

(Continued)

TABLE 5.27 (Continued)

Catchment Area, ha	50.00	100.00	150.00	200.00	250.00	300.00
Shear Friction Factor F	3.33	3.01	2.76	2.96	2.79	2.66
Total Volume of Gabion Boxes, m^3	18.90	21.00	21.90	24.90	25.50	26.40
C. WING WALLS (ABUTMENTS) ON BOTH BANKS						
Length in U/S $L_{U/S\ ABUT}$, m	1.00	1.00	1.00	1.00	1.00	1.00
Length in D/S $L_{D/S\ ABUT}$, m	4.50	4.50	4.50	4.50	4.50	4.50
Top Level including Free Board, m	100.80	101.00	101.30	101.30	101.50	101.80
Foundation Level, m	97.50	97.50	97.50	97.50	97.50	97.50
Total Height H_{ABUT}, m	3.30	3.50	3.80	3.80	4.00	4.30
Base Width of Abutment B_{ABUT}, m	2.50	2.50	2.50	2.50	2.50	2.50
Size of Gabion Box (L×B×H), m						
Bottom Layer 1	2.50 × 1.00 × 1.00	2.50 × 1.00 × 1.00	2.50 × 1.00 × 1.00	2.50 × 1.00 × 1.00	2.50 × 1.00 × 1.00	2.50 × 1.00 × 1.00
Layer 2	2.00 × 1.00 × 1.00	2.00 × 1.00 × 1.00	2.00 × 1.00 × 1.00	2.00 × 1.00 × 1.00	2.00 × 1.00 × 1.00	2.00 × 1.00 × 1.00
Layer 3	1.00 × 1.00 × 1.00	1.00 × 1.00 × 1.00	1.00 × 1.00 × 1.00	1.00 × 1.00 × 1.00	1.00 × 1.00 × 1.00	1.00 × 1.00 × 1.00
Layer 4	1.00 × 1.00 × 0.30	1.00 × 1.00 × 0.50	1.00 × 1.00 × 0.80	1.00 × 1.00 × 0.80	1.00 × 1.00 × 1.00	1.00 × 1.00 × 1.30
Pressure at Toe, t/m^2	8.09	8.97	10.41	10.41	11.43	13.15
Pressure at Heel, t/m^2	2.87	2.67	2.27	2.27	1.93	1.24
Factor of Safety against Overturning	5.78	5.14	4.38	4.38	3.98	3.45
Shear Friction Factor F	5.72	5.20	4.55	4.55	4.19	3.73
Total Volume of Gabion Boxes, m^3	63.80	66.00	69.30	69.30	71.50	74.80

(Continued)

TABLE 5.27 (Continued)

Catchment Area, ha	50.00	100.00	150.00	200.00	250.00	300.00
D. D/S WING WALLS ON BOTH BANKS						
Length in D/S $L_{D/S\ WING}$, m	4.20	4.20	4.20	4.20	4.20	4.20
Top Level including Free Board, m	98.80	99.00	99.30	99.30	99.50	99.80
Foundation Level, m	97.50	97.50	97.50	97.50	97.50	97.50
Total Height $H_{D/S\ WING}$, m	1.30	1.50	1.80	1.80	2.00	2.30
Base Width of D/S Wing Wall $B_{D/S\ WING}$, m	1.00	1.00	1.50	1.50	1.50	1.50
Size of Gabion Box (L×B×H), m						
Bottom Layer 1	1.00 × 1.00 × 1.00	1.00 × 1.00 × 1.00	1.50 × 1.00 × 1.00	1.50 × 1.00 × 1.00	1.50 × 1.00 × 1.00	1.50 × 1.00 × 1.00
Layer 2	1.00 × 1.00 × 0.30	1.00 × 1.00 × 0.50	1.00 × 1.00 × 0.80	1.00 × 1.00 × 0.80	1.00 × 1.00 × 1.00	1.00 × 1.00 × 1.00
Layer 3						1.00 × 1.00 × 0.30
Pressure at Toe, t/m²	3.16	4.11	4.12	4.12	4.94	6.32
Pressure at Heel, t/m²	1.00	0.69	1.74	1.74	1.60	1.21
Factor of Safety against Overturning	5.76	4.21	6.87	6.87	5.50	4.20
Shear Friction Factor F	11.83	8.93	9.81	9.81	8.05	6.33
Total Volume of Gabion Boxes, m³	10.92	12.60	19.32	19.32	21.00	23.52

TABLE 5.28
Design Results for Gabion Check Dam with Vertical U/S Face Height = 2.50 m and Stream Width L = 20.0 m

Catchment Area, ha	50.00	100.00	150.00	200.00	250.00	300.00
A. WEIR						
Span of Weir W, m	16.00	16.00	16.00	16.00	16.00	16.00

(Continued)

TABLE 5.28 (Continued)

Catchment Area, ha	50.00	100.00	150.00	200.00	250.00	300.00
Flood Discharge Q, m^3/s	6.54	11.00	14.91	18.50	21.87	25.07
Head Over Crest, m	0.37	0.52	0.64	0.73	0.82	0.90
Top Level, m	100.00	100.00	100.00	100.00	100.00	100.00
Foundation Level, m	97.50	97.50	97.50	97.50	97.50	97.50
Total Height of Weir H_{WEIR}, m	2.50	2.50	2.50	2.50	2.50	2.50
Bottom Width of Weir B_{WEIR}, m	3.00	3.50	3.50	3.50	3.50	3.50
Size of Gabion Box (L×B×H), m						
Bottom Layer 1	3.00 × 1.00 × 1.00	3.50 × 1.00 × 1.00	3.50 × 1.00 × 1.00	3.50 × 1.00 × 1.00	3.50 × 1.00 × 1.00	3.50 × 1.00 × 1.00
Layer 2	2.50 × 1.00 × 1.00	2.50 × 1.00 × 1.00	2.50 × 1.00 × 1.00	2.50 × 1.00 × 1.00	2.50 × 1.00 × 1.00	2.50 × 1.00 × 1.00
Layer 3	2.00 × 1.00 × 0.50	1.50 × 1.00 × 0.50	1.50 × 1.00 × 0.50	1.50 × 1.00 × 0.50	1.50 × 1.00 × 0.50	1.50 × 1.00 × 0.50
Pressure at Toe, t/m^2	4.71	3.45	3.60	3.72	3.83	3.93
Pressure at Heel, t/m^2	0.05	0.52	0.44	0.37	0.31	0.25
Factor of Safety against Overturning	1.64	1.63	1.62	1.60	1.59	1.58
Shear Friction Factor F	3.76	3.89	3.67	3.50	3.35	3.23
Total Volume of Gabion Boxes, m^3	104.00	108.00	108.00	108.00	108.00	108.00
Length of D/S Apron Floor $L_{D/S\ APRON}$, m	5.60	5.20	5.20	5.20	5.20	5.20
B. NON-OVER FLOW PORTION						
Width, m	4.00	4.00	4.00	4.00	4.00	4.00
Top Level including Free Board, m	100.80	101.00	101.30	101.30	101.30	101.30
Foundation Level, m	97.50	97.50	97.50	97.50	97.50	97.50
Total Height of Non-Over Flow H_{NOF}, m	3.30	3.50	3.80	3.80	3.80	3.80
Bottom Width of Non-Over Flow B_{NOF}, m	3.00	3.00	3.00	3.00	3.00	3.00

(Continued)

TABLE 5.28 (Continued)

Catchment Area, ha	**50.00**	**100.00**	**150.00**	**200.00**	**250.00**	**300.00**
Size of Gabion Box (L×B×H), m						
Bottom Layer 1	3.00 × 1.00 × 1.00	3.00 × 1.00 × 1.00	3.00 × 1.00 × 1.00	3.00 × 1.00 × 1.00	3.00 × 1.00 × 1.00	3.00 × 1.00 × 1.00
Layer 2	2.00 × 1.00 × 1.00	2.00 × 1.00 × 1.00	2.00 × 1.00 × 1.00	2.00 × 1.00 × 1.00	2.00 × 1.00 × 1.00	2.50 × 1.00 × 1.00
Layer 3	1.00 × 1.00 × 1.00	1.00 × 1.00 × 1.00	1.00 × 1.00 × 1.00	1.00 × 1.00 × 1.00	1.50 × 1.00 × 1.00	2.00 × 1.00 × 1.00
Layer 4	1.00 × 1.00 × 0.30	1.00 × 1.00 × 0.50	1.00 × 1.00 × 0.80	1.00 × 1.00 × 0.80	1.00 × 1.00 × 0.80	1.50 × 1.00 × 0.80
Pressure at Toe, t/m^2	3.64	4.05	4.31	4.67	5.20	6.72
Pressure at Heel, t/m^2	0.58	0.39	0.45	0.09	0.09	0.06
Factor of Safety against Overturning	1.62	1.58	1.60	1.54	1.57	1.70
Shear Friction Factor F	3.53	3.18	2.97	2.78	2.67	2.68
Total Volume of Gabion Boxes, m^3	25.20	26.00	27.20	27.20	29.20	34.80
C. WING WALLS (ABUTMENTS) ON BOTH BANKS						
Length in U/S $L_{U/S\ ABUT}$, m	1.00	1.00	1.00	1.00	1.00	1.00
Length in D/S $L_{D/S\ ABUT}$, m	4.00	4.00	4.00	4.00	4.00	4.00
Top Level including Free Board, m	100.80	101.00	101.30	101.30	101.30	101.30
Foundation Level, m	97.50	97.50	97.50	97.50	97.50	97.50
Total Height H_{ABUT}, m	3.30	3.50	3.80	3.80	3.80	3.80
Base Width of Abutment B_{ABUT}, m	2.00	2.00	2.50	2.50	2.50	2.50
Size of Gabion Box (L×B×H), m						
Bottom Layer 1	2.00 × 1.00 × 1.00	2.00 × 1.00 × 1.00	2.50 × 1.00 × 1.00	2.50 × 1.00 × 1.00	2.50 × 1.00 × 1.00	2.50 × 1.00 × 1.00
Layer 2	1.00 × 1.00 × 1.00	1.00 × 1.00 × 1.00	2.00 × 1.00 × 1.00	2.00 × 1.00 × 1.00	2.00 × 1.00 × 1.00	2.00 × 1.00 × 1.00
Layer 3	1.00 × 1.00 × 1.00	1.00 × 1.00 × 1.00	1.00 × 1.00 × 1.00	1.00 × 1.00 × 1.00	1.00 × 1.00 × 1.00	1.00 × 1.00 × 1.00
Layer 4	1.00 × 1.00 × 0.30	1.00 × 1.00 × 0.50	1.00 × 1.00 × 0.80	1.00 × 1.00 × 0.80	1.00 × 1.00 × 0.80	1.00 × 1.00 × 0.80

(Continued)

TABLE 5.28 (Continued)

Catchment Area, ha	50.00	100.00	150.00	200.00	250.00	300.00
Pressure at T oe, t/m^2	9.69	10.88	10.41	10.41	10.41	10.41
Pressure at Heel, t/m^2	1.33	0.82	2.27	2.27	2.27	2.27
Factor of Safety against Overturning	3.73	3.32	4.38	4.38	4.38	4.38
Shear Friction Factor F	4.59	4.17	4.55	4.55	4.55	4.55
Total Volume of Gabion Boxes, m^3	43.00	45.00	63.00	63.00	63.00	63.00
D. D/S WING WALLS ON BOTH BANKS						
Length in D/S $L_{D/S\ WING}$, m	4.60	4.20	4.20	4.20	4.20	4.20
Top Level including Free Board, m	98.80	99.00	99.30	99.30	99.30	99.30
Foundation Level, m	97.50	97.50	97.50	97.50	97.50	97.50
Total Height $H_{D/S\ WING}$, m	1.30	1.50	1.80	1.80	1.80	1.80
Base Width of D/S Wing Wall $B_{D/S\ WING}$, m	1.00	1.00	1.50	1.50	1.50	1.50
Size of Gabion Box (L×B×H), m						
Bottom Layer 1	1.00 × 1.00 × 1.00	1.00 × 1.00 × 1.00	1.50 × 1.00 × 1.00	1.50 × 1.00 × 1.00	1.50 × 1.00 × 1.00	1.50 × 1.00 × 1.00
Layer 2	1.00 × 1.00 × 0.30	1.00 × 1.00 × 0.50	1.00 × 1.00 × 0.80	1.00 × 1.00 × 0.80	1.00 × 1.00 × 0.80	1.00 × 1.00 × 0.80
Pressure at Toe, t/m^2	3.16	4.11	4.12	4.12	4.12	4.12
Pressure at Heel, t/m^2	1.00	0.69	1.74	1.74	1.74	1.74
Factor of Safety against Overturning	5.76	4.21	6.87	6.87	6.87	6.87
Shear Friction Factor F	11.83	8.93	9.81	9.81	9.81	9.81
Total Volume of Gabion Boxes, m^3	11.96	12.60	19.32	19.32	19.32	19.32

5.4.3 Section with Steps on Both Faces

The tables developed after the rigorous design of the different components for 2.50 m high gabion check dams for different combinations are presented from Table 5.29 to Table 5.32.

TABLE 5.29
Design Results for Gabion Check Dam with Steps on Both Faces Height = 2.50 m and Stream Width L = 5.0 m

Catchment Area, ha	50.00	100.00
A. WEIR		
Span of Weir W, m	4.00	4.00
Flood Discharge Q, m^3/s	6.54	11.00
Head Over Crest, m	0.92	1.31
Top Level, m	100.00	100.00
Foundation Level, m	97.50	97.50
Total Height of Weir H_{WEIR}, m	2.50	2.50
Bottom Width of Weir B_{WEIR}, m	4.50	4.50
Size of Gabion Box (L×B×H), m		
Bottom Layer 1	4.50 × 1.00 × 1.00	4.50 × 1.00 × 1.00
Layer 2	3.50 × 1.00 × 1.00	4.00 × 1.00 × 1.00
Layer 3	3.00 × 1.00 × 0.50	3.50 × 1.00 × 0.50
Pressure at Toe, t/m^2	5.38	6.30
Pressure at Heel, t/m^2	0.09	0.07
Factor of Safety against Overturning	1.84	1.92
Shear Friction Factor F	4.32	3.79
Total Volume of Gabion Boxes, m^3	38.00	41.00
Length of D/S Apron Floor $L_{D/S\ APRON}$, m	5.00	5.00
B. NON-OVER FLOW PORTION		
Length, m	1.00	1.00
Top Level including Free Board, m	101.50	101.80
Foundation Level, m	97.50	97.50
Total Height of Non-Over Flow H_{NOF}, m	4.00	4.30
Bottom Width of Non-Over Flow B_{NOF}, m	4.00	4.00
Size of Gabion Box (L×B×H), m		
Bottom Layer 1	4.00 × 1.00 × 1.00	4.00 × 1.00 × 1.00
Layer 2	3.00 × 1.00 × 1.00	3.00 × 1.00 × 1.00
Layer 3	2.00 × 1.00 × 1.00	2.00 × 1.00 × 1.00
Layer 4	1.00 × 1.00 × 1.00	1.00 × 1.00 × 1.30
Pressure at Toe, t/m^2	6.06	7.16

(Continued)

TABLE 5.29 (Continued)

Catchment Area, ha	50.00	100.00
Pressure at Heel, t/m^2	0.58	0.05
Factor of Safety against Overturning	1.93	1.82
Shear Friction Factor F	3.50	2.84
Total Volume of Gabion Boxes, m^3	10.00	10.30
C. WING WALLS (ABUTMENTS) ON BOTH BANKS		
Length in U/S $L_{U/S\ ABUT}$, m	4.00	4.00
Length in D/S $L_{D/S\ ABUT}$, m	2.00	2.00
Top Level including Free Board, m	101.50	101.80
Foundation Level, m	97.50	97.50
Total Height H_{ABUT}, m	4.00	4.30
Base Width of Abutment B_{ABUT}, m	2.50	2.50
Size of Gabion Box (L×B×H), m		
Bottom Layer 1	2.50 × 1.00 × 1.00	2.50 × 1.00 × 1.00
Layer 2	2.00 × 1.00 × 1.00	2.00 × 1.00 × 1.00
Layer 3	1.00 × 1.00 × 1.00	1.00 × 1.00 × 1.00
Layer 4	1.00 × 1.00 × 1.00	1.00 × 1.00 × 1.30
Pressure at Toe, t/m^2	11.43	13.15
Pressure at Heel, t/m^2	1.93	1.24
Factor of Safety against Overturning	3.98	3.45
Shear Friction Factor F	4.19	3.73
Total Volume of Gabion Boxes, m^3	78.00	81.60
D. D/S WING WALLS ON BOTH BANKS		
Length in D/S $L_{D/S\ WING}$, m	3.00	3.00
Top Level including Free Board, m	99.50	99.80
Foundation Level, m	97.50	97.50
Total Height $H_{D/S\ WING}$, m	2.00	2.30
Base Width of D/S Wing Wall $B_{D/S\ WING}$, m	1.50	1.50
Size of Gabion Box (L×B×H), m		
Bottom Layer 1	1.50 × 1.00 × 1.00	1.50 × 1.00 × 1.00
Layer 2	1.00 × 1.00 × 1.00	1.00 × 1.00 × 1.00
Layer 3		1.00 × 1.00 × 0.30
Pressure at Toe, t/m^2	4.94	6.32
Pressure at Heel, t/m^2	1.60	1.21
Factor of Safety against Overturning	5.50	4.20
Shear Friction Factor F	8.05	6.33
Total Volume of Gabion Boxes, m^3	15.00	16.80

TABLE 5.30
Design Results for Gabion Check Dam with Steps on Both Faces Height = 2.50 m and Stream Width L = 10.0 m

Catchment Area, ha	50.00	100.00	150.00	200.00
A. WEIR				
Span of Weir W, m	8.00	8.00	8.00	8.00
Flood Discharge Q, m^3/s	6.54	11.00	14.91	18.50
Head Over Crest, m	0.58	0.82	1.01	1.16
Top Level, m	100.00	100.00	100.00	100.00
Foundation Level, m	97.50	97.50	97.50	97.50
Total Height of Weir H_{WEIR}, m	2.50	2.50	2.50	2.50
Bottom Width of Weir B_{WEIR}, m	4.00	4.50	4.50	4.50
Size of Gabion Box (L×B×H), m				
Bottom Layer 1	4.00 × 1.00 × 1.00	4.50 × 1.00 × 1.00	4.50 × 1.00 × 1.00	4.50 × 1.00 × 1.00
Layer 2	3.50 × 1.00 × 1.00	3.50 × 1.00 × 1.00	3.50 × 1.00 × 1.00	4.00 × 1.00 × 1.00
Layer 3	3.00 × 1.00 × 0.50	3.00 × 1.00 × 0.50	3.00 × 1.00 × 0.50	3.50 × 1.00 × 0.50
Pressure at Toe, t/m^2	5.44	5.21	5.52	6.07
Pressure at Heel, t/m^2	0.09	0.16	0.02	0.15
Factor of Safety against Overturning	1.84	1.85	1.83	1.93
Shear Friction Factor F	4.56	4.51	4.17	4.00
Total Volume of Gabion Boxes, m^3	72.00	76.00	76.00	82.00
Length of D/S Apron Floor $L_{D/S\ APRON}$, m	5.40	5.00	5.00	5.00
B. NON-OVER FLOW PORTION				
Length, m	2.00	2.00	2.00	2.00
Top Level including Free Board, m	101.00	101.30	101.50	101.80
Foundation Level, m	97.50	97.50	97.50	97.50
Total Height of Non-Over Flow H_{NOF}, m	3.50	3.80	4.00	4.30
Bottom Width of Non-Over Flow B_{NOF}, m	3.50	3.50	4.00	4.00
Size of Gabion Box (L×B×H), m				
Bottom Layer 1	3.50 × 1.00 × 1.00	3.50 × 1.00 × 1.00	4.00 × 1.00 × 1.00	4.00 × 1.00 × 1.00
Layer 2	3.00 × 1.00 × 1.00	3.00 × 1.00 × 1.00	3.00 × 1.00 × 1.00	3.00 × 1.00 × 1.00
Layer 3	2.00 × 1.00 × 1.00	2.00 × 1.00 × 1.00	2.00 × 1.00 × 1.00	2.00 × 1.00 × 1.00
Layer 4	1.00 × 1.00 × 0.50	1.50 × 1.00 × 0.80	1.00 × 1.00 × 1.00	1.00 × 1.00 × 1.30
Pressure at Toe, t/m^2	6.19	7.14	6.25	6.74
Pressure at Heel, t/m^2	0.21	0.04	0.46	0.34
Factor of Safety against Overturning	1.88	1.87	1.90	1.88
Shear Friction Factor F	3.82	3.33	3.33	3.08

(Continued)

TABLE 5.30 (Continued)

Catchment Area, ha	**50.00**	**100.00**	**150.00**	**200.00**
Total Volume of Gabion Boxes, m^3	18.00	19.40	20.00	20.60
C. WING WALLS (ABUTMENTS) ON BOTH BANKS				
Length in U/S $L_{U/S\ ABUT}$, m	3.50	3.50	4.00	4.00
Length in D/S $L_{D/S\ ABUT}$, m	2.00	2.00	2.00	2.00
Top Level including Free Board, m	101.00	101.30	101.50	101.80
Foundation Level, m	97.50	97.50	97.50	97.50
Total Height H_{ABUT}, m	3.50	3.80	4.00	4.30
Base Width of Abutment B_{ABUT}, m	2.00	2.50	2.50	2.50
Size of Gabion Box (L×B×H), m				
Bottom Layer 1	2.00 × 1.00 × 1.00	2.50 × 1.00 × 1.00	2.50 × 1.00 × 1.00	2.50 × 1.00 × 1.00
Layer 2	1.00 × 1.00 × 1.00	2.00 × 1.00 × 1.00	2.00 × 1.00 × 1.00	2.00 × 1.00 × 1.00
Layer 3	1.00 × 1.00 × 1.00	1.00 × 1.00 × 1.00	1.00 × 1.00 × 1.00	1.00 × 1.00 × 1.00
Layer 4	1.00 × 1.00 × 0.50	1.00 × 1.00 × 0.80	1.00 × 1.00 × 1.00	1.00 × 1.00 × 1.30
Pressure at Toe, t/m^2	10.88	10.41	11.43	13.15
Pressure at Heel, t/m^2	0.82	2.27	1.93	1.24
Factor of Safety against Overturning	3.32	4.38	3.98	3.45
Shear Friction Factor F	4.17	4.55	4.19	3.73
Total Volume of Gabion Boxes, m^3	49.50	69.30	78.00	81.60
D. D/S WING WALLS ON BOTH BANKS				
Length in D/S $L_{D/S\ WING}$, m	3.40	3.00	3.00	3.00
Top Level including Free Board, m	99.00	99.30	99.50	99.80
Foundation Level, m	97.50	97.50	97.50	97.50
Total Height $H_{D/S\ WING}$, m	1.50	1.80	2.00	2.30
Base Width of D/S Wing Wall $B_{D/S\ WING}$, m	1.00	1.50	1.50	1.50
Size of Gabion Box (L×B×H), m				
Bottom Layer 1	1.00 × 1.00 × 1.00	1.50 × 1.00 × 1.00	1.50 × 1.00 × 1.00	1.50 × 1.00 × 1.00
Layer 2	1.00 × 1.00 × 0.50	1.00 × 1.00 × 0.80	1.00 × 1.00 × 1.00	1.00 × 1.00 × 1.00
Layer 3				1.00 × 1.00 × 0.30
Pressure at Toe, t/m^2	4.11	4.12	4.94	6.32
Pressure at Heel, t/m^2	0.69	1.74	1.60	1.21
Factor of Safety against Overturning	4.21	6.87	5.50	4.20
Shear Friction Factor F	8.93	9.81	8.05	6.33
Total Volume of Gabion Boxes, m^3	10.20	13.80	15.00	16.80

TABLE 5.31
Design Results for Gabion Check Dam with Steps on Both Faces Height = 2.50 m and Stream Width L = 15.0 m

Catchment Area, ha	50.00	100.00	150.00	200.00	250.00	300.00
A. WEIR						
Span of Weir W, m	12.00	12.00	12.00	12.00	12.00	12.00
Flood Discharge Q, m^3/s	6.54	11.00	14.91	18.50	21.87	25.07
Head Over Crest, m	0.44	0.63	0.77	0.89	0.99	1.09
Top Level, m	100.00	100.00	100.00	100.00	100.00	100.00
Foundation Level, m	97.50	97.50	97.50	97.50	97.50	97.50
Total Height of Weir H_{WEIR}, m	2.50	2.50	2.50	2.50	2.50	2.50
Bottom Width of Weir B_{WEIR}, m	4.00	4.00	4.00	4.00	4.00	4.50
Size of Gabion Box (L×B×H), m						
Bottom Layer 1	4.00 × 1.00 × 1.00	4.00 × 1.00 × 1.00	4.00 × 1.00 × 1.00	4.00 × 1.00 × 1.00	4.00 × 1.00 × 1.00	4.50 × 1.00 × 1.00
Layer 2	3.00 × 1.00 × 1.00	3.50 × 1.00 × 1.00	3.50 × 1.00 × 1.00	3.50 × 1.00 × 1.00	3.50 × 1.00 × 1.00	4.00 × 1.00 × 1.00
Layer 3	2.00 × 1.00 × 0.50	3.00 × 1.00 × 0.50	3.50 × 1.00 × 0.50	3.50 × 1.00 × 0.50	3.50 × 1.00 × 0.50	3.50 × 1.00 × 0.50
Pressure at Toe, t/m^2	4.67	5.53	5.82	6.03	6.21	5.95
Pressure at Heel, t/m^2	0.03	0.05	0.16	0.09	0.04	0.19
Factor of Safety against Overturning	1.73	1.84	1.89	1.89	1.88	1.93
Shear Friction Factor F	4.78	4.46	4.21	4.00	3.83	4.11
Total Volume of Gabion Boxes, m^3	96.00	108.00	111.00	111.00	111.00	123.00
Length of D/S Apron Floor $L_{D/S\ APRON}$, m	5.40	5.40	5.40	5.40	5.40	5.00
B. NON-OVER FLOW PORTION						
Length, m	3.00	3.00	3.00	3.00	3.00	3.00
Top Level including Free Board, m	100.80	101.00	101.30	101.30	101.50	101.80

(Continued)

TABLE 5.31 (Continued)

Catchment Area, ha	**50.00**	**100.00**	**150.00**	**200.00**	**250.00**	**300.00**
Foundation Level, m	97.50	97.50	97.50	97.50	97.50	97.50
Total Height of Non-Over Flow H_{NOF}, m	3.30	3.50	3.80	3.80	4.00	4.30
Bottom Width of Non-Over Flow B_{NOF}, m	3.50	3.50	3.50	3.50	3.50	3.50
Size of Gabion Box (L×B×H), m						
Bottom Layer 1	3.50 × 1.00 × 1.00	3.50 × 1.00 × 1.00	3.50 × 1.00 × 1.00	3.50 × 1.00 × 1.00	3.50 × 1.00 × 1.00	3.50 × 1.00 × 1.00
Layer 2	3.00 × 1.00 × 1.00	3.00 × 1.00 × 1.00	3.00 × 1.00 × 1.00	3.00 × 1.00 × 1.00	3.00 × 1.00 × 1.00	3.00 × 1.00 × 1.00
Layer 3	2.00 × 1.00 × 1.00	2.00 × 1.00 × 1.00	2.00 × 1.00 × 1.00	2.50 × 1.00 × 1.00	2.50 × 1.00 × 1.00	2.50 × 1.00 × 1.00
Layer 4	1.00 × 1.00 × 0.30	1.00 × 1.00 × 0.50	1.00 × 1.00 × 0.80	2.00 × 1.00 × 0.80	2.50 × 1.00 × 1.00	2.50 × 1.00 × 1.30
Pressure at Toe, t/m^2	5.80	6.31	6.80	7.81	8.58	9.27
Pressure at Heel, t/m^2	0.31	0.13	0.02	0.02	0.03	0.06
Factor of Safety against Overturning	1.89	1.86	1.85	1.93	1.98	2.02
Shear Friction Factor F	4.17	3.70	3.41	3.27	3.14	3.03
Total Volume of Gabion Boxes, m^3	26.40	27.00	27.90	31.80	34.50	36.75
C. WING WALLS (ABUTMENTS) ON BOTH BANKS						
Length in U/S $L_{U/S\ ABUT}$, m	3.50	3.50	3.50	3.50	3.50	3.50
Length in D/S $L_{D/S\ ABUT}$, m	2.00	2.00	2.00	2.00	2.00	2.00
Top Level including Free Board, m	100.80	101.00	101.30	101.30	101.50	101.80
Foundation Level, m	97.50	97.50	97.50	97.50	97.50	97.50
Total Height H_{ABUT}, m	3.30	3.50	3.80	3.80	4.00	4.30
Base Width of Abutment B_{ABUT}, m	2.00	2.00	2.50	2.50	2.50	2.50

(Continued)

TABLE 5.31 (Continued)

Catchment Area, ha	50.00	100.00	150.00	200.00	250.00	300.00
Size of Gabion Box (L×B×H), m						
Bottom Layer 1	2.00 × 1.00 × 1.00	2.00 × 1.00 × 1.00	2.50 × 1.00 × 1.00	2.50 × 1.00 × 1.00	2.50 × 1.00 × 1.00	2.50 × 1.00 × 1.00
Layer 2	1.00 × 1.00 × 1.00	1 × 1.0 × 1	2.00 × 1.00 × 1.00	2.00 × 1.00 × 1.00	2.00 × 1.00 × 1.00	2.00 × 1.00 × 1.00
Layer 3	1.00 × 1.00 × 1.00	1.00 × 1.00 × 1.00	1.00 × 1.00 × 1.00	1.00 × 1.00 × 1.00	1.00 × 1.00 × 1.00	1.00 × 1.00 × 1.00
Layer 4	1.00 × 1.00 × 0.30	1.00 × 1.00 × 0.50	1.00 × 1.00 × 0.80	1.00 × 1.00 × 0.80	1.00 × 1.00 × 1.00	1.00 × 1.00 × 1.30
Pressure at Toe, t/m^2	9.69	10.88	10.41	10.41	11.43	13.15
Pressure at Heel, t/m^2	1.33	0.82	2.27	2.27	1.93	1.24
Factor of Safety against Overturning	3.73	3.32	4.38	4.38	3.98	3.45
Shear Friction Factor F	4.59	4.17	4.55	4.55	4.19	3.73
Total Volume of Gabion Boxes, m^3	47.30	49.50	69.30	69.30	71.50	74.80
D. D/S WING WALLS ON BOTH BANKS						
Length in D/S $L_{D/S\ WING}$, m	3.40	3.40	3.40	3.40	3.40	3.00
Top Level including Free Board, m	98.80	99.00	99.30	99.30	99.50	99.80
Foundation Level, m	97.50	97.50	97.50	97.50	97.50	97.50
Total Height $H_{D/S\ WING}$, m	1.30	1.50	1.80	1.80	2.00	2.30
Base Width of D/S Wing Wall $B_{D/S\ WING}$, m	1.00	1.00	1.50	1.50	1.50	1.50
Size of Gabion Box (L×B×H), m						
Bottom Layer 1	1.00 × 1.00 × 1.00	1.00 × 1.00 × 1.00	1.50 × 1.00 × 1.00	1.50 × 1.00 × 1.00	1.50 × 1.00 × 1.00	1.50 × 1.00 × 1.00
Layer 2	1.00 × 1.00 × 0.30	1.00 × 1.00 × 0.50	1.00 × 1.00 × 0.80	1.00 × 1.00 × 0.80	1.00 × 1.00 × 1.00	1.00 × 1.00 × 1.00
Layer 3						1.00 × 1.00 × 0.30
Pressure at Toe, t/m^2	3.16	4.11	4.12	4.12	4.94	6.32

(Continued)

TABLE 5.31 (Continued)

Catchment Area, ha	50.00	100.00	150.00	200.00	250.00	300.00
Pressure at Heel, t/m^2	1.00	0.69	1.74	1.74	1.60	1.21
Factor of Safety against Overturning	5.76	4.21	6.87	6.87	5.50	4.20
Shear Friction Factor F	11.83	8.93	9.81	9.81	8.05	6.33
Total Volume of Gabion Boxes, m^3	8.84	10.20	15.64	15.64	17.00	16.80

TABLE 5.32
Design Results for Gabion Check Dam with Steps on Both Faces Height = 2.50 m and Stream Width L = 20.0 m

Catchment Area, ha	50.00	100.00	150.00	200.00	250.00	300.00
A. WEIR						
Span of Weir W, m	16.00	16.00	16.00	16.00	16.00	16.00
Flood Discharge Q, m^3/s	6.54	11.00	14.91	18.50	21.87	25.07
Head Over Crest, m	0.37	0.52	0.64	0.73	0.82	0.90
Top Level, m	100.00	100.00	100.00	100.00	100.00	100.00
Foundation Level, m	97.50	97.50	97.50	97.50	97.50	97.50
Total Height of Weir H_{WEIR}, m	2.50	2.50	2.50	2.50	2.50	2.50
Bottom Width of Weir B_{WEIR}, m	4.00	4.00	4.00	4.50	4.50	4.50
Size of Gabion Box (L×B×H), m						
Bottom Layer 1	4.00 × 1.00 × 1.00	4.00 × 1.00 × 1.00	4.00 × 1.00 × 1.00	4.50 × 1.00 × 1.00	4.50 × 1.00 × 1.00	4.50 × 1.00 × 1.00
Layer 2	3.00 × 1.00 × 1.00	3.00 × 1.00 × 1.00	3.50 × 1.00 × 1.00	3.50 × 1.00 × 1.00	3.50 × 1.00 × 1.00	3.50 × 1.00 × 1.00
Layer 3	2.00 × 1.00 × 0.50	2.50 × 1.00 × 0.50	3.00 × 1.00 × 0.50	2.50 × 1.00 × 0.50	2.50 × 1.00 × 0.50	3.00 × 1.00 × 0.50
Pressure at Toe, t/m^2	4.52	4.94	5.54	4.96	5.11	5.34
Pressure at Heel, t/m^2	0.12	0.03	0.05	0.10	0.02	0.11
Factor of Safety against Overturning	1.74	1.76	1.84	1.80	1.79	1.84
Shear Friction Factor F	5.00	4.63	4.44	4.66	4.48	4.36
Total Volume of Gabion Boxes, m^3	128.00	132.00	144.00	148.00	148.00	152.00

(Continued)

TABLE 5.32 (Continued)

Catchment Area, ha	50.00	100.00	150.00	200.00	250.00	300.00
Length of D/S Apron Floor $L_{D/S\ APRON}$, m	5.40	5.40	5.40	5.00	5.00	5.00
B. NON-OVER FLOW PORTION						
Length, m	4.00	4.00	4.00	4.00	4.00	4.00
Top Level including Free Board, m	100.80	101.00	101.30	101.30	101.30	101.30
Foundation Level, m	97.50	97.50	97.50	97.50	97.50	97.50
Total Height of Non-Over Flow H_{NOF}, m	3.30	3.50	3.80	3.80	3.80	3.80
Bottom Width of Non-Over Flow B_{NOF}, m	3.50	3.50	3.50	3.50	3.50	3.50
Size of Gabion Box (L×B×H), m						
Bottom Layer 1	3.50 × 1.00 × 1.00	3.50 × 1.00 × 1.00	3.50 × 1.00 × 1.00	3.50 × 1.00 × 1.00	3.50 × 1.00 × 1.00	3.50 × 1.00 × 1.00
Layer 2	3.00 × 1.00 × 1.00	3.00 × 1.00 × 1.00	3.00 × 1.00 × 1.00	3.00 × 1.00 × 1.00	3.00 × 1.00 × 1.00	3.00 × 1.00 × 1.00
Layer 3	2.00 × 1.00 × 1.00	2.00 × 1.00 × 1.00	2.00 × 1.00 × 1.00	2.00 × 1.00 × 1.00	2.00 × 1.00 × 1.00	2.50 × 1.00 × 1.00
Layer 4	1.00 × 1.00 × 0.30	1.00 × 1.00 × 0.50	1.00 × 1.00 × 0.80	1.00 × 1.00 × 0.80	1.50 × 1.00 × 0.80	2.50 × 1.00 × 0.80
Pressure at Toe, t/m²	5.64	6.06	6.46	6.69	7.13	8.07
Pressure at Heel, t/m²	0.44	0.30	0.26	0.11	0.04	0.06
Factor of Safety against Overturning	1.93	1.90	1.91	1.87	1.87	1.96
Shear Friction Factor F	4.41	3.99	3.72	3.49	3.34	3.28
Total Volume of Gabion Boxes, m³	35.20	36.00	37.20	37.20	38.80	44.00
C. WING WALLS (ABUTMENTS) ON BOTH BANKS						
Length in U/S $L_{U/S\ ABUT}$, m	3.50	3.50	3.50	3.50	3.50	3.50
Length in D/S $L_{D/S\ ABUT}$, m	2.00	2.00	2.00	2.00	2.00	2.00
Top Level including Free Board, m	100.80	101.00	101.30	101.30	101.30	101.30
Foundation Level, m	97.50	97.50	97.50	97.50	97.50	97.50
Total Height H_{ABUT}, m	3.30	3.50	3.80	3.80	3.80	3.80
Base Width of Abutment B_{ABUT}, m	2.00	2.00	2.50	2.50	2.50	2.50
Size of Gabion Box (L×B×H), m						
Bottom Layer 1	2.00 × 1.00 × 1.00	2.00 × 1.00 × 1.00	2.50 × 1.00 × 1.00	2.50 × 1.00 × 1.00	2.50 × 1.00 × 1.00	2.50 × 1.00 × 1.00
Layer 2	1.00 × 1.00 × 1.00	1.00 × 1.00 × 1.00	2.00 × 1.00 × 1.00	2.00 × 1.00 × 1.00	2.00 × 1.00 × 1.00	2.00 × 1.00 × 1.00
Layer 3	1.00 × 1.00 × 1.00	1.00 × 1.00 × 1.00	1.00 × 1.00 × 1.00	1.00 × 1.00 × 1.00	1.00 × 1.00 × 1.00	1.00 × 1.00 × 1.00
Layer 4	1.00 × 1.00 × 0.30	1.00 × 1.00 × 0.50	1.00 × 1.00 × 0.80	1.00 × 1.00 × 0.80	1.00 × 1.00 × 0.80	1.00 × 1.00 × 0.80

(Continued)

TABLE 5.32 (Continued)

Catchment Area, ha	50.00	100.00	150.00	200.00	250.00	300.00
Pressure at Toe, t/m^2	9.69	10.88	10.41	10.41	10.41	10.41
Pressure at Heel, t/m^2	1.33	0.82	2.27	2.27	2.27	2.27
Factor of Safety against Overturning	3.73	3.32	4.38	4.38	4.38	4.38
Shear Friction Factor F	4.59	4.17	4.55	4.55	4.55	4.55
Total Volume of Gabion Boxes, m^3	47.30	49.50	69.30	69.30	69.30	69.30
D. D/S WING WALLS ON BOTH BANKS						
Length in D/S $L_{D/S\ WING}$, m	3.40	3.40	3.40	3.00	3.00	3.00
Top Level including Free Board, m	98.80	99.00	99.30	99.30	99.30	99.30
Foundation Level, m	97.50	97.50	97.50	97.50	97.50	97.50
Total Height $H_{D/S\ WING}$, m	1.30	1.50	1.80	1.80	1.80	1.80
Base Width of D/S Wing Wall $B_{D/S\ WING}$, m	1.00	1.00	1.50	1.50	1.50	1.50
Size of Gabion Box (L×B×H), m						
Bottom Layer 1	1.00 × 1.00 × 1.00	1.00 × 1.00 × 1.00	1.50 × 1.00 × 1.00	1.50 × 1.00 × 1.00	1.50 × 1.00 × 1.00	1.50 × 1.00 × 1.00
Layer 2	1.00 × 1.00 × 0.30	1.00 × 1.00 × 0.50	1.00 × 1.00 × 0.80	1.00 × 1.00 × 0.80	1.00 × 1.00 × 0.80	1.00 × 1.00 × 0.80
Pressure at Toe, t/m^2	3.16	4.11	4.12	4.12	4.12	4.12
Pressure at Heel, t/m^2	1.00	0.69	1.74	1.74	1.74	1.74
Factor of Safety against Overturning	5.76	4.21	6.87	6.87	6.87	6.87
Shear Friction Factor F	11.83	8.93	9.81	9.81	9.81	9.81
Total Volume of Gabion Boxes, m^3	8.84	10.20	15.64	13.80	13.80	13.80

5.5 DESIGN OF 3.00 M HIGH GABION CHECK DAMS

5.5.1 Section with Vertical D/S Face

The tables developed after the rigorous design of the different components for 3.00 m high gabion check dams for different combinations are presented from Table 5.33 to Table 5.36.

TABLE 5.33
Design Results for Gabion Check Dam with Vertical D/S Face Height = 3.00 m and Stream Width L = 5.0 m

Catchment Area, ha	50.00	100.00
A. WEIR		
Span of Weir W, m	4.00	4.00
Flood Discharge Q, m^3/s	6.54	11.00
Head Over Crest, m	0.92	1.31
Top Level, m	100.00	100.00
Foundation Level, m	97.00	97.00
Total Height of Weir H_{WEIR}, m	3.00	3.00
Bottom Width of Weir B_{WEIR}, m	5.00	5.50
Size of Gabion Box (L×B×H), m		
Bottom Layer 1	5.00 × 1.00 × 1.00	5.50 × 1.00 × 1.00
Layer 2	5.00 × 1.00 × 1.00	5.00 × 1.00 × 1.00
Layer 3	4.50 × 1.00 × 1.00	5.00 × 1.00 × 1.00
Pressure at Toe, t/m^2	7.45	7.70
Pressure at Heel, t/m^2	0.14	0.27
Factor of Safety against Overturning	1.96	2.02
Shear Friction Factor F	3.87	3.72
Total Volume of Gabion Boxes, m^3	58.00	62.00
Length of D/S Apron Floor $L_{D/S\ APRON}$, m	6.50	6.50
B. NON-OVER FLOW PORTION		
Length, m	1.00	1.00
Top Level including Free Board, m	101.50	101.80
Foundation Level, m	97.00	97.00
Total Height of Non-Over Flow H_{NOF}, m	4.50	4.80
Bottom Width of Non-Over Flow B_{NOF}, m	5.00	5.50
Size of Gabion Box (L×B×H), m		
Bottom Layer 1	5.00 × 1.00 × 1.00	5.50 × 1.00 × 1.00
Layer 2	4.50 × 1.00 × 1.00	5.00 × 1.00 × 1.00
Layer 3	4.00 × 1.00 × 1.00	4.50 × 1.00 × 1.00
Layer 4	4.00 × 1.00 × 1.50	4.50 × 1.00 × 1.80
Pressure at Toe, t/m^2	10.46	11.22
Pressure at Heel, t/m^2	0.03	0.29
Factor of Safety against Overturning	2.22	2.33
Shear Friction Factor F	3.81	3.54
Total Volume of Gabion Boxes, m^3	19.50	23.10

(Continued)

TABLE 5.33 (Continued)

Catchment Area, ha	50.00	100.00
C. WING WALLS (ABUTMENTS) ON BOTH BANKS		
Length in U/S $L_{U/S\ ABUT}$, m	5.50	5.50
Length in D/S $L_{D/S\ ABUT}$, m	2.00	2.00
Top Level including Free Board, m	101.50	101.80
Foundation Level, m	97.00	97.00
Total Height H_{ABUT}, m	4.50	4.80
Base Width of Abutment B_{ABUT}, m	3.00	3.00
Size of Gabion Box (L×B×H), m		
Bottom Layer 1	3.00 × 1.00 × 1.00	3.00 × 1.00 × 1.00
Layer 2	2.00 × 1.00 × 1.00	2.00 × 1.00 × 1.00
Layer 3	1.00 × 1.00 × 1.00	1.00 × 1.00 × 1.00
Layer 4	1.00 × 1.00 × 1.50	1.00 × 1.00 × 1.80
Pressure at Toe, t/m^2	12.22	13.79
Pressure at Heel, t/m^2	2.98	2.45
Factor of Safety against Overturning	4.57	4.03
Shear Friction Factor F	4.18	3.78
Total Volume of Gabion Boxes, m3	112.50	117.00
D. D/S WING WALLS ON BOTH BANKS		
Length in D/S LD/S WING, m	4.50	4.50
Top Level including Free Board, m	99.00	99.30
Foundation Level, m	97.00	97.00
Total Height HD/S WING, m	2.00	2.30
Base Width of D/S Wing Wall BD/S WING, m	1.50	1.50
Size of Gabion Box (L×B×H), m		
Bottom Layer 1	1.50 × 1.00 × 1.00	1.50 × 1.00 × 1.00
Layer 2	1.00 × 1.00 × 1.00	1.00 × 1.00 × 1.00
Layer 3		1.00 × 1.00 × 0.30
Pressure at Toe, t/m2	4.94	6.32
Pressure at Heel, t/m2	1.60	1.21
Factor of Safety against Overturning	5.50	4.20
Shear Friction Factor F	8.05	6.33
Total Volume of Gabion Boxes, m3	22.50	25.20

TABLE 5.34
Design Results for Gabion Check Dam with Vertical D/S Face Height = 3.00 m and Stream Width L = 10.0 m

Catchment Area, ha	50.00	100.00	150.00	200.00
A. WEIR				
Span of Weir W, m	8.00	8.00	8.00	8.00
Flood Discharge Q, m^3/s	6.54	11.00	14.91	18.50
Head Over Crest, m	0.58	0.82	1.01	1.16
Top Level, m	100.00	100.00	100.00	100.00
Foundation Level, m	97.00	97.00	97.00	97.00
Total Height of Weir H_{WEIR}, m	3.00	3.00	3.00	3.00
Bottom Width of Weir B_{WEIR}, m	5.00	5.00	5.00	5.00
Size of Gabion Box (L×B×H), m				
Bottom Layer 1	5.00 × 1.00 × 1.00	5.00 × 1.00 × 1.00	5.00 × 1.00 × 1.00	5.00 × 1.00 × 1.00
Layer 2	4.50 × 1.00 × 1.00	5.00 × 1.00 × 1.00	5.00 × 1.00 × 1.00	5.00 × 1.00 × 1.00
Layer 3	4.50 × 1.00 × 1.00	4.50 × 1.00 × 1.00	4.50 × 1.00 × 1.00	4.50 × 1.00 × 1.00
Pressure at Toe, t/m^2	7.01	7.29	7.59	7.83
Pressure at Heel, t/m^2	0.05	0.18	0.10	0.05
Factor of Safety against Overturning	1.92	1.96	1.95	1.95
Shear Friction Factor F	4.42	4.02	3.75	3.56
Total Volume of Gabion Boxes, m^3	112.00	116.00	116.00	116.00
Length of D/S Apron Floor $L_{D/S\ APRON}$, m	7.00	7.00	7.00	7.00
B. NON-OVER FLOW PORTION				
Length, m	2.00	2.00	2.00	2.00
Top Level including Free Board, m	101.00	101.30	101.50	101.80
Foundation Level, m	97.00	97.00	97.00	97.00
Total Height of Non-Over Flow H_{NOF}, m	4.00	4.30	4.50	4.80
Bottom Width of Non-Over Flow B_{NOF}, m	5.00	5.00	5.00	5.00
Size of Gabion Box (L×B×H), m				
Bottom Layer 1	5.00 × 1.00 × 1.00	5.00 × 1.00 × 1.00	5.00 × 1.00 × 1.00	5.00 × 1.00 × 1.00
Layer 2	4.50 × 1.00 × 1.00	4.50 × 1.00 × 1.00	4.50 × 1.00 × 1.00	5 × 1.0 × 1
Layer 3	4.00 × 1.00 × 1.00	4.00 × 1.00 × 1.00	4.50 × 1.00 × 1.00	4.00 × 1.00 × 1.00
Layer 4	4.00 × 1.00 × 1.00	4.00 × 1.00 × 1.30	4.00 × 1.00 × 1.50	4.00 × 1.00 × 1.80
Pressure at Toe, t/m^2	8.88	9.87	10.55	11.46
Pressure at Heel, t/m^2	0.17	0.07	0.09	0.03
Factor of Safety against Overturning	2.16	2.19	2.22	2.26
Shear Friction Factor F	4.44	3.97	3.66	3.46

(Continued)

TABLE 5.34 (Continued)

Catchment Area, ha	50.00	100.00	150.00	200.00
Total Volume of Gabion Boxes, m³	35.00	37.40	40.00	42.40
C. WING WALLS (ABUTMENTS) ON BOTH BANKS				
Length in U/S $L_{U/S\ ABUT}$, m	5.00	5.00	5.00	5.00
Length in D/S $L_{D/S\ ABUT}$, m	2.00	2.00	2.00	2.00
Top Level including Free Board, m	101.00	101.30	101.50	101.80
Foundation Level, m	97.00	97.00	97.00	97.00
Total Height H_{ABUT}, m	4.00	4.30	4.50	4.80
Base Width of Abutment B_{ABUT}, m	2.50	2.50	2.50	3.00
Size of Gabion Box (L×B×H), m				
Bottom Layer 1	2.50 × 1.00 × 1.00	2.50 × 1.00 × 1.00	2.50 × 1.00 × 1.00	3.00 × 1.00 × 1.00
Layer 2	2.00 × 1.00 × 1.00	2.00 × 1.00 × 1.00	2.00 × 1.00 × 1.00	2.00 × 1.00 × 1.00
Layer 3	1.00 × 1.00 × 1.00	1.00 × 1.00 × 1.00	1.00 × 1.00 × 1.00	1.00 × 1.00 × 1.00
Layer 4	1.00 × 1.00 × 1.00	1.00 × 1.00 × 1.30	1.00 × 1.00 × 1.50	1.00 × 1.00 × 1.80
Pressure at Toe, t/m²	11.43	13.15	14.41	13.79
Pressure at Heel, t/m²	1.93	1.24	0.67	2.45
Factor of Safety against Overturning	3.98	3.45	3.15	4.03
Shear Friction Factor F	4.19	3.73	3.47	3.78
Total Volume of Gabion Boxes, m³	91.00	95.20	98.00	109.20
D. D/S WING WALLS ON BOTH BANKS				
Length in D/S $L_{D/S\ WING}$, m	5.00	5.00	5.00	5.00
Top Level including Free Board, m	98.50	98.80	99.00	99.30
Foundation Level, m	97.00	97.00	97.00	97.00
Total Height $H_{D/S\ WING}$, m	1.50	1.80	2.00	2.30
Base Width of D/S Wing Wall $B_{D/S\ WING}$, m	1.00	1.50	1.50	1.50
Size of Gabion Box (L×B×H), m				
Bottom Layer 1	1.00 × 1.00 × 1.00	1.50 × 1.00 × 1.00	1.50 × 1.00 × 1.00	1.50 × 1.00 × 1.00
Layer 2	1.00 × 1.00 × 0.50	1.00 × 1.00 × 0.80	1.00 × 1.00 × 1.00	1.00 × 1.00 × 1.00
Layer 3				1.00 × 1.00 × 0.30
Pressure at Toe, t/m²	4.11	4.12	4.94	6.32
Pressure at Heel, t/m²	0.69	1.74	1.60	1.21
Factor of Safety against Overturning	4.21	6.87	5.50	4.20
Shear Friction Factor F	8.93	9.81	8.05	6.33
Total Volume of Gabion Boxes, m³	15.00	23.00	25.00	28.00

TABLE 5.35
Design Results for Gabion Check Dam with Vertical D/S Face Height = 3.00 m and Stream Width L = 15.0 m

Catchment Area, ha	50.00	100.00	150.00	200.00	250.00	300.00
A. WEIR						
Span of Weir W, m	12.00	12.00	12.00	12.00	12.00	12.00
Flood Discharge Q, m^3/s	6.54	11.00	14.91	18.50	21.87	25.07
Head Over Crest, m	0.44	0.63	0.77	0.89	0.99	1.09
Top Level, m	100.00	100.00	100.00	100.00	100.00	100.00
Foundation Level, m	97.00	97.00	97.00	97.00	97.00	97.00
Total Height of Weir H_{WEIR}, m	3.00	3.00	3.00	3.00	3.00	3.00
Bottom Width of Weir B_{WEIR}, m	5.00	5.00	5.00	5.00	5.00	5.00
Size of Gabion Box (L×B×H), m						
Bottom Layer 1	5.00 × 1.00 × 1.00	5.00 × 1.00 × 1.00	5.00 × 1.00 × 1.00	5.00 × 1.00 × 1.00	5.00 × 1.00 × 1.00	5.00 × 1.00 × 1.00
Layer 2	4.50 × 1.00 × 1.00	4.50 × 1.00 × 1.00	5.00 × 1.00 × 1.00	5.00 × 1.00 × 1.00	5.00 × 1.00 × 1.00	5.00 × 1.00 × 1.00
Layer 3	4.50 × 1.00 × 1.00	4.50 × 1.00 × 1.00	4.50 × 1.00 × 1.00	4.50 × 1.00 × 1.00	4.50 × 1.00 × 1.00	4.50 × 1.00 × 1.00
Pressure at Toe, t/m^2	6.79	7.08	7.21	7.40	7.56	7.71
Pressure at Heel, t/m^2	0.10	0.03	0.20	0.15	0.11	0.07
Factor of Safety against Overturning	1.93	1.92	1.96	1.96	1.95	1.95
Shear Friction Factor F	4.71	4.33	4.10	3.92	3.77	3.64
Total Volume of Gabion Boxes, m^3	168.00	168.00	174.00	174.00	174.00	174.00
Length of D/S Apron Floor $L_{D/S\ APRON}$, m	7.00	7.00	7.00	7.00	7.00	7.00
B. NON-OVER FLOW PORTION						
Length, m	3.00	3.00	3.00	3.00	3.00	3.00
Top Level including Free Board, m	100.80	101.00	101.30	101.30	101.50	101.80
Foundation Level, m	97.00	97.00	97.00	97.00	97.00	97.00

(Continued)

TABLE 5.35 (Continued)

Catchment Area, ha	50.00	100.00	150.00	200.00	250.00	300.00
Total Height of Non-Over Flow H_{NOF}, m	3.80	4.00	4.30	4.30	4.50	4.80
Bottom Width of Non-Over Flow B_{NOF}, m	5.00	5.00	5.00	5.00	5.00	5.00
Size of Gabion Box (L×B×H), m						
Bottom Layer 1	5.00 × 1.00 × 1.00	5.00 × 1.00 × 1.00	5.00 × 1.00 × 1.00	5.00 × 1.00 × 1.00	5.00 × 1.00 × 1.00	5.00 × 1.00 × 1.00
Layer 2	4.50 × 1.00 × 1.00	4.50 × 1.00 × 1.00	4.50 × 1.00 × 1.00	5.00 × 1.00 × 1.00	5.00 × 1.00 × 1.00	5.00 × 1.00 × 1.00
Layer 3	4.00 × 1.00 × 1.00	4.00 × 1.00 × 1.00	4.00 × 1.00 × 1.00	4.00 × 1.00 × 1.00	4.00 × 1.00 × 1.00	4.00 × 1.00 × 1.00
Layer 4	3.50 × 1.00 × 0.80	4.00 × 1.00 × 1.00	4.00 × 1.00 × 1.30	4.00 × 1.00 × 1.30	4.00 × 1.00 × 1.50	4.00 × 1.00 × 1.80
Pressure at Toe, t/m^2	8.31	8.94	9.78	9.89	10.50	11.29
Pressure at Heel, t/m^2	0.02	0.14	0.13	0.19	0.14	0.16
Factor of Safety against Overturning	2.08	2.15	2.21	2.20	2.23	2.29
Shear Friction Factor F	4.73	4.32	4.09	3.84	3.69	3.59
Total Volume of Gabion Boxes, m^3	48.90	52.50	56.10	57.60	60.00	63.60
C. WING WALLS (ABUTMENTS) ON BOTH BANKS						
Length in U/S $L_{U/S\ ABUT}$, m	5.00	5.00	5.00	5.00	5.00	5.00
Length in D/S $L_{D/S\ ABUT}$, m	2.00	2.00	2.00	2.00	2.00	2.00
Top Level including Free Board, m	100.80	101.00	101.30	101.30	101.50	101.80
Foundation Level, m	97.00	97.00	97.00	97.00	97.00	97.00
Total Height H_{ABUT}, m	3.80	4.00	4.30	4.30	4.50	4.80
Base Width of Abutment B_{ABUT}, m	2.50	2.50	2.50	2.50	2.50	3.00

(Continued)

TABLE 5.35 (Continued)

Catchment Area, ha	**50.00**	**100.00**	**150.00**	**200.00**	**250.00**	**300.00**
Size of Gabion Box (L×B×H), m						
Bottom Layer 1	2.50 × 1.00 × 1.00	2.50 × 1.00 × 1.00	2.50 × 1.00 × 1.00	2.50 × 1.00 × 1.00	2.50 × 1.00 × 1.00	3.00 × 1.00 × 1.00
Layer 2	2.00 × 1.00 × 1.00	2.00 × 1.00 × 1.00	2.00 × 1.00 × 1.00	2.00 × 1.00 × 1.00	2.00 × 1.00 × 1.00	2.00 × 1.00 × 1.00
Layer 3	1.00 × 1.00 × 1.00	1.00 × 1.00 × 1.00	1.00 × 1.00 × 1.00	1.00 × 1.00 × 1.00	1.00 × 1.00 × 1.00	1.00 × 1.00 × 1.00
Layer 4	1.00 × 1.00 × 0.80	1.00 × 1.00 × 1.00	1.00 × 1.00 × 1.30	1.00 × 1.00 × 1.30	1.00 × 1.00 × 1.50	1.00 × 1.00 × 1.80
Pressure at Toe, t/m	10.41	11.43	13.15	13.15	14.41	13.79
Pressure at Heel, t/m^2	2.27	1.93	1.24	1.24	0.67	2.45
Factor of Safety against Overturning	4.38	3.98	3.45	3.45	3.15	4.03
Shear Friction Factor F	4.55	4.19	3.73	3.73	3.47	3.78
Total Volume of Gabion Boxes, m^3	88.20	91.00	95.20	95.20	98.00	109.20
D. D/S WING WALLS ON BOTH BANKS						
Length in D/S $L_{D/S\ WING}$, m	5.00	5.00	5.00	5.00	5.00	5.00
Top Level including Free Board, m	98.30	98.50	98.80	98.80	99.00	99.30
Foundation Level, m	97.00	97.00	97.00	97.00	97.00	97.00
Total Height $H_{D/S\ WING}$, m	1.30	1.50	1.80	1.80	2.00	2.30
Base Width of D/S Wing Wall $B_{D/S\ WING}$, m	1.00	1.00	1.50	1.50	1.50	1.50
Size of Gabion Box (L×B×H), m						
Bottom Layer 1	1.00 × 1.00 × 1.00	1.00 × 1.00 × 1.00	1.50 × 1.00 × 1.00	1.50 × 1.00 × 1.00	1.50 × 1.00 × 1.00	1.50 × 1.00 × 1.00
Layer 2	1.00 × 1.00 × 0.30	1.00 × 1.00 × 0.50	1.00 × 1.00 × 0.80	1.00 × 1.00 × 0.80	1.00 × 1.00 × 1.00	1.00 × 1.00 1.00
Layer 3						1.00 × 1.00 × 0.30
Pressure at Toe, t/m^2	3.16	4.11	4.12	4.12	4.94	6.32

(Continued)

TABLE 5.35 (Continued)

Catchment Area, ha	50.00	100.00	150.00	200.00	250.00	300.00
Pressure at Heel, t/m^2	1.00	0.69	1.74	1.74	1.60	1.21
Factor of Safety against Overturning	5.76	4.21	6.87	6.87	5.50	4.20
Shear Friction Factor F	11.83	8.93	9.81	9.81	8.05	6.33
Total Volume of Gabion Boxes, m^3	13.00	15.00	23.00	23.00	25.00	28.00

TABLE 5.36
Design Results for Gabion Check Dam with Vertical D/S Face Height = 3.00 m and Stream Width L = 20.0 m

Catchment Area, ha	50.00	100.00	150.00	200.00	250.00	300.00
A. WEIR						
Span of Weir W, m	16.00	16.00	16.00	16.00	16.00	16.00
Flood Discharge Q, m^3/s	6.54	11.00	14.91	18.50	21.87	25.07
Head Over Crest, m	0.37	0.52	0.64	0.73	0.82	0.90
Top Level, m	100.00	100.00	100.00	100.00	100.00	100.00
Foundation Level, m	97.00	97.00	97.00	97.00	97.00	97.00
Total Height of Weir H_{WEIR}, m	3.00	3.00	3.00	3.00	3.00	3.00
Bottom Width of Weir B_{WEIR}, m	5.00	5.00	5.00	5.00	5.00	5.00
Size of Gabion Box (L×B×H), m						
Bottom Layer 1	5.00 × 1.00 × 1.00	5.00 × 1.00 × 1.00	5.00 × 1.00 × 1.00	5.00 × 1.00 × 1.00	5.00 × 1.00 × 1.00	5.00 × 1.00 × 1.00
Layer 2	4.50 × 1.00 × 1.00	4.50 × 1.00 × 1.00	4.50 × 1.00 × 1.00	5.00 × 1.00 × 1.00	5.00 × 1.00 × 1.00	5.00 × 1.00 × 1.00
Layer 3	4.50 × 1.00 × 1.00	4.50 × 1.00 × 1.00	4.50 × 1.00 × 1.00	4.50 × 1.00 × 1.00	4.50 × 1.00 × 1.00	4.50 × 1.00 × 1.00
Pressure at Toe, t/m^2	6.67	6.91	7.10	7.15	7.29	7.41

(Continued)

TABLE 5.36 (Continued)

Catchment Area, ha	50.00	100.00	150.00	200.00	250.00	300.00
Pressure at Heel, t/m^2	0.13	0.07	0.03	0.21	0.18	0.15
Factor of Safety against Overturning	1.93	1.92	1.92	1.96	1.96	1.96
Shear Friction Factor F	4.89	4.55	4.32	4.17	4.02	3.90
Total Volume of Gabion Boxes, m^3	224.00	224.00	224.00	232.00	232.00	232.00
Length of D/S Apron Floor $L_{D/S\ APRON}$, m	7.00	7.00	7.00	7.00	7.00	7.00
B. NON-OVER FLOW PORTION						
Length, m	4.00	4.00	4.00	4.00	4.00	4.00
Top Level including Free Board, m	100.80	101.00	101.30	101.30	101.30	101.30
Foundation Level, m	97.00	97.00	97.00	97.00	97.00	97.00
Total Height of Non-Over Flow H_{NOF}, m	3.80	4.00	4.30	4.30	4.30	4.30
Bottom Width of Non-Over Flow B_{NOF}, m	5.00	5.00	5.00	5.00	5.00	5.00
Size of Gabion Box (L×B×H), m						
Bottom Layer 1	5.00 × 1.00 × 1.00	5.00 × 1.00 × 1.00	5.00 × 1.00 × 1.00	5.00 × 1.00 × 1.00	5.00 × 1.00 × 1.00	5.00 × 1.00 × 1.00
Layer 2	4.50 × 1.00 × 1.00	4.50 × 1.00 × 1.00	4.50 × 1.00 × 1.00	4.50 × 1.00 × 1.00	4.50 × 1.00 × 1.00	4.50 × 1.00 × 1.00
Layer 3	4.00 × 1.00 × 1.00	4.00 × 1.00 × 1.00	4.00 × 1.00 × 1.00	4.00 × 1.00 × 1.00	4.00 × 1.00 × 1.00	4.50 × 1.00 × 1.00
Layer 4	3.50 × 1.00 × 0.80	4.00 × 1.00 × 1.00	4.00 × 1.00 × 1.30	4.00 × 1.00 × 1.30	4.00 × 1.00 × 1.30	4.00 × 1.00 × 1.30
Pressure at Toe, t/m^2	8.22	8.78	9.58	9.72	9.86	9.94
Pressure at Heel, t/m^2	0.05	0.25	0.27	0.17	0.07	0.15
Factor of Safety against Overturning	2.10	2.19	2.26	2.22	2.19	2.20
Shear Friction Factor F	4.96	4.62	4.41	4.17	3.97	3.82

(Continued)

TABLE 5.36 (Continued)

Catchment Area, ha	**50.00**	**100.00**	**150.00**	**200.00**	**250.00**	**300.00**
Total Volume of Gabion Boxes, m^3	65.20	70.00	74.80	74.80	74.80	76.80
C. WING WALLS (ABUTMENTS) ON BOTH BANKS						
Length in U/S $L_{U/S\ ABUT}$, m	5.00	5.00	5.00	5.00	5.00	5.00
Length in D/S $L_{D/S\ ABUT}$, m	2.00	2.00	2.00	2.00	2.00	2.00
Top Level including Free Board, m	100.80	101.00	101.30	101.30	101.30	101.30
Foundation Level, m	97.00	97.00	97.00	97.00	97.00	97.00
Total Height H_{ABUT}, m	3.80	4.00	4.30	4.30	4.30	4.30
Base Width of Abutment B_{ABUT}, m	2.50	2.50	2.50	2.50	2.50	2.50
Size of Gabion Box (L×B×H), m						
Bottom Layer 1	2.50 × 1.00 × 1.00	2.50 × 1.00 × 1.00	2.50 × 1.00 × 1.00	2.50 × 1.00 × 1.00	2.50 × 1.00 × 1.00	2.50 × 1.00 × 1.00
Layer 2	2.00 × 1.00 × 1.00	2.00 × 1.00 × 1.00	2.00 × 1.00 × 1.00	2.00 × 1.00 × 1.00	2.00 × 1.00 × 1.00	2.00 × 1.00 × 1.00
Layer 3	1.00 × 1.00 × 1.00	1.00 × 1.00 × 1.00	1.00 × 1.00 × 1.00	1.00 × 1.00 × 1.00	1.00 × 1.00 × 1.00	1.00 × 1.00 × 1.00
Layer 4	1.00 × 1.00 × 0.80	1.00 × 1.00 × 1.00	1.00 × 1.00 × 1.30	1.00 × 1.00 × 1.30	1.00 × 1.00 × 1.30	1.00 × 1.00 × 1.30
Pressure at Toe, t/m^2	10.41	11.43	13.15	13.15	13.15	13.15
Pressure at Heel, t/m^2	2.27	1.93	1.24	1.24	1.24	1.24
Factor of Safety against Overturning	4.38	3.98	3.45	3.45	3.45	3.45
Shear Friction Factor F	4.55	4.19	3.73	3.73	3.73	3.73
Total Volume of Gabion Boxes, m^3	88.20	91.00	95.20	95.20	95.20	95.20
D. D/S WING WALLS ON BOTH BANKS						
Length in D/S $L_{D/S\ WING}$, m	5.00	5.00	5.00	5.00	5.00	5.00

(Continued)

TABLE 5.36 (Continued)

Catchment Area, ha	**50.00**	**100.00**	**150.00**	**200.00**	**250.00**	**300.00**
Top Level including Free Board, m	98.30	98.50	98.80	98.80	98.80	98.80
Foundation Level, m	97.00	97.00	97.00	97.00	97.00	97.00
Total Height $H_{D/S\ WING}$, m	1.30	1.50	1.80	1.80	1.80	1.80
Base Width of D/S Wing Wall $B_{D/S\ WING}$, m	1.00	1.00	1.50	1.50	1.50	1.50
Size of Gabion Box (L×B×H), m						
Bottom Layer 1	1.00 × 1.00 × 1.00	1.00 × 1.00 × 1.00	1.50 × 1.00 × 1.00	1.50 × 1.00 × 1.00	1.50 × 1.00 × 1.00	1.50 × 1.00 × 1.00
Layer 2	1.00 × 1.00 × 0.30	1.00 × 1.00 × 0.50	1.00 × 1.00 × 0.80	1.00 × 1.00 × 0.80	1.00 × 1.00 × 0.80	1.00 × 1.00 × 0.80
Pressure at Toe, t/m²	3.16	4.11	4.12	4.12	4.12	4.12
Pressure at Heel, t/m²	1.00	0.69	1.74	1.74	1.74	1.74
Factor of Safety against Overturning	5.76	4.21	6.87	6.87	6.87	6.87
Shear Friction Factor F	11.83	8.93	9.81	9.81	9.81	9.81
Total Volume of Gabion Boxes, m³	13.00	15.00	23.00	23.00	23.00	23.00

5.5.2 Section with Vertical U/S Face

The tables developed after the rigorous design of the different components for 3.00 m high gabion check dams for different combinations are presented from Table 5.37 to Table 5.40.

TABLE 5.37
Design Results for Gabion Check Dam with Vertical U/S Face Height = 3.00 m and Stream Width L = 5.0 m

Catchment Area, ha	50.00	100.00
A. WEIR		
Span of Weir W, m	4.00	4.00
Flood Discharge Q, m^3/s	6.54	11.00
Head Over Crest, m	0.92	1.31
Top Level, m	100.00	100.00
Foundation Level, m	97.00	97.00
Total Height of Weir H_{WEIR}, m	3.00	3.00
Bottom Width of Weir B_{WEIR}, m	4.00	4.00
Size of Gabion Box (L×B×H), m		
Bottom Layer 1	4.00 × 1.00 × 1.00	4.00 × 1.00 × 1.00
Layer 2	3.00 × 1.00 × 1.00	3.00 × 1.00 × 1.00
Layer 3	2.00 × 1.00 × 1.00	2.00 × 1.00 × 1.00
Pressure at Toe, t/m^2	4.51	5.07
Pressure at Heel, t/m^2	0.30	0.01
Factor of Safety against Overturning	1.55	1.51
Shear Friction Factor F	2.80	2.44
Total Volume of Gabion Boxes, m^3	36.00	36.00
Length of D/S Apron Floor $L_{D/S\ APRON}$, m	6.40	6.40
B. NON-OVER FLOW PORTION		
Width, m	1.00	1.00
Top Level including Free Board, m	101.50	101.80
Foundation Level, m	97.00	97.00
Total Height of Non-Over Flow H_{NOF}, m	4.50	4.80
Bottom Width of Non-Over Flow B_{NOF}, m	4.00	4.00
Size of Gabion Box (L×B×H), m		
Bottom Layer 1	4.00 × 1.00 × 1.00	4.00 × 1.00 × 1.00
Layer 2	3.00 × 1.00 × 1.00	3.50 × 1.00 × 1.00
Layer 3	2.00 × 1.00 × 1.00	2.50 × 1.00 × 1.00
Layer 4	1.00 × 1.00 × 1.50	1.50 × 1.00 × 1.80
Pressure at Toe, t/m^2	4.83	6.91
Pressure at Heel, t/m^2	0.57	0.25
Factor of Safety against Overturning	1.59	1.64
Shear Friction Factor F	2.57	2.24
Total Volume of Gabion Boxes, m^3	10.50	12.70
C. WING WALLS (ABUTMENTS) ON BOTH BANKS		
Length in U/S $L_{U/S\ ABUT}$, m	1.00	1.00

(Continued)

TABLE 5.37 (Continued)

Catchment Area, ha	50.00	100.00
Length in D/S $L_{D/S\ ABUT}$, m	5.00	5.00
Top Level including Free Board, m	101.50	101.80
Foundation Level, m	97.00	97.00
Total Height H_{ABUT}, m	4.50	4.80
Base Width of Abutment B_{ABUT}, m	2.50	3.00
Size of Gabion Box (L×B×H), m		
Bottom Layer 1	2.50 × 1.00 × 1.00	3.00 × 1.00 × 1.00
Layer 2	2.00 × 1.00 × 1.00	2.00 × 1.00 × 1.00
Layer 3	1.00 × 1.00 × 1.00	1.00 × 1.00 × 1.00
Layer 4	1.00 × 1.00 × 1.50	1.00 × 1.00 × 1.80
Pressure at Toe, t/m^2	14.41	13.79
Pressure at Heel, t/m^2	0.67	2.45
Factor of Safety against Overturning	3.15	4.03
Shear Friction Factor F	3.47	3.78
Total Volume of Gabion Boxes, m^3	84.00	93.60
D. D/S WING WALLS ON BOTH BANKS		
Length in D/S $L_{D/S\ WING}$, m	5.40	5.40
Top Level including Free Board, m	99.00	99.30
Foundation Level, m	97.00	97.00
Total Height $H_{D/S\ WING}$, m	2.00	2.30
Base Width of D/S Wing Wall $B_{D/S\ WING}$, m	1.50	1.50
Size of Gabion Box (L×B×H), m		
Bottom Layer 1	1.50 × 1.00 × 1.00	1.50 × 1.00 × 1.00
Layer 2	1.00 × 1.00 × 1.00	1.00 × 1.00 × 1.00
Layer 3		1.00 × 1.00 × 0.30
Pressure at Toe, t/m^2	4.94	6.32
Pressure at Heel, t/m^2	1.60	1.21
Factor of Safety against Overturning	5.50	4.20
Shear Friction Factor F	8.05	6.33
Total Volume of Gabion Boxes, m^3	27.00	30.24

TABLE 5.38
Design Results for Gabion Check Dam with Vertical U/S Face Height = 3.00 m and Stream Width L = 10.0 m

Catchment Area, ha	50.00	100.00	150.00	200.00
A. WEIR				
Span of Weir W, m	8.00	8.00	8.00	8.00
Flood Discharge Q, m^3/s	6.54	11.00	14.91	18.50
Head Over Crest, m	0.58	0.82	1.01	1.16
Top Level, m	100.00	100.00	100.00	100.00
Foundation Level, m	97.00	97.00	97.00	97.00

(Continued)

TABLE 5.38 (Continued)

Catchment Area, ha	50.00	100.00	150.00	200.00
Total Height of Weir H_{WEIR}, m	3.00	3.00	3.00	3.00
Bottom Width of Weir B_{WEIR}, m	4.00	4.00	4.00	4.00
Size of Gabion Box (L×B×H), m				
Bottom Layer 1	4.00 × 1.00 × 1.00	4.00 × 1.00 × 1.00	4.00 × 1.00 × 1.00	4.00 × 1.00 × 1.00
Layer 2	3.00 × 1.00 × 1.00	3.00 × 1.00 × 1.00	3.00 × 1.00 × 1.00	3.00 × 1.00 × 1.00
Layer 3	2.00 × 1.00 × 1.00	2.00 × 1.00 × 1.00	2.00 × 1.00 × 1.00	2.00 × 1.00 × 1.00
Pressure at Toe, t/m²	4.03	4.37	4.64	4.86
Pressure at Heel, t/m²	0.56	0.38	0.23	0.12
Factor of Safety against Overturning	1.59	1.56	1.54	1.53
Shear Friction Factor F	3.23	2.92	2.71	2.57
Total Volume of Gabion Boxes, m³	72.00	72.00	72.00	72.00
Length of D/S Apron Floor $L_{D/S\ APRON}$, m	6.40	6.40	6.40	6.40
B. NON-OVER FLOW PORTION				
Width, m	2.00	2.00	2.00	2.00
Top Level including Free Board, m	101.00	101.30	101.50	101.80
Foundation Level, m	97.00	97.00	97.00	97.00
Total Height of Non-Over Flow H_{NOF}, m	4.00	4.30	4.50	4.80
Bottom Width of Non-Over Flow B_{NOF}, m	4.00	4.00	4.00	4.00
Size of Gabion Box (L×B×H), m				
Bottom Layer 1	4.00 × 1.00 × 1.00	4.00 × 1.00 × 1.00	4.00 × 1.00 × 1.00	4.00 × 1.00 × 1.00
Layer 2	3.00 × 1.00 × 1.00	3.00 × 1.00 × 1.00	3.00 × 1.00 × 1.00	3.00 × 1.00 × 1.00
Layer 3	2.00 × 1.00 × 1.00	2.00 × 1.00 × 1.00	2.00 × 1.00 × 1.00	2.00 × 1.00 × 1.00
Layer 4	1.00 × 1.00 × 1.00	1.00 × 1.00 × 1.30	1.00 × 1.00 × 1.50	1.00 × 1.00 × 1.80
Pressure at Toe, t/m²	4.04	4.58	5.11	5.54
Pressure at Heel, t/m²	0.96	0.66	0.29	0.10
Factor of Safety against Overturning	1.66	1.61	1.55	1.52
Shear Friction Factor F	3.08	2.70	2.45	2.28
Total Volume of Gabion Boxes, m³	20.00	20.60	21.00	21.60
C. WING WALLS (ABUTMENTS) ON BOTH BANKS				
Length in U/S $L_{U/S\ ABUT}$, m	1.00	1.00	1.00	1.00
Length in D/S $L_{D/S\ ABUT}$, m	5.00	5.00	5.00	5.00
Top Level including Free Board, m	101.00	101.30	101.50	101.80
Foundation Level, m	97.00	97.00	97.00	97.00

(Continued)

TABLE 5.38 (Continued)

Catchment Area, ha	50.00	100.00	150.00	200.00
Total Height H_{ABUT}, m	4.00	4.30	4.50	4.80
Base Width of Abutment B_{ABUT}, m	2.50	2.50	2.50	3.00
Size of Gabion Box (L×B×H), m				
Bottom Layer 1	2.50 × 1.00 × 1.00	2.50 × 1.00 × 1.00	2.50 × 1.00 × 1.00	3.00 × 1.00 × 1.00
Layer 2	2.00 × 1.00 × 1.00	2.00 × 1.00 × 1.00	2.00 × 1.00 × 1.00	2.00 × 1.00 × 1.00
Layer 3	1.00 × 1.00 × 1.00	1.00 × 1.00 × 1.00	1.00 × 1.00 × 1.00	1.00 × 1.00 × 1.00
Layer 4	1.00 × 1.00 × 1.00	1.00 × 1.00 × 1.30	1.00 × 1.00 × 1.50	1.00 × 1.00 × 1.80
Pressure at Toe, t/m^2	11.43	13.15	14.41	13.79
Pressure at Heel, t/m^2	1.93	1.24	0.67	2.45
Factor of Safety against Overturning	3.98	3.45	3.15	4.03
Shear Friction Factor F	4.19	3.73	3.47	3.78
Total Volume of Gabion Boxes, m^3	78.00	81.60	84.00	93.60
D. D/S WING WALLS ON BOTH BANKS				
Length in D/S $L_{D/S\ WING}$, m	5.40	5.40	5.40	5.40
Top Level including Free Board, m	98.50	98.80	99.00	99.30
Foundation Level, m	97.00	97.00	97.00	97.00
Total Height $H_{D/S\ WING}$, m	1.50	1.80	2.00	2.30
Base Width of D/S Wing Wall $B_{D/S\ WING}$, m	1.00	1.50	1.50	1.50
Size of Gabion Box (L×B×H), m				
Bottom Layer 1	1.00 × 1.00 × 1.00	1.50 × 1.00 × 1.00	1.50 × 1.00 × 1.00	1.50 × 1.00 × 1.00
Layer 2	1.00 × 1.00 × 0.50	1.00 × 1.00 × 0.80	1.00 × 1.00 × 1.00	1.00 × 1.00 × 1.00
Layer 3				0.50 × 1.00 × 0.30
Pressure at Toe, t/m^2	4.11	4.12	4.94	6.34
Pressure at Heel, t/m^2	0.69	1.74	1.60	1.23
Factor of Safety against Overturning	4.21	6.87	5.50	4.22
Shear Friction Factor F	8.93	9.81	8.05	6.34
Total Volume of Gabion Boxes, m^3	16.20	24.84	27.00	28.62

TABLE 5.39
Design Results for Gabion Check Dam with Vertical U/S Face Height = 3.00 m and Stream Width L = 15.0 m

Catchment Area, ha	50.00	100.00	150.00	200.00	250.00	300.00
A. WEIR						
Span of Weir W, m	12.00	12.00	12.00	12.00	12.00	12.00

(Continued)

TABLE 5.39 (Continued)

Catchment Area, ha	**50.00**	**100.00**	**150.00**	**200.00**	**250.00**	**300.00**
Flood Discharge Q, m^3/s	6.54	11.00	14.91	18.50	21.87	25.07
Head Over Crest, m	0.44	0.63	0.77	0.89	0.99	1.09
Top Level, m	100.00	100.00	100.00	100.00	100.00	100.00
Foundation Level, m	97.00	97.00	97.00	97.00	97.00	97.00
Total Height of Weir H_{WEIR}, m	3.00	3.00	3.00	3.00	3.00	3.00
Bottom Width of Weir B_{WEIR}, m	4.00	4.00	4.00	4.00	4.00	4.00
Size of Gabion Box (L×B×H), m						
Bottom Layer 1	4.00 × 1.00 × 1.00	4.00 × 1.00 × 1.00	4.00 × 1.00 × 1.00	4.00 × 1.00 × 1.00	4.00 × 1.00 × 1.00	4.00 × 1.00 × 1.00
Layer 2	3.00 × 1.00 × 1.00	3.00 × 1.00 × 1.00	3.00 × 1.00 × 1.00	3.00 × 1.00 × 1.00	3.00 × 1.00 × 1.00	3.00 × 1.00 × 1.00
Layer 3	2.00 × 1.00 × 1.00	2.00 × 1.00 × 1.00	2.00 × 1.00 × 1.00	2.00 × 1.00 × 1.00	2.00 × 1.00 × 1.00	2.00 × 1.00 × 1.00
Pressure at Toe, t/m^2	3.82	4.09	4.29	4.47	4.62	4.75
Pressure at Heel, t/m^2	0.67	0.53	0.42	0.33	0.25	0.17
Factor of Safety against Overturning	1.61	1.59	1.57	1.56	1.55	1.53
Shear Friction Factor F	3.45	3.16	2.98	2.84	2.73	2.63
Total Volume of Gabion Boxes, m^3	108.00	108.00	108.00	108.00	108.00	108.00
Length of D/S Apron Floor $L_{D/S\ APRON}$, m	6.40	6.40	6.40	6.40	6.40	6.40
B. NON-OVER FLOW PORTION						
Width, m	3.00	3.00	3.00	3.00	3.00	3.00
Top Level including Free Board, m	100.80	101.00	101.30	101.30	101.50	101.80
Foundation Level, m	97.00	97.00	97.00	97.00	97.00	97.00
Total Height of Non-Over Flow H_{NOF}, m	3.80	4.00	4.30	4.30	4.50	4.80

(Continued)

TABLE 5.39 (Continued)

Catchment Area, ha	50.00	100.00	150.00	200.00	250.00	300.00
Bottom Width of Non-Over Flow B_{NOF}, m	3.50	3.50	3.50	3.50	3.50	3.50
Size of Gabion Box (L×B×H), m						
Bottom Layer 1	3.50 × 1.00 × 1.00	3.50 × 1.00 × 1.00	3.50 × 1.00 × 1.00	3.50 × 1.00 × 1.00	3.50 × 1.00 × 1.00	3.50 × 1.00 × 1.00
Layer 2	2.50 × 1.00 × 1.00	2.50 × 1.00 × 1.00	2.50 × 1.00 × 1.00	2.50 × 1.00 × 1.00	2.50 × 1.00 × 1.00	2.50 × 1.00 × 1.00
Layer 3	1.50 × 1.00 × 1.00	2.00 × 1.00 × 1.00	2.00 × 1.00 × 1.00	2.50 × 1.00 × 1.00	2.50 × 1.00 × 1.00	2.00 × 1.00 × 1.00
Layer 4	1.00 × 1.00 × 0.80	1.00 × 1.00 × 1.00	1.00 × 1.00 × 1.30	1.50 × 1.00 × 1.30	2.00 × 1.00 × 1.50	1.50 × 1.00 × 1.80
Pressure at Toe, t/m^2	4.32	5.10	5.48	6.47	7.18	6.70
Pressure at Heel, t/m^2	0.27	0.13	0.02	0.09	0.33	0.09
Factor of Safety against Overturning	1.51	1.53	1.52	1.59	1.67	1.57
Shear Friction Factor F	2.90	2.65	2.46	2.39	2.33	2.16
Total Volume of Gabion Boxes, m^3	24.90	27.00	27.90	31.35	34.50	32.10
C. WING WALLS (ABUTMENTS) ON BOTH BANKS						
Length in U/S $L_{U/S\ ABUT}$, m	1.00	1.00	1.00	1.00	1.00	1.00
Length in D/S $L_{D/S\ ABUT}$, m	4.50	4.50	4.50	4.50	4.50	4.50
Top Level including Free Board, m	100.80	101.00	101.30	101.30	101.50	101.80
Foundation Level, m	97.00	97.00	97.00	97.00	97.00	97.00
Total Height H_{ABUT}, m	3.80	4.00	4.30	4.30	4.50	4.80
Base Width of Abutment B_{ABUT}, m	2.50	2.50	2.50	2.50	2.50	3.00
Size of Gabion Box (L×B×H), m						
Bottom Layer 1	2.50 × 1.00 × 1.00	2.50 × 1.00 × 1.00	2.50 × 1.00 × 1.00	2.50 × 1.00 × 1.00	2.50 × 1.00 × 1.00	3.00 × 1.00 × 1.00
Layer 2	2.00 × 1.00 × 1.00	2.00 × 1.00 × 1.00	2.00 × 1.00 × 1.00	2.00 × 1.00 × 1.00	2.00 × 1.00 × 1.00	2.00 × 1.00 × 1.00

(Continued)

TABLE 5.39 (Continued)

Catchment Area, ha	**50.00**	**100.00**	**150.00**	**200.00**	**250.00**	**300.00**
Layer 3	1.00 × 1.00 × 1.00	1.00 × 1.00 × 1.00	1.00 × 1.00 × 1.00	1.00 × 1.00 × 1.00	1.00 × 1.00 × 1.00	1.00 × 1.00 × 1.00
Layer 4	1.00 × 1.00 × 0.80	1.00 × 1.00 × 1.00	1.00 × 1.00 × 1.30	1.00 × 1.00 × 1.30	1.00 × 1.00 × 1.50	1.00 × 1.00 × 1.80
Pressure at Toe, t/m^2	10.41	11.43	13.15	13.15	14.41	13.79
Pressure at Heel, t/m^2	2.27	1.93	1.24	1.24	0.67	2.45
Factor of Safety against Overturning	4.38	3.98	3.45	3.45	3.15	4.03
Shear Friction Factor F	4.55	4.19	3.73	3.73	3.47	3.78
Total Volume of Gabion Boxes, m^3	69.30	71.50	74.80	74.80	77.00	85.80
D. D/S WING WALLS ON BOTH BANKS						
Length in D/S $L_{D/S\ WING}$, m	5.40	5.40	5.40	5.40	5.40	5.40
Top Level including Free Board, m	98.30	98.50	98.80	98.80	99.00	99.30
Foundation Level, m	97.00	97.00	97.00	97.00	97.00	97.00
Total Height $H_{D/S\ WING}$, m	1.30	1.50	1.80	1.80	2.00	2.30
Base Width of D/S Wing Wall $B_{D/S\ WING}$, m	1.00	1.00	1.50	1.50	1.50	1.50
Size of Gabion Box (L×B×H), m						
Bottom Layer 1	1.00 × 1.00 × 1.00	1.00 × 1.00 × 1.00	1.50 × 1.00 × 1.00	1.50 × 1.00 × 1.00	1.50 × 1.00 × 1.00	1.50 × 1.00 × 1.00
Layer 2	1.00 × 1.00 × 0.30	1.00 × 1.00 × 0.50	1.00 × 1.00 × 0.80	1.00 × 1.00 × 0.80	1.00 × 1.00 × 1.00	1.00 × 1.00 × 1.00
Layer 3						0.50 × 1.00 × 0.30
Pressure at Toe, t/m^2	3.16	4.11	4.12	4.12	4.94	6.34
Pressure at Heel, t/m^2	1.00	0.69	1.74	1.74	1.60	1.23
Factor of Safety against Overturning	5.76	4.21	6.87	6.87	5.50	4.22

(Continued)

TABLE 5.39 (Continued)

Catchment Area, ha	50.00	100.00	150.00	200.00	250.00	300.00
Shear Friction Factor F	11.83	8.93	9.81	9.81	8.05	6.34
Total Volume of Gabion Boxes, m^3	14.04	16.20	24.84	24.84	27.00	28.62

TABLE 5.40
Design Results for Gabion Check Dam with Vertical U/S Face Height = 3.00 m and Stream Width L = 20.0 m

Catchment Area, ha	50.00	100.00	150.00	200.00	250.00	300.00
A. WEIR						
Span of Weir W, m	16.00	16.00	16.00	16.00	16.00	16.00
Flood Discharge Q, m^3/s	6.54	11.00	14.91	18.50	21.87	25.07
Head Over Crest, m	0.37	0.52	0.64	0.73	0.82	0.90
Top Level, m	100.00	100.00	100.00	100.00	100.00	100.00
Foundation Level, m	97.00	97.00	97.00	97.00	97.00	97.00
Total Height of Weir H_{WEIR}, m	3.00	3.00	3.00	3.00	3.00	3.00
Bottom Width of Weir B_{WEIR}, m	3.50	3.50	4.00	4.00	4.00	4.00
Size of Gabion Box (L×B×H), m						
Bottom Layer 1	3.50 × 1.00 × 1.00	3.50 × 1.00 × 1.00	4.00 × 1.00 × 1.00	4.00 × 1.00 × 1.00	4.00 × 1.00 × 1.00	4.00 × 1.00 × 1.00
Layer 2	2.50 × 1.00 × 1.00	2.50 × 1.00 × 1.00	3.00 × 1.00 × 1.00	3.00 × 1.00 × 1.00	3.00 × 1.00 × 1.00	3.00 × 1.00 × 1.00
Layer 3	2.00 × 1.00 × 1.00	2.50 × 1.00 × 1.00	2.00 × 1.00 × 1.00	2.00 × 1.00 × 1.00	2.00 × 1.00 × 1.00	2.00 × 1.00 × 1.00
Pressure at Toe, t/m^2	4.43	5.22	4.10	4.24	4.37	4.48
Pressure at Heel, t/m^2	0.16	0.04	0.52	0.45	0.38	0.32

(Continued)

TABLE 5.40 (Continued)

Catchment Area, ha	**50.00**	**100.00**	**150.00**	**200.00**	**250.00**	**300.00**
Factor of Safety against Overturning	1.53	1.57	1.59	1.58	1.57	1.56
Shear Friction Factor F	3.16	2.99	3.16	3.03	2.92	2.83
Total Volume of Gabion Boxes, m^3	128.00	136.00	144.00	144.00	144.00	144.00
Length of D/S Apron Floor $L_{D/S\ APRON}$, m	6.80	6.80	6.40	6.40	6.40	6.40
B. NON-OVER FLOW PORTION						
Width, m	4.00	4.00	4.00	4.00	4.00	4.00
Top Level including Free Board, m	100.80	101.00	101.30	101.30	101.30	101.30
Foundation Level, m	97.00	97.00	97.00	97.00	97.00	97.00
Total Height of Non-Over Flow H_{NOF}, m	3.80	4.00	4.30	4.30	4.30	4.30
Bottom Width of Non-Over Flow B_{NOF}, m	3.50	3.50	3.50	3.50	3.50	3.50
Size of Gabion Box (L×B×H), m						
Bottom Layer 1	3.50 × 1.00 × 1.00	3.50 × 1.00 × 1.00	3.50 × 1.00 × 1.00	3.50 × 1.00 × 1.00	3.50 × 1.00 × 1.00	3.50 × 1.00 × 1.00
Layer 2	2.50 × 1.00 × 1.00	2.50 × 1.00 × 1.00	2.50 × 1.00 × 1.00	2.50 × 1.00 × 1.00	2.50 × 1.00 × 1.00	2.50 × 1.00 × 1.00
Layer 3	1.50 × 1.00 × 1.00	1.50 × 1.00 × 1.00	1.50 × 1.00 × 1.00	1.50 × 1.00 × 1.00	1.50 × 1.00 × 1.00	2.00 × 1.00 × 1.00
Layer 4	1.00 × 1.00 × 0.80	1.00 × 1.00 × 1.00	1.00 × 1.00 × 1.30	1.50 × 1.00 × 1.30	1.50 × 1.00 × 1.30	1.50 × 1.00 × 1.30
Pressure at Toe, t/m^2	4.09	4.48	4.77	5.17	5.51	6.07
Pressure at Heel, t/m^2	0.49	0.30	0.28	0.47	0.13	0.03
Factor of Safety against Overturning	1.55	1.52	1.52	1.58	1.53	1.55

(Continued)

TABLE 5.40 (Continued)

Catchment Area, ha	50.00	100.00	150.00	200.00	250.00	300.00
Shear Friction Factor F	3.05	2.78	2.62	2.52	2.40	2.34
Total Volume of Gabion Boxes, m^3	33.20	34.00	35.20	37.80	37.80	39.80
C. WING WALLS (ABUTMENTS) ON BOTH BANKS						
Length in U/S $L_{U/S\ ABUT}$, m	1.00	1.00	1.00	1.00	1.00	1.00
Length in D/S $L_{D/S\ ABUT}$, m	4.50	4.50	4.50	4.50	4.50	4.50
Top Level including Free Board, m	100.80	101.00	101.30	101.30	101.30	101.30
Foundation Level, m	97.00	97.00	97.00	97.00	97.00	97.00
Total Height H_{ABUT}, m	3.80	4.00	4.30	4.30	4.30	4.30
Base Width of Abutment B_{ABUT}, m	2.50	2.50	2.50	2.50	2.50	2.50
Size of Gabion Box (L×B×H), m						
Bottom Layer 1	2.50 × 1.00 × 1.00	2.50 × 1.00 × 1.00	2.50 × 1.00 × 1.00	2.50 × 1.00 × 1.00	2.50 × 1.00 × 1.00	2.50 × 1.00 × 1.00
Layer 2	2.00 × 1.00 × 1.00	2.00 × 1.00 × 1.00	2.00 × 1.00 × 1.00	2.00 × 1.00 × 1.00	2.00 × 1.00 × 1.00	2.00 × 1.00 × 1.00
Layer 3	1.00 × 1.00 × 1.00	1.00 × 1.00 × 1.00	1.00 × 1.00 × 1.00	1.00 × 1.00 × 1.00	1.00 × 1.00 × 1.00	1.00 × 1.00 × 1.00
Layer 4	1.00 × 1.00 × 0.80	1.00 × 1.00 × 1.00	1.00 × 1.00 × 1.30	1.00 × 1.00 × 1.30	1.00 × 1.00 × 1.30	1.00 × 1.00 × 1.30
Pressure at Toe, t/m^2	10.41	11.43	13.15	13.15	13.15	13.15
Pressure at Heel, t/m^2	2.27	1.93	1.24	1.24	1.24	1.24
Factor of Safety against Overturning	4.38	3.98	3.45	3.45	3.45	3.45
Shear Friction Factor F	4.55	4.19	3.73	3.73	3.73	3.73

(Continued)

TABLE 5.40 (Continued)

Catchment Area, ha	**50.00**	**100.00**	**150.00**	**200.00**	**250.00**	**300.00**
Total Volume of Gabion Boxes, m^3	69.30	71.50	74.80	74.80	74.80	74.80
D. D/S WING WALLS ON BOTH BANKS						
Length in D/S $L_{D/S\ WING}$, m	5.80	5.80	5.40	5.40	5.40	5.40
Top Level including Free Board, m	98.30	98.50	98.80	98.80	98.80	98.80
Foundation Level, m	97.00	97.00	97.00	97.00	97.00	97.00
Total Height $H_{D/S\ WING}$, m	1.30	1.50	1.80	1.80	1.80	1.80
Base Width of D/S Wing Wall $B_{D/S\ WING}$, m	1.00	1.00	1.50	1.50	1.50	1.50
Size of Gabion Box (L×B×H), m						
Bottom Layer 1	1.00 × 1.00 × 1.00	1.00 × 1.00 × 1.00	1.50 × 1.00 × 1.00	1.50 × 1.00 × 1.00	1.50 × 1.00 × 1.00	1.50 × 1.00 × 1.00
Layer 2	1.00 × 1.00 × 0.30	1.00 × 1.00 × 0.50	1.00 × 1.00 × 0.80	1.00 × 1.00 × 0.80	1.00 × 1.00 × 0.80	1.00 × 1.00 × 0.80
Pressure at Toe, t/m^2	3.16	4.11	4.12	4.12	4.12	4.12
Pressure at Heel, t/m^2	1.00	0.69	1.74	1.74	1.74	1.74
Factor of Safety against Overturning	5.76	4.21	6.87	6.87	6.87	6.87
Shear Friction Factor F	11.83	8.93	9.81	9.81	9.81	9.81
Total Volume of Gabion Boxes, m^3	15.08	17.40	24.84	24.84	24.84	24.84

5.5.3 Section with Steps on Both Faces

The tables developed after the rigorous design of the different components for 3.00 m high gabion check dams for different combinations are presented from Table 5.41 to Table 5.44.

TABLE 5.41
Design Results for Gabion Check Dam with Steps on Both Faces Height = 3.00 m and Stream Width L = 5.0 m

Catchment Area, ha	50.00	100.00
A. WEIR		
Span of Weir W, m	4.00	4.00
Flood Discharge Q, m^3/s	6.54	11.00
Head Over Crest, m	0.92	1.31
Top Level, m	100.00	100.00
Foundation Level, m	97.00	97.00
Total Height of Weir H_{WEIR}, m	3.00	3.00
Bottom Width of Weir B_{WEIR}, m	5.00	5.50
Size of Gabion Box (L×B×H), m		
Bottom Layer 1	5.00 × 1.00 × 1.00	5.50 × 1.00 × 1.00
Layer 2	4.00 × 1.00 × 1.00	4.50 × 1.00 × 1.00
Layer 3	3.50 × 1.00 × 1.00	4.00 × 1.00 × 1.00
Pressure at Toe, t/m^2	6.34	6.68
Pressure at Heel, t/m^2	0.02	0.19
Factor of Safety against Overturning	1.79	1.88
Shear Friction Factor F	3.71	3.58
Total Volume of Gabion Boxes, m^3	50.00	56.00
Length of D/S Apron Floor $L_{D/S\ APRON}$, m	5.90	5.90
B. NON-OVER FLOW PORTION		
Length, m	1.00	1.00
Top Level including Free Board, m	101.50	101.80
Foundation Level, m	97.00	97.00
Total Height of Non-Over Flow H_{NOF}, m	4.50	4.80
Bottom Width of Non-Over Flow B_{NOF}, m	4.50	4.50
Size of Gabion Box (L×B×H), m		
Bottom Layer 1	4.50 × 1.00 × 1.00	4.50 × 1.00 × 1.00
Layer 2	3.50 × 1.00 × 1.00	3.50 × 1.00 × 1.00
Layer 3	2.50 × 1.00 × 1.00	2.50 × 1.00 × 1.00
Layer 4	1.50 × 1.00 × 1.50	2.00 × 1.00 × 1.80
Pressure at Toe, t/m^2	6.93	8.41
Pressure at Heel, t/m^2	0.49	0.09
Factor of Safety against Overturning	1.86	1.83
Shear Friction Factor F	3.10	2.63
Total Volume of Gabion Boxes, m^3	12.75	14.10
C. WING WALLS (ABUTMENTS) ON BOTH BANKS		
Length in U/S $L_{U/S\ ABUT}$, m	4.50	4.50
Length in D/S $L_{D/S\ ABUT}$, m	2.00	2.00
Top Level including Free Board, m	101.50	101.80
Foundation Level, m	97.00	97.00
Total Height H_{ABUT}, m	4.50	4.80
Base Width of Abutment B_{ABUT}, m	2.50	3.00
Size of Gabion Box (L×B×H), m		
Bottom Layer 1	2.50 × 1.00 × 1.00	3.00 × 1.00 × 1.00

(Continued)

TABLE 5.41 (Continued)

Catchment Area, ha	50.00	100.00
Layer 2	2.00 × 1.00 × 1.00	2.00 × 1.00 × 1.00
Layer 3	1.00 × 1.00 × 1.00	1.00 × 1.00 × 1.00
Layer 4	1.00 × 1.00 × 1.50	1.00 × 1.00 × 1.80
Pressure at Toe, t/m^2	14.41	13.79
Pressure at Heel, t/m^2	0.67	2.45
Factor of Safety against Overturning	3.15	4.03
Shear Friction Factor F	3.47	3.78
Total Volume of Gabion Boxes, m^3	91.00	101.40
D. D/S WING WALLS ON BOTH BANKS		
Length in D/S $L_{D/S\ WING}$, m	3.90	3.90
Top Level including Free Board, m	99.00	99.30
Foundation Level, m	97.00	97.00
Total Height $H_{D/S\ WING}$, m	2.00	2.30
Base Width of D/S Wing Wall $B_{D/S\ WING}$, m	1.50	1.50
Size of Gabion Box (L×B×H), m		
Bottom Layer 1	1.50 × 1.00 × 1.00	1.50 × 1.00 × 1.00
Layer 2	1.00 × 1.00 × 1.00	1.00 × 1.00 × 1.00
Layer 3		1.00 × 1.00 × 0.30
Pressure at Toe, t/m^2	4.94	6.32
Pressure at Heel, t/m^2	1.60	1.21
Factor of Safety against Overturning	5.50	4.20
Shear Friction Factor F	8.05	6.33
Total Volume of Gabion Boxes, m^3	19.50	21.84

TABLE 5.42
Design Results for Gabion Check Dam with Steps on Both Faces Height = 3.00 m and Stream Width L = 10.0 m

Catchment Area, ha	50.00	100.00	150.00	200.00
A. WEIR				
Span of Weir W, m	8.00	8.00	8.00	8.00
Flood Discharge Q, m^3/s	6.54	11.00	14.91	18.50
Head Over Crest, m	0.58	0.82	1.01	1.16
Top Level, m	100.00	100.00	100.00	100.00
Foundation Level, m	97.00	97.00	97.00	97.00
Total Height of Weir H_{WEIR}, m	3.00	3.00	3.00	3.00
Bottom Width of Weir B_{WEIR}, m	5.00	5.00	5.00	5.00
Size of Gabion Box (L×B×H), m				
Bottom Layer 1	5.00 × 1.00 × 1.00	5.00 × 1.00 × 1.00	5.00 × 1.00 × 1.00	5.00 × 1.00 × 1.00
Layer 2	4.00 × 1.00 × 1.00	4.00 × 1.00 × 1.00	4.00 × 1.00 × 1.00	4.00 × 1.00 × 1.00

(Continued)

TABLE 5.42 (Continued)

Catchment Area, ha	50.00	100.00	150.00	200.00
Layer 3	3.00 × 1.00 × 1.00	3.50 × 1.00 × 1.00	4.00 × 1.00 × 1.00	4.00 × 1.00 × 1.00
Pressure at Toe, t/m²	5.54	6.16	6.64	6.90
Pressure at Heel, t/m²	0.20	0.11	0.16	0.06
Factor of Safety against Overturning	1.77	1.80	1.85	1.84
Shear Friction Factor F	4.22	3.85	3.63	3.45
Total Volume of Gabion Boxes, m³	96.00	100.00	104.00	104.00
Length of D/S Apron Floor $L_{D/S\ APRON}$, m	6.30	6.30	6.30	6.30
B. NON-OVER FLOW PORTION				
Length, m	2.00	2.00	2.00	2.00
Top Level including Free Board, m	101.00	101.30	101.50	101.80
Foundation Level, m	97.00	97.00	97.00	97.00
Total Height of Non-Over Flow H_{NOF}, m	4.00	4.30	4.50	4.80
Bottom Width of Non-Over Flow B_{NOF}, m	4.00	4.00	4.50	4.50
Size of Gabion Box (L×B×H), m				
Bottom Layer 1	4.00 × 1.00 × 1.00	4.00 × 1.00 × 1.00	4.50 × 1.00 × 1.00	4.50 × 1.00 × 1.00
Layer 2	3.00 × 1.00 × 1.00	3.00 × 1.00 × 1.00	3.50 × 1.00 × 1.00	3.50 × 1.00 ×1.00
Layer 3	2.00 ×1.00 × 1.00	2.50 × 1.00 × 1.00	2.50 × 1.00 × 1.00	2.50 × 1.00 × 1.00
Layer 4	1.50 × 1.00 × 1.00	2.50 × 1.00 × 1.30	1.50 × 1.00 × 1.50	1.50 × 1.00 × 1.80
Pressure at Toe, t/m²	6.51	8.12	7.12	7.69
Pressure at Heel, t/m²	0.04	0.05	0.36	0.23
Factor of Safety against Overturning	1.73	1.84	1.83	1.82
Shear Friction Factor F	3.26	3.00	2.97	2.78
Total Volume of Gabion Boxes, m³	21.00	25.50	25.50	26.40
C. WING WALLS (ABUTMENTS) ON BOTH BANKS				
Length in U/S $L_{U/S\ ABUT}$, m	4.00	4.00	4.50	4.50
Length in D/S $L_{D/S\ ABUT}$, m	2.00	2.00	2.00	2.00
Top Level including Free Board, m	101.00	101.30	101.50	101.80
Foundation Level, m	97.00	97.00	97.00	97.00
Total Height H_{ABUT}, m	4.00	4.30	4.50	4.80
Base Width of Abutment B_{ABUT}, m	2.50	2.50	2.50	3.00
Size of Gabion Box (L×B×H), m				
Bottom Layer 1	2.50 × 1.00 × 1.00	2.50 × 1.00 × 1.00	2.50 × 1.00 × 1.00	3.00 × 1.00 × 1.00

(Continued)

TABLE 5.42 (Continued)

Catchment Area, ha	50.00	100.00	150.00	200.00
Layer 2	2.00 × 1.00 × 1.00	2.00 × 1.00 × 1.00	2.00 × 1.00 × 1.00	2.00 × 1.00 × 1.00
Layer 3	1.00 × 1.00 × 1.00	1.00 × 1.00 × 1.00	1.00 × 1.00 × 1.00	1.00 × 1.00 × 1.00
Layer 4	1.00 × 1.00 × 1.00	1.00 × 1.00 × 1.30	1.00 × 1.00 × 1.50	1.00 × 1.00 × 1.80
Pressure at Toe, t/m²	11.43	13.15	14.41	13.79
Pressure at Heel, t/m²	1.93	1.24	0.67	2.45
Factor of Safety against Overturning	3.98	3.45	3.15	4.03
Shear Friction Factor F	4.19	3.73	3.47	3.78
Total Volume of Gabion Boxes, m³	78.00	81.60	91.00	101.40
D. D/S WING WALLS ON BOTH BANKS				
Length in D/S $L_{D/S\ WING}$, m	4.30	4.30	4.30	4.30
Top Level including Free Board, m	98.50	98.80	99.00	99.30
Foundation Level, m	97.00	97.00	97.00	97.00
Total Height $H_{D/S\ WING}$, m	1.50	1.80	2.00	2.30
Base Width of D/S Wing Wall $B_{D/S\ WING}$, m	1.00	1.50	1.50	1.50
Size of Gabion Box (L×B×H), m				
Bottom Layer 1	1.00 × 1.00 × 1.00	1.50 × 1.00 × 1.00	1.50 × 1.00 × 1.00	1.50 × 1.00 × 1.00
Layer 2	1 × 1.0 × 0.5	1 × 1.0 × 0.8	1.00 × 1.00 × 1.00	1.00 × 1.00 × 1.00
Layer 3				1.00 × 1.00 × 0.30
Pressure at Toe, t/m²	4.11	4.12	4.94	6.32
Pressure at Heel, t/m²	0.69	1.74	1.60	1.21
Factor of Safety against Overturning	4.21	6.87	5.50	4.20
Shear Friction Factor F	8.93	9.81	8.05	6.33
Total Volume of Gabion Boxes, m³	12.90	19.78	21.50	24.08

TABLE 5.43
Design Results for Gabion Check Dam with Steps on Both Faces Height = 3.00 m and Stream Width L = 15.0 m

Catchment Area, ha	50.00	100.00	150.00	200.00	250.00	300.00
A. WEIR						
Span of Weir W, m	12.00	12.00	12.00	12.00	12.00	12.00

(Continued)

TABLE 5.43 (Continued)

Catchment Area, ha	50.00	100.00	150.00	200.00	250.00	300.00
Flood Discharge Q, m^3/s	6.54	11.00	14.91	18.50	21.87	25.07
Head Over Crest, m	0.44	0.63	0.77	0.89	0.99	1.09
Top Level, m	100.00	100.00	100.00	100.00	100.00	100.00
Foundation Level, m	97.00	97.00	97.00	97.00	97.00	97.00
Total Height of Weir H_{WEIR}, m	3.00	3.00	3.00	3.00	3.00	3.00
Bottom Width of Weir B_{WEIR}, m	4.50	5.00	5.00	5.00	5.00	5.00
Size of Gabion Box (L×B×H), m						
Bottom Layer 1	4.50 × 1.00 × 1.00	5.00 × 1.00 × 1.00	5.00 × 1.00 × 1.00	5.00 × 1.00 × 1.00	5.00 × 1.00 × 1.00	5.00 × 1.00 × 1.00
Layer 2	3.50 × 1.00 × 1.00	4.00 × 1.00 × 1.00	4.00 × 1.00 × 1.00	4.00 × 1.00 × 1.00	4.50 × 1.00 × 1.00	4.50 × 1.00 × 1.00
Layer 3	3.00 × 1.00 × 1.00	3.00 × 1.00 × 1.00	3.00 × 1.00 × 1.00	3.50 × 1.00 × 1.00	4.00 × 1.00 × 1.00	4.00 × 1.00 × 1.00
Pressure at Toe, t/m^2	5.73	5.63	5.89	6.28	6.85	7.01
Pressure at Heel, t/m^2	0.04	0.15	0.01	0.05	0.15	0.09
Factor of Safety against Overturning	1.73	1.77	1.75	1.79	1.88	1.87
Shear Friction Factor F	4.07	4.13	3.89	3.75	3.69	3.56
Total Volume of Gabion Boxes, m^3	132.00	144.00	144.00	150.00	162.00	162.00
Length of D/S Apron Floor $L_{D/S\ APRON}$, m	6.80	6.30	6.30	6.30	6.30	6.30
B. NON-OVER FLOW PORTION						
Length, m	3.00	3.00	3.00	3.00	3.00	3.00
Top Level including Free Board, m	100.80	101.00	101.30	101.30	101.50	101.80
Foundation Level, m	97.00	97.00	97.00	97.00	97.00	97.00

(Continued)

TABLE 5.43 (Continued)

Catchment Area, ha	50.00	100.00	150.00	200.00	250.00	300.00
Total Height of Non-Over Flow H_{NOF}, m	3.80	4.00	4.30	4.30	4.50	4.80
Bottom Width of Non-Over Flow B_{NOF}, m	4.00	4.00	4.00	4.50	4.50	4.50
Size of Gabion Box (L×B×H), m						
Bottom Layer 1	4.00 × 1.00 × 1.00	4.00 × 1.00 × 1.00	4.00 × 1.00 × 1.00	4.50 × 1.00 × 1.00	4.50 × 1.00 × 1.00	4.50 × 1.00 × 1.00
Layer 2	3.00 × 1.00 × 1.00	3.00 × 1.00 × 1.00	3.00 × 1.00 × 1.00	3.50 × 1.00 × 1.00	3.50 × 1.00 × 1.00	3.50 × 1.00 × 1.00
Layer 3	2.00 × 1.00 × 1.00	2.50 × 1.00 × 1.00	2.50 × 1.00 × 1.00	2.50 × 1.00 × 1.00	2.50 × 1.00 × 1.00	2.50 × 1.00 × 1.00
Layer 4	1.00 × 1.00 × 0.80	1.50 × 1.00 × 1.00	2.00 × 1.00 × 1.30	1.50 × 1.00 × 1.30	1.50 × 1.00 × 1.50	1.50 × 1.00 × 1.80
Pressure at Toe, t/m^2	5.90	6.85	7.67	6.75	7.10	7.48
Pressure at Heel, t/m^2	0.06	0.01	0.07	0.43	0.37	0.37
Factor of Safety against Overturning	1.70	1.75	1.82	1.83	1.83	1.85
Shear Friction Factor F	3.48	3.21	3.04	3.14	2.99	2.88
Total Volume of Gabion Boxes, m^3	29.40	33.00	36.30	37.35	38.25	39.60
C. WING WALLS (ABUTMENTS) ON BOTH BANKS						
Length in U/S $L_{U/S\ ABUT}$, m	4.00	4.00	4.00	4.50	4.50	4.50
Length in D/S $L_{D/S\ ABUT}$, m	2.00	2.00	2.00	2.00	2.00	2.00
Top Level including Free Board, m	100.80	101.00	101.30	101.30	101.50	101.80
Foundation Level, m	97.00	97.00	97.00	97.00	97.00	97.00
Total Height H_{ABUT}, m	3.80	4.00	4.30	4.30	4.50	4.80
Base Width of Abutment B_{ABUT}, m	2.50	2.50	2.50	2.50	2.50	3.00

(Continued)

TABLE 5.43 (Continued)

Catchment Area, ha	**50.00**	**100.00**	**150.00**	**200.00**	**250.00**	**300.00**
Size of Gabion Box (L×B×H), m						
Bottom Layer 1	2.50 × 1.00 × 1.00	2.50 × 1.00 × 1.00	2.50 × 1.00 × 1.00	2.50 × 1.00 × 1.00	2.50 × 1.00 × 1.00	3.00 × 1.00 × 1.00
Layer 2	2.00 × 1.00 × 1.00	2.00 × 1.00 × 1.00	2.00 × 1.00 × 1.00	2.00 × 1.00 × 1.00	2.00 × 1.00 × 1.00	2.00 × 1.00 × 1.00
Layer 3	1.00 × 1.00 × 1.00	1.00 × 1.00 × 1.00	1.00 × 1.00 × 1.00	1.00 × 1.00 × 1.00	1.00 × 1.00 × 1.00	1.00 × 1.00 × 1.00
Layer 4	1.00 × 1.00 × 0.80	1.00 × 1.00 × 1.00	1.00 × 1.00 × 1.30	1.00 × 1.00 × 1.30	1.00 × 1.00 × 1.50	1.00 × 1.00 × 1.80
Pressure at Toe, t/m^2	10.41	11.43	13.15	13.15	14.41	13.79
Pressure at Heel, t/m^2	2.27	1.93	1.24	1.24	0.67	2.45
Factor of Safety against Overturning	4.38	3.98	3.45	3.45	3.15	4.03
Shear Friction Factor F	4.55	4.19	3.73	3.73	3.47	3.78
Total Volume of Gabion Boxes, m^3	75.60	78.00	81.60	88.40	91.00	101.40
D. D/S WING WALLS ON BOTH BANKS						
Length in D/S $L_{D/S\ WING}$, m	4.80	4.30	4.30	4.30	4.30	4.30
Top Level including Free Board, m	98.30	98.50	98.80	98.80	99.00	99.30
Foundation Level, m	97.00	97.00	97.00	97.00	97.00	97.00
Total Height $H_{D/S\ WING}$, m	1.30	1.50	1.80	1.80	2.00	2.30
Base Width of D/S Wing Wall $B_{D/S\ WING}$, m	1.00	1.00	1.50	1.50	1.50	1.50
Size of Gabion Box (L×B×H), m						
Bottom Layer 1	1.00 × 1.00 × 1.00	1.00 × 1.00 × 1.00	1.50 × 1.00 × 1.00	1.50 × 1.00 × 1.00	1.50 × 1.00 × 1.00	1.50 × 1.00 × 1.00
Layer 2	1.00 × 1.00 × 0.30	1.00 × 1.00 × 0.50	1.00 × 1.00 × 0.80	1.00 × 1.00 × 0.80	1.00 × 1.00 × 1.00	1.00 × 1.00 × 1.00
Layer 3						1.00 × 1.00 × 0.30
Pressure at Toe, t/m^2	3.16	4.11	4.12	4.12	4.94	6.32

(Continued)

TABLE 5.43 (Continued)

Catchment Area, ha	50.00	100.00	150.00	200.00	250.00	300.00
Pressure at Heel, t/m^2	1.00	0.69	1.74	1.74	1.60	1.21
Factor of Safety against Overturning	5.76	4.21	6.87	6.87	5.50	4.20
Shear Friction Factor F	11.83	8.93	9.81	9.81	8.05	6.33
Total Volume of Gabion Boxes, m^3	12.48	12.90	19.78	19.78	21.50	24.08

TABLE 5.44
Design Results for Gabion Check Dam with Steps on Both Faces Height = 3.00 m and Stream Width L = 20.0 m

Catchment Area, ha	50.00	100.00	150.00	200.00	250.00	300.00
A. WEIR						
Span of Weir W, m	16.00	16.00	16.00	16.00	16.00	16.00
Flood Discharge Q, m^3/s	6.54	11.00	14.91	18.50	21.87	25.07
Head Over Crest, m	0.37	0.52	0.64	0.73	0.82	0.90
Top Level, m	100.00	100.00	100.00	100.00	100.00	100.00
Foundation Level, m	97.00	97.00	97.00	97.00	97.00	97.00
Total Height of Weir H_{WEIR}, m	3.00	3.00	3.00	3.00	3.00	3.00
Bottom Width of Weir B_{WEIR}, m	4.50	4.50	5.00	5.00	5.00	5.00
Size of Gabion Box (L×B×H), m						
Bottom Layer 1	4.50 × 1.00 × 1.00	4.50 × 1.00 × 1.00	5.00 × 1.00 × 1.00	5.00 × 1.00 × 1.00	5.00 × 1.00 × 1.00	5.00 × 1.00 × 1.00
Layer 2	3.50 × 1.00 × 1.00	4.00 × 1.00 × 1.00	4.00 × 1.00 × 1.00	4.00 × 1.00 × 1.00	4.00 × 1.00 × 1.00	4.00 × 1.00 × 1.00
Layer 3	2.50 × 1.00 × 1.00	3.50 × 1.00 × 1.00	3 × 1.0 × 1	3 × 1.0 × 1	3.50 × 1.00 × 1.00	3.50 × 1.00 × 1.00
Pressure at Toe, t/m^2	5.35	6.35	5.64	5.82	6.16	6.30

(Continued)

TABLE 5.44 (Continued)

Catchment Area, ha	50.00	100.00	150.00	200.00	250.00	300.00
Pressure at Heel, t/m^2	0.05	0.05	0.15	0.05	0.11	0.04
Factor of Safety against Overturning	1.70	1.80	1.76	1.75	1.80	1.79
Shear Friction Factor F	4.18	4.01	4.12	3.95	3.86	3.74
Total Volume of Gabion Boxes, m^3	168.00	192.00	192.00	192.00	200.00	200.00
Length of D/S Apron Floor $L_{D/S\ APRON}$, m	6.80	6.80	6.30	6.30	6.30	6.30
B. NON-OVER FLOW PORTION						
Length, m	4.00	4.00	4.00	4.00	4.00	4.00
Top Level including Free Board, m	100.80	101.00	101.30	101.30	101.30	101.30
Foundation Level, m	97.00	97.00	97.00	97.00	97.00	97.00
Total Height of Non-Over Flow H_{NOF}, m	3.80	4.00	4.30	4.30	4.30	4.30
Bottom Width of Non-Over Flow B_{NOF}, m	4.00	4.00	4.00	4.00	4.00	4.50
Size of Gabion Box (L×B×H), m						
Bottom Layer 1	4.00 × 1.00 × 1.00	4.00 × 1.00 × 1.00	4.00 × 1.00 × 1.00	4.00 × 1.00 × 1.00	4.00 × 1.00 × 1.00	4.50 × 1.00 × 1.00
Layer 2	3.00 × 1.00 × 1.00	3.00 × 1.00 × 1.00	3.00 × 1.00 × 1.00	3.00 × 1.00 × 1.00	3.00 × 1.00 × 1.00	3.50 × 1.00 × 1.00
Layer 3	2.00 × 1.00 × 1.00	2.00 × 1.00 × 1.00	2.00 × 1.00 × 1.00	2.00 × 1.00 × 1.00	2.50 × 1.00 × 1.00	2.50 × 1.00 × 1.00
Layer 4	1.00 × 1.00 × 0.80	1.00 × 1.00 × 1.00	1.50 × 1.00 × 1.30	2.00 × 1.00 × 1.30	2.50 × 1.00 × 1.30	1.50 × 1.00 × 1.30
Pressure at Toe, t/m^2	5.75	6.15	6.82	7.34	8.11	6.76
Pressure at Heel, t/m^2	0.15	0.04	0.13	0.10	0.06	0.42
Factor of Safety against Overturning	1.71	1.70	1.77	1.80	1.85	1.83

(Continued)

TABLE 5.44 (Continued)

Catchment Area, ha	50.00	100.00	150.00	200.00	250.00	300.00
Shear Friction Factor F	3.66	3.35	3.20	3.08	3.00	3.12
Total Volume of Gabion Boxes, m^3	39.20	40.00	43.80	46.40	51.00	49.80
C. WING WALLS (ABUTMENTS) ON BOTH BANKS						
Length in U/S $L_{U/S\ ABUT}$, m	4.00	4.00	4.00	4.00	4.00	4.50
Length in D/S $L_{D/S\ ABUT}$, m	2.00	2.00	2.00	2.00	2.00	2.00
Top Level including Free Board, m	100.80	101.00	101.30	101.30	101.30	101.30
Foundation Level, m	97.00	97.00	97.00	97.00	97.00	97.00
Total Height H_{ABUT}, m	3.80	4.00	4.30	4.30	4.30	4.30
Base Width of Abutment B_{ABUT}, m	2.50	2.50	2.50	2.50	2.50	2.50
Size of Gabion Box (L×B×H), m						
Bottom Layer 1	2.50 × 1.00 × 1.00	2.50 × 1.00 × 1.00	2.50 × 1.00 × 1.00	2.50 × 1.00 × 1.00	2.50 × 1.00 × 1.00	2.50 × 1.00 × 1.00
Layer 2	2.00 × 1.00 × 1.00	2.00 × 1.00 × 1.00	2.00 × 1.00 × 1.00	2.00 × 1.00 × 1.00	2.00 × 1.00 × 1.00	2.00 × 1.00 × 1.00
Layer 3	1.00 × 1.00 × 1.00	1.00 × 1.00 × 1.00	1.00 × 1.00 × 1.00	1.00 × 1.00 × 1.00	1.00 × 1.00 × 1.00	1.00 × 1.00 × 1.00
Layer 4	1.00 × 1.00 × 0.80	1.00 × 1.00 × 1.00	1.00 × 1.00 × 1.30	1.00 × 1.00 × 1.30	1.00 × 1.00 × 1.30	1.00 × 1.00 × 1.30
Pressure at Toe, t/m^2	10.41	11.43	13.15	13.15	13.15	13.15
Pressure at Heel, t/m^2	2.27	1.93	1.24	1.24	1.24	1.24
Factor of Safety against Overturning	4.38	3.98	3.45	3.45	3.45	3.45
Shear Friction Factor F	4.55	4.19	3.73	3.73	3.73	3.73
Total Volume of Gabion Boxes, m^3	75.60	78.00	81.60	81.60	81.60	88.40

(Continued)

TABLE 5.44 (Continued)

Catchment Area, ha	50.00	100.00	150.00	200.00	250.00	300.00
D. D/S WING WALLS ON BOTH BANKS						
Length in D/S $L_{D/S\ WING}$, m	4.80	4.80	4.30	4.30	4.30	4.30
Top Level including Free Board, m	98.30	98.50	98.80	98.80	98.80	98.80
Foundation Level, m	97.00	97.00	97.00	97.00	97.00	97.00
Total Height $H_{D/S\ WING}$, m	1.30	1.50	1.80	1.80	1.80	1.80
Base Width of D/S Wing Wall $B_{D/S\ WING}$, m	1.00	1.00	1.50	1.50	1.50	1.50
Size of Gabion Box (L×B×H), m						
Bottom Layer 1	1.00 × 1.00 × 1.00	1.00 × 1.00 × 1.00	1.50 × 1.00 × 1.00	1.50 × 1.00 × 1.00	1.50 × 1.00 × 1.00	1.50 × 1.00 × 1.00
Layer 2	1.00 × 1.00 × 0.30	1.00 × 1.00 × 0.50	1.00 × 1.00 × 0.80	1.00 × 1.00 × 0.80	1.00 × 1.00 × 0.80	1.00 × 1.00 × 0.80
Pressure at Toe, t/m^2	3.16	4.11	4.12	4.12	4.12	4.12
Pressure at Heel, t/m^2	1.00	0.69	1.74	1.74	1.74	1.74
Factor of Safety against Overturning	5.76	4.21	6.87	6.87	6.87	6.87
Shear Friction Factor F	11.83	8.93	9.81	9.81	9.81	9.81
Total Volume of Gabion Boxes, m^3	12.48	14.40	19.78	19.78	19.78	19.78

5.6 APPLICATION OF DESIGN TABLES

Steps to carry out the design of gabion check dams include the following:

Step 1: For the chosen shape of the cross section, given the value of the height of a check dam H and the width of the stream, select the appropriate table.

Step 2: From the appropriate table, for the known catchment area, select the appropriate column of the table and read the size of gabion boxes from section A for the weir section, top level, foundation level and factor of safety against overturning, base pressures at toe and heel, length of downstream apron for check dams greater than 1.50 m high, total volume of gabion boxes, etc.

Step 3: Read the size of gabion boxes for the non-over flow section from section B, top level, foundation level and factor of safety against overturning, base pressures at toe and pressure at heel, total volume of gabion boxes, etc.

Step 4: Read the size of gabion boxes for wing walls (abutments) from section C for check dams greater than 1.50 m high, length of wing walls to be provided, top level, foundation level and factor of safety against overturning, base pressures at toe and pressure at heel, total volume of gabion boxes, etc.

Step 5: Read the size of gabion boxes for downstream wing walls from section D for check dams greater than 1.50 m high, top level, foundation level and factor of safety against overturning, base pressures at toe and pressure at heel, total volume of gabion boxes, etc.

Step 6: Engineering drawing for the gabion check dam can be prepared by selecting the appropriate figure (from 9.1 to 9.5) on the basis of the design data obtained from Step 2 to Step 5.

Step 7: Specifications and general guidelines to be followed during the construction of gabion check dams can be read from Figure 9.6.

5.7 DESIGN EXAMPLES

To illustrate the use of tables, the following examples have been chosen:

5.7.1 Example 1

Design a gabion check dam for the following parameters:

i. Shape of the gabion section: vertical D/S face
ii. Height of a check dam H = 2.50 m
iii. Width of stream/gadhera = 10.0 m
iv. Catchment area = 150 ha

Solution:

Step 1: For a chosen shape of the cross section as vertical D/S face, the given value of the height of a check dam as 2.50 m and width of stream as 10.0 m, select Table 5.22.

Step 2: From Table 5.22, for the known catchment area, select the appropriate column of the table. The values given in section A for the design of gabion weir are as follows:

A. WEIR	
Span of Weir W, m	8.00
Flood Discharge Q, m^3/s	14.91
Head Over Crest, m	1.01
Top Level, m	100.00
Foundation Level, m	97.50

Total Height of Weir H_{WEIR}, m	2.50
Bottom Width of Weir B_{WEIR}, m	4.00
Size of Gabion Box (L×B×H), m	
Bottom Layer 1	4.00 × 1.00 × 1.00
Layer 2	4.00 × 1.00 × 1.00
Layer 3	4.00 × 1.00 × 0.50
Pressure at Toe, t/m^2	6.49
Pressure at Heel, t/m^2	0.35
Factor of Safety against Overturning	2.01
Shear Friction Factor F	3.88
Total Volume of Gabion Boxes, m^3	80.00
Length of D/S Apron Floor $L_{D/S\ APRON}$, m	6.00

Step 3: From Table 5.22, read the values given in section B for the design of the non-over flow section:

Length, m	2.00
Top Level including Free Board, m	101.50
Foundation Level, m	97.50
Total Height of Non-Over Flow H_{NOF}, m	4.00
Bottom Width of Non-Over Flow B_{NOF}, m	4.00
Size of Gabion Box (L×B×H), m	
Bottom Layer 1	4.00 × 1.00 × 1.00
Layer 2	4.00 × 1.00 × 1.00
Layer 3	3.50 × 1.00 × 1.00
Layer 4	3.50 × 1.00 × 1.00
Pressure at Toe, t/m^2	9.69
Pressure at Heel, t/m^2	0.21
Factor of Safety against Overturning	2.27
Shear Friction Factor F	3.71
Total Volume of Gabion Boxes, m^3	30.00

Step 4: From Table 5.22, read the values given in section C for the design of wing walls (abutments)

Length in U/S $L_{U/S\ ABUT}$, m	4.00
Length in D/S $L_{D/S\ ABUT}$, m	2.00
Top Level including Free Board, m	101.50
Foundation Level, m	97.50
Total Height H_{ABUT}, m	4.00

Base Width of Abutment B_{ABUT}, m	2.50
Size of Gabion Box (L×B×H), m	
Bottom Layer 1	2.50 × 1.00 × 1.00
Layer 2	2.00 × 1.00 × 1.00
Layer 3	1.00 × 1.00 × 1.00
Layer 4	1.00 × 1.00 × 1.00
Pressure at Toe, t/m²	11.43
Pressure at Heel, t/m²	1.93
Factor of Safety against Overturning	3.98
Shear Friction Factor F	4.19
Total Volume of Gabion Boxes, m³	78.00

Step 5: From Table 5.22, read the values given in section D for the design of D/S wing walls:

Length in D/S $L_{D/S\ WING}$, m	4.00
Top Level including Free Board, m	99.50
Foundation Level, m	97.50
Total Height $H_{D/S\ WING}$, m	2.00
Base Width of D/S Wing Wall $B_{D/S\ WING}$, m	1.50
Size of Gabion Box (L×B×H), m	
Bottom Layer 1	1.50 × 1.00 × 1.00
Layer 2	1.00 × 1.00 × 1.00
Pressure at Toe, t/m²	4.94
Pressure at Heel, t/m²	1.60
Factor of Safety against Overturning	5.50
Shear Friction Factor F	8.05
Total Volume of Gabion Boxes, m³	20.00

Step 6: Based on the design data obtained from Step 2 to Step 5, an engineering drawing for the 2.50 m high gabion check dam with vertical D/S face can be prepared with the help of Figure 9.3.

Step 7: Specifications and general guidelines for the construction of gabion check dams can be read from Figure 9.6.

5.7.2 Example 2

Design a gabion check dam for the following parameters:

i. Shape of gabion section: vertical U/S face
ii. All the other data are the same, as given in example 1.

Solution:

Step 1: For a chosen shape of the cross section as vertical U/S face, the given value of the height of a check dam as 2.50 m and width of stream as 10.0 m, select Table 5.26.

Step 2: From Table 5.26, for the known catchment area, select the appropriate column of the table. The values given in section A for the design of gabion weir are as follows:

Span of Weir W, m	8.00
Flood Discharge Q, m^3/s	14.91
Head Over Crest, m	1.01
Top Level, m	100.00
Foundation Level, m	97.50
Total Height of Weir H_{WEIR}, m	2.50
Bottom Width of Weir B_{WEIR}, m	3.50
Size of Gabion Box (L×B×H), m	
Bottom Layer 1	3.50 × 1.00 × 1.00
Layer 2	2.50 × 1.00 × 1.00
Layer 3	1.50 × 1.00 × 0.50
Pressure at Toe, t/m^2	4.07
Pressure at Heel, t/m^2	0.18
Factor of Safety against Overturning	1.57
Shear Friction Factor F	3.09
Total Volume of Gabion Boxes, m^3	54.00
Length of D/S Apron Floor $L_{D/S\ APRON}$, m	5.20

Step 3: From Table 5.26, read the values given in section B for the design of the non-over flow section:

Length, m	2.00
Top Level including Free Board, m	101.50
Foundation Level, m	97.50
Total Height of Non-Over Flow H_{NOF}, m	4.00
Bottom Width of Non-Over Flow B_{NOF}, m	3.50
Size of Gabion Box (L×B×H), m	
Bottom Layer 1	3.50 × 1.00 × 1.00
Layer 2	3.00 × 1.00 × 1.00
Layer 3	2.00 × 1.00 × 1.00
Layer 4	1.00 × 1.00 × 1.00
Pressure at Toe, t/m^2	5.48
Pressure at Heel, t/m^2	0.70
Factor of Safety against Overturning	1.78

Shear Friction Factor F	2.86
Total Volume of Gabion Boxes, m^3	19.00

Step 4: Since other parameters are the same, the design of wing walls (abutments) is the same as shown in example 1.

Step 5: Since other parameters are the same, the design of D/S wing walls is the same as shown in example 1.

Step 6: Based on the design data obtained from Step 2 to Step 5, an engineering drawing for the 2.50 m high gabion check dam with vertical U/S face can be prepared with the help of Figure 9.4.

Step 7: Specifications and general guidelines for the construction of gabion check dams can be read from Figure 9.6.

5.7.3 Example 3

Design a gabion check dam for the following parameters:

i. Shape of gabion section: steps on both faces
ii. All the other data are the same, as given in example 1.

Solution:

Step 1: For a chosen shape of the cross section as steps on both faces, the given value of the height of a check dam as 2.50 m and width of stream as 10.0 m, select Table 5.30.

Step 2: From Table 5.30, for the known catchment area, select the appropriate column of the table. The values given in section A for the design of the gabion weir are as follows:

Span of Weir W, m	8.00
Flood Discharge Q, m^3/s	14.91
Head Over Crest, m	1.01
Top Level, m	100.00
Foundation Level, m	97.50
Total Height of Weir H_{WEIR}, m	2.50
Bottom Width of Weir B_{WEIR}, m	4.50
Size of Gabion Box (L×B×H), m	
Bottom Layer 1	4.50 × 1.00 × 1.00
Layer 2	3.50 × 1.00 × 1.00
Layer 3	3.00 × 1.00 × 0.50
Pressure at Toe, t/m^2	5.52
Pressure at Heel, t/m^2	0.02
Factor of Safety against Overturning	1.83
Shear Friction Factor F	4.17

Total Volume of Gabion Boxes, m^3	76.00
Length of D/S Apron Floor $L_{D/S\ APRON}$, m	5.00

Step 3: From Table 5.30, read the values given in section B for the design of the non-over flow section:

Length, m	2.00
Top Level including Free Board, m	101.50
Foundation Level, m	97.50
Total Height of Non-Over Flow H_{NOF}, m	4.00
Bottom Width of Non-Over Flow B_{NOF}, m	4.00
Size of Gabion Box (L×B×H), m	
Bottom Layer 1	4.00 × 1.00 × 1.00
Layer 2	3.00 × 1.00 × 1.00
Layer 3	2.00 × 1.00 × 1.00
Layer 4	1.00 × 1.00 × 1.00
Pressure at Toe, t/m^2	6.25
Pressure at Heel, t/m^2	0.46
Factor of Safety against Overturning	1.90
Shear Friction Factor F	3.33
Total Volume of Gabion Boxes, m^3	20.00

Step 4: Since other parameters are the same, the design of the wing walls (abutments) is the same as shown in example 1.

Step 5: Since other parameters are the same, the design of D/S wing walls is the same as shown in example 1.

Step 6: Based on the design data obtained from Step 2 to Step 5, an engineering drawing for the 2.50 m high gabion check dam with steps on both faces can be prepared with the help of Figure 9.5.

Step 7: Specifications and general guidelines for the construction of gabion check dams can be read from Figure 9.6.

REFERENCE

IS:6512 (2019) *Indian Standard Criteria for Design of Solid Gravity Dams*, IS:6512, Bureau of Indian Standards, New Delhi.

6 Tables for the Quick Design of Masonry Check Dams

6.1 PARAMETERS CONSIDERED FOR THE DEVELOPMENT OF TABLES

Masonry check dams can be constructed with different shapes having vertical U/S face or vertical D/S face or slopes on both faces. The following forces are considered, in general, in the design of masonry check dams:

- Self weight of wall;
- Water pressure;
- Uplift pressures; and
- Active earth pressures in the design of abutment and wing walls.

The following parameters were varied for developing the design tables for masonry check dams:

Height of a Check Dam Weir	:	1.50 m, 2.00 m, 2.50 m, and 3.00 m
Catchment Area[1]	:	50 ha, 100 ha, 150 ha, 200 ha, 250 ha, and 300 ha
Width of Stream/Gadhera	:	5.0 m, 10.0 m, 15.0 m, and 20.0 m
Shape of a Check Dam Section	:	• Vertical U/S Face • Vertical D/S Face • Slopes on both Faces
Unit Weight of Water, γ_{water}	:	1.00 t/m^3 or 9.81 kN/m^3
Unit Weight of Masonry, $\gamma_{Masonry}$	:	2.30 t/m^3 or 22.56 kN/m^3
Effect of Water for Uplift Pressure	:	100% of Total Height
Coefficient of Friction between the Masonry and Soil/Rock Mass μ	:	0.50

[1]For small width of streams, the design of check dams has been carried out by limiting the head over crest to about 1.30 m.

6.1.1 Typical Cross Sections Adopted

The generalised cross sections of masonry check dam weir and non-over flow for heights varying from 1.50 m to 3.00 m for carrying out the analysis for different shapes are depicted in Figure 6.1 to Figure 6.3.

DOI: 10.1201/9781003605621-6

6.1.1.1 Section with Vertical U/S Face

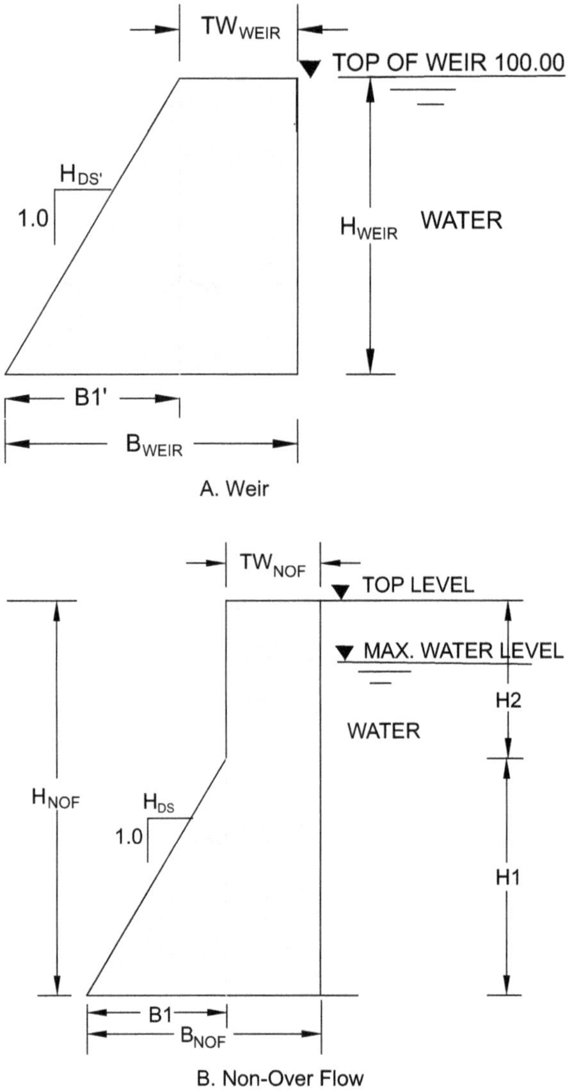

FIGURE 6.1A, 6.1B Typical Cross Sections of Masonry Weir and Non-Over Flow with Vertical U/S Face.

6.1.1.2 Section with Vertical D/S Face

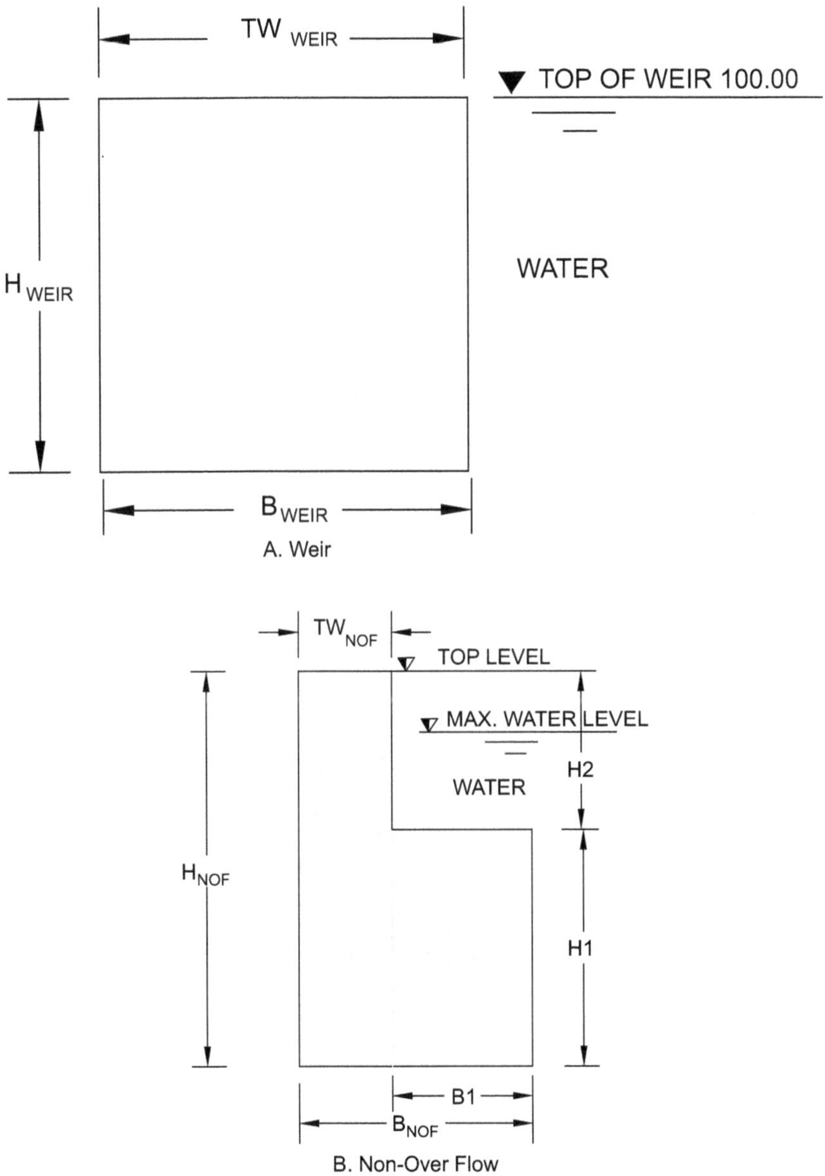

FIGURE 6.2A, 6.2B Typical Cross Sections of Masonry Weir and Non-Over Flow with Vertical D/S Face.

6.1.1.3 Section with Slopes on Both Faces

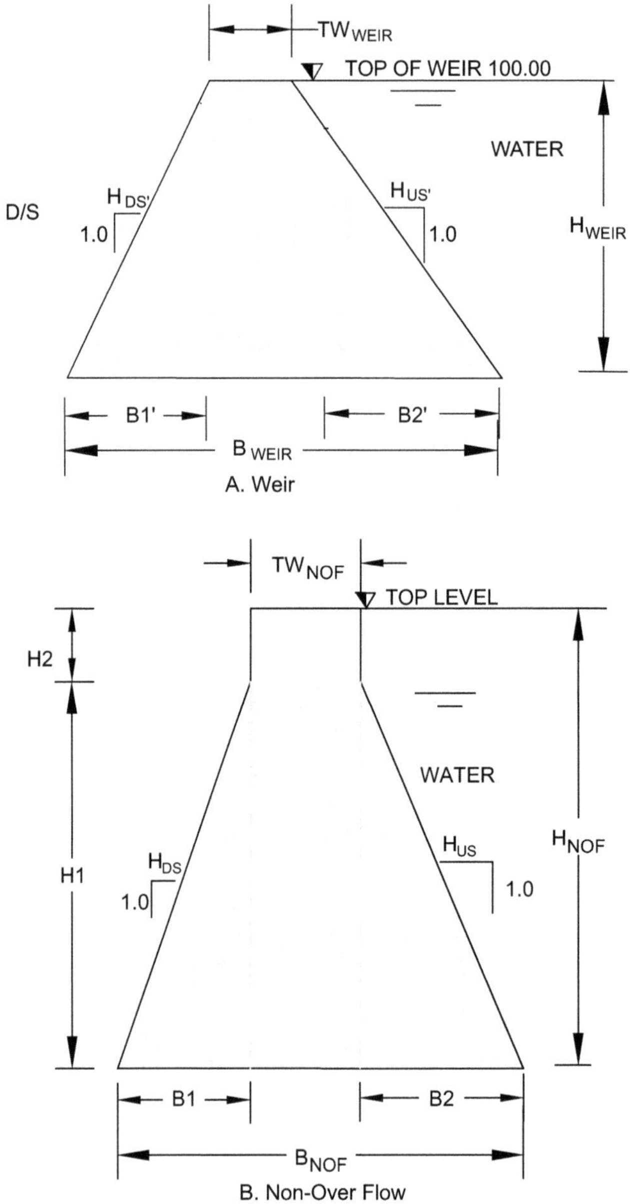

FIGURE 6.3A, 6.3B Typical Cross Sections of Masonry Weir and Non-Over Flow with Slopes on Both Faces.

6.1.2 Design Approach

For each masonry check dam of specified height, the stability analysis was carried out by the conventional gravity approach (IS:6512, 2019), discussed in Chapter 3 on design philosophy, for six sets of catchment areas and four sets of stream widths for each shape to develop the design tables. For a small width of streams, the design of masonry check dams has been limited to the head over crest to about 1.30 m.

The base width of the masonry check dam (non-over flow and overflow sections) has been so proportioned in the analysis that there is nearly zero stress near the heel under static loading; the factor of safety against overturning is equal to or more than 1.50.

6.2 DESIGN OF 1.50 M HIGH MASONRY CHECK DAMS

6.2.1 Section with Vertical U/S Face

The tables developed after the rigorous design of the different components for 1.50 m high masonry check dams for different combinations are presented from Table 6.1 to Table 6.4.

TABLE 6.1
Design Results for Masonry Check Dam with Vertical U/S Face Height = 1.50 m and Stream Width L = 5.0 m

Catchment Area, ha	50.00	100.00
A. WEIR		
Span of Weir W, m	4.00	4.00
Flood Discharge Q, m^3/s	6.54	11.00
Head Over Crest, m	0.92	1.31
Top Level, m	100.00	100.00
Foundation Level, m	98.50	98.50
Height of Weir H_{WEIR}, m	1.50	1.50
Top Width of Weir TW_{WEIR}, mm	500	500
Base Width of Weir B_{WEIR}, mm	1820	2000
D/S Slope – H_{DS}: 1.0(V)	0.880	1.000
Width B1', mm	1320	1500
Pressure at Toe, t/m^2	3.22	3.20
Pressure at Heel, t/m^2	0.02	0.05
Factor of Safety against Overturning	1.53	1.52
Shear Friction Factor F	3.10	2.82
Total Volume of Masonry, m^3	6.96	7.50
B. NON-OVER FLOW PORTION		
Length, m	1.00	1.00

(Continued)

TABLE 6.1 (Continued)

Top Level including Free Board, m	101.50	102.00
Foundation Level, m	98.50	98.50
Height of Non-Over Flow H_{NOF}, m	3.00	3.50
Top Width of Non-Over Flow TW_{NOF}, mm	500	500
Base Width of Non-Over Flow B_{NOF}, mm	1620	1960
D/S Slope – H_{DS}: 1.0(V)	0.540	0.560
Width B1, mm	1120	1460
Height H1, mm	2074	2607
Height H2, mm	926	893
Pressure at Toe, t/m^2	6.03	7.02
Pressure at Heel, t/m^2	0.03	0.04
Factor of Safety against Overturning	1.65	1.69
Shear Friction Factor F	2.52	2.26
Total Volume of Masonry, m^3	2.66	3.65

TABLE 6.2
Design Results for Masonry Check Dam with Vertical U/S Face Height = 1.50 m and Stream Width L = 10.0 m

Catchment Area, ha	50.00	100.00	150.00	200.00
A. WEIR				
Span of Weir W, m	8.00	8.00	8.00	8.00
Flood Discharge Q, m^3/s	6.54	11.00	14.91	18.50
Head Over Crest, m	0.58	0.82	1.01	1.16
Top Level, m	100.00	100.00	100.00	100.00
Foundation Level, m	98.50	98.50	98.50	98.50
Height of Weir H_{WEIR}, m	1.50	1.50	1.50	1.50
Top Width of Weir TW_{WEIR}, mm	500	500	500	500
Base Width of Weir B_{WEIR}, mm	1640	1770	1860	1920
D/S Slope – $H_{DS'}$: 1.0(V)	0.760	0.846	0.906	0.946
Width B1', mm	1140	1270	1360	1420
Pressure at Toe, t/m^2	3.22	3.21	3.22	3.25
Pressure at Heel, t/m^2	0.02	0.02	0.02	0.01
Factor of Safety against Overturning	1.54	1.53	1.52	1.52
Shear Friction Factor F	3.42	3.19	3.02	2.90
Total Volume of Masonry, m^3	12.84	13.62	14.16	14.52
B. NON-OVER FLOW PORTION				
Length, m	2.00	2.00	2.00	2.00
Top Level including Free Board, m	101.10	101.40	101.60	101.80
Foundation Level, m	98.50	98.50	98.50	98.50
Height of Non-Over Flow H_{NOF}, m	2.60	2.90	3.10	3.30

(Continued)

TABLE 6.2 (Continued)

Catchment Area, ha	**50.00**	**100.00**	**150.00**	**200.00**
Top Width of Non-Over Flow TW_{NOF}, mm	500	500	500	500
Base Width of Non-Over Flow B_{NOF}, mm	1370	1550	1700	1820
D/S Slope – H_{DS}: 1.0(V)	0.527	0.534	0.548	0.551
Width B1, mm	870	1050	1200	1320
Height H1, mm	1651	1964	2188	2393
Height H2, mm	949	936	912	907
Pressure at Toe, t/m^2	5.22	5.84	6.15	6.61
Pressure at Heel, t/m^2	0.06	0.03	0.10	0.07
Factor of Safety against Overturning	1.63	1.65	1.67	1.68
Shear Friction Factor F	2.83	2.58	2.48	2.35
Total Volume of Masonry, m^3	4.04	4.96	5.72	6.47

TABLE 6.3
Design Results for Masonry Check Dam with Vertical U/S Face Height = 1.50 m and Stream Width L = 15.0 m

Catchment Area, ha	**50.00**	**100.00**	**150.00**	**200.00**	**250.00**	**300.00**
A. WEIR						
Span of Weir W, m	12.00	12.00	12.00	12.00	12.00	12.00
Flood Discharge Q, m^3/s	6.54	11.00	14.91	18.50	21.87	25.07
Head Over Crest, m	0.44	0.63	0.77	0.89	0.99	1.09
Top Level, m	100.00	100.00	100.00	100.00	100.00	100.00
Foundation Level, m	98.50	98.50	98.50	98.50	98.50	98.50
Height of Weir H_{WEIR}, m	1.50	1.50	1.50	1.50	1.50	1.50
Top Width of Weir TW_{WEIR}, mm	500	500	500	500	500	500
Base Width of Weir B_{WEIR}, mm	1560	1670	1740	1810	1850	1900
D/S Slope – $H_{DS'}$: 1.0(V)	0.706	0.780	0.826	0.873	0.900	0.933
Width B1', mm	1060	1170	1240	1310	1350	1400
Pressure at Toe, t/m^2	3.24	3.22	3.23	3.21	3.22	3.21
Pressure at Heel, t/m^2	0.01	0.01	0.01	0.02	0.02	0.03
Factor of Safety against Overturning	1.54	1.54	1.53	1.53	1.52	1.53
Shear Friction Factor F	3.57	3.36	3.22	3.13	3.03	2.96
Total Volume of Masonry, m^3	18.54	19.53	20.16	20.79	21.15	21.60
B. NON-OVER FLOW PORTION						
Length, m	3.00	3.00	3.00	3.00	3.00	3.00

(Continued)

TABLE 6.3 (Continued)

Catchment Area, ha	50.00	100.00	150.00	200.00	250.00	300.00
Top Level including Free Board, m	101.00	101.20	101.40	101.50	101.60	101.70
Foundation Level, m	98.50	98.50	98.50	98.50	98.50	98.50
Height of Non-Over Flow H_{NOF}, m	2.50	2.70	2.90	3.00	3.10	3.20
Top Width of Non-Over Flow TW_{NOF}, mm	500	500	500	500	500	500
Base Width of Non-Over Flow B_{NOF}, mm	1300	1430	1550	1620	1700	1760
D/S Slope – H_{DS}: 1.0(V)	0.520	0.529	0.534	0.540	0.548	0.550
Width B1, mm	800	930	1050	1120	1200	1260
Height H1, mm	1538	1755	1964	2074	2188	2291
Height H2, mm	962	945	936	926	912	909
Pressure at Toe, t/m^2	5.08	5.40	5.84	6.03	6.15	6.38
Pressure at Heel, t/m^2	0.02	0.08	0.03	0.03	0.10	0.09
Factor of Safety against Overturning	1.62	1.64	1.65	1.65	1.67	1.68
Shear Friction Factor F	2.91	2.74	2.58	2.52	2.48	2.41
Total Volume of Masonry, m^3	5.60	6.51	7.44	7.99	8.57	9.15

TABLE 6.4
Design Results for Masonry Check Dam with Vertical U/S Face Height = 1.50 m and Stream Width L = 20.0 m

Catchment Area, ha	50.00	100.00	150.00	200.00	250.00	300.00
A. WEIR						
Span of Weir W, m	16.00	16.00	16.00	16.00	16.00	16.00
Flood Discharge Q, m^3/s	6.54	11.00	14.91	18.50	21.87	25.07
Head Over Crest, m	0.37	0.52	0.64	0.73	0.82	0.90
Top Level, m	100.00	100.00	100.00	100.00	100.00	100.00
Foundation Level, m	98.50	98.50	98.50	98.50	98.50	98.50
Height of Weir H_{WEIR}, m	1.50	1.50	1.50	1.50	1.50	1.50
Top Width of Weir TW_{WEIR}, mm	500	500	500	500	500	500
Base Width of Weir B_{WEIR}, mm	1520	1610	1680	1720	1770	1810

(Continued)

TABLE 6.4 (Continued)

Catchment Area, ha	50.00	100.00	150.00	200.00	250.00	300.00
D/S Slope – $H_{DS'}$: 1.0(V)	0.680	0.740	0.786	0.813	0.846	0.873
Width B1', mm	1020	1110	1180	1220	1270	1310
Pressure at Toe, t/m^2	3.23	3.20	3.19	3.23	3.21	3.22
Pressure at Heel, t/m^2	0.02	0.03	0.04	0.01	0.02	0.01
Factor of Safety against Overturning	1.55	1.54	1.54	1.53	1.53	1.53
Shear Friction Factor F	3.66	3.49	3.36	3.26	3.19	3.12
Total Volume of Masonry, m^3	24.24	25.32	26.16	26.64	27.24	27.72
B. NON-OVER FLOW PORTION						
Length, m	4.00	4.00	4.00	4.00	4.00	4.00
Top Level including Free Board, m	100.90	101.10	101.20	101.30	101.40	101.50
Foundation Level, m	98.50	98.50	98.50	98.50	98.50	98.50
Height of Non-Over Flow H_{NOF}, m	2.40	2.60	2.70	2.80	2.90	3.00
Top Width of Non-Over Flow TW_{NOF}, mm	500	500	500	500	500	500
Base Width of Non-Over Flow B_{NOF}, mm	1240	1370	1430	1500	1550	1620
D/S Slope – H_{DS}: 1.0(V)	0.517	0.527	0.529	0.535	0.534	0.540
Width B1, mm	740	870	930	1000	1050	1120
Height H1, mm	1433	1651	1755	1865	1964	2074
Height H2, mm	967	949	945	935	936	926
Pressure at Toe, t/m^2	4.90	5.22	5.40	5.57	5.84	6.03
Pressure at Heel, t/m^2	0.02	0.06	0.08	0.09	0.03	0.03
Factor of Safety against Overturning	1.62	1.63	1.64	1.65	1.65	1.65
Shear Friction Factor F	3.01	2.83	2.74	2.67	2.58	2.52
Total Volume of Masonry, m^3	6.93	8.09	8.67	9.32	9.92	10.66

6.2.2 Section with Vertical D/S Face

The tables developed after the rigorous design of the different components for 1.50 m high masonry check dams for different combinations are presented from Table 6.5 to Table 6.8.

TABLE 6.5
Design Results for Masonry Check Dam with Vertical D/S Face Height = 1.50 m and Stream Width L = 5.0 m

Catchment Area, ha	50.00	100.00
A. WEIR		
Span of Weir W, m	4.00	4.00
Flood Discharge Q, m^3/s	6.54	11.00
Head Over Crest, m	0.92	1.31
Top Level, m	100.00	100.00
Foundation Level, m	98.50	98.50
Height of Weir H_{WEIR}, m	1.50	1.50
Top Width of Weir TW_{WEIR}, mm	1910	2010
Base Width of Weir B_{WEIR}, mm	1910	2010
Pressure at Toe, t/m^2	6.63	7.13
Pressure at Heel, t/m^2	0.00	0.01
Factor of Safety against Overturning	2.13	2.15
Shear Friction Factor F	3.71	3.29
Total Volume of Masonry, m^3	11.46	12.06
B. NON-OVER FLOW PORTION		
Length, m	1.00	1.00
Top Level including Free Board, m	101.50	102.00
Foundation Level, m	98.50	98.50
Height of Non-Over Flow H_{NOF}, m	3.00	3.50
Height H1, mm	2400	2700
Height H2, mm	600	800
Top Width of Non-Over Flow TW_{NOF}, mm	1000	1000
Base Width of Non-Over Flow B_{NOF}, mm	2380	2450
Width B1, mm	1380	1450
Pressure at Toe, t/m^2	10.81	12.67
Pressure at Heel, t/m^2	0.01	0.10
Factor of Safety against Overturning	2.67	2.55
Shear Friction Factor F	4.33	3.35
Total Volume of Masonry, m^3	6.31	7.41

TABLE 6.6
Design Results for Masonry Check Dam with Vertical D/S Face Height = 1.50 m and Stream Width L = 10.0 m

Catchment Area, ha	50.00	100.00	150.00	200.00
A. WEIR				
Span of Weir W, m	8.00	8.00	8.00	8.00
Flood Discharge Q, m^3/s	6.54	11.00	14.91	18.50
Head Over Crest, m	0.58	0.82	1.01	1.16
Top Level, m	100.00	100.00	100.00	100.00
Foundation Level, m	98.50	98.50	98.50	98.50

(Continued)

TABLE 6.6 (Continued)

Catchment Area, ha	50.00	100.00	150.00	200.00
Height of Weir H_{WEIR}, m	1.50	1.50	1.50	1.50
Top Width of Weir TW_{WEIR}, mm	1800	1890	1940	1980
Base Width of Weir B_{WEIR}, mm	1800	1890	1940	1980
Pressure at Toe, t/m^2	6.17	6.47	6.73	6.92
Pressure at Heel, t/m^2	0.00	0.02	0.01	0.02
Factor of Safety against Overturning	2.12	2.13	2.14	2.14
Shear Friction Factor F	4.20	3.86	3.61	3.44
Total Volume of Masonry, m^3	21.60	22.68	23.28	23.76
B. NON-OVER FLOW PORTION				
Length, m	2.00	2.00	2.00	2.00
Top Level including Free Board, m	101.10	101.40	101.60	101.80
Foundation Level, m	98.50	98.50	98.50	98.50
Height of Non-Over Flow H_{NOF}, m	2.60	2.90	3.10	3.30
Height H1, mm	2080	2320	2480	2640
Height H2, mm	520	580	620	660
Top Width of Non-Over Flow TW_{NOF}, mm	1000	1000	1000	1000
Base Width of Non-Over Flow B_{NOF}, mm	2200	2340	2420	2510
Width B1, mm	1200	1340	1420	1510
Pressure at Toe, t/m^2	9.18	10.39	11.23	12.03
Pressure at Heel, t/m^2	0.00	0.01	0.00	0.02
Factor of Safety against Overturning	2.55	2.65	2.70	2.75
Shear Friction Factor F	5.21	4.53	4.14	3.84
Total Volume of Masonry, m^3	10.19	12.02	13.24	14.57

TABLE 6.7
Design Results for Masonry Check Dam with Vertical D/S Face Height = 1.50 m and Stream Width L = 15.0 m

Catchment Area, ha	50.00	100.00	150.00	200.00	250.00	300.00
A. WEIR						
Span of Weir W, m	12.00	12.00	12.00	12.00	12.00	12.00
Flood Discharge Q, m^3/s	6.54	11.00	14.91	18.50	21.87	25.07
Head Over Crest, m	0.44	0.63	0.77	0.89	0.99	1.09
Top Level, m	100.00	100.00	100.00	100.00	100.00	100.00
Foundation Level, m	98.50	98.50	98.50	98.50	98.50	98.50
Height of Weir H_{WEIR}, m	1.50	1.50	1.50	1.50	1.50	1.50
Top Width of Weir TW_{WEIR}, mm	1740	1820	1870	1910	1930	1960
Base Width of Weir B_{WEIR}, mm	1740	1820	1870	1910	1930	1960
Pressure at Toe, t/m^2	5.99	6.23	6.42	6.57	6.72	6.84
Pressure at Heel, t/m^2	0.00	0.01	0.01	0.02	0.00	0.01
Factor of Safety against Overturning	2.11	2.12	2.13	2.14	2.13	2.14
Shear Friction Factor F	4.44	4.12	3.92	3.76	3.62	3.51
Total Volume of Masonry, m^3	31.32	32.76	33.66	34.38	34.74	35.28

(Continued)

TABLE 6.7 (Continued)

Catchment Area, ha	50.00	100.00	150.00	200.00	250.00	300.00
B. NON-OVER FLOW PORTION						
Length, m	3.00	3.00	3.00	3.00	3.00	3.00
Top Level including Free Board, m	101.00	101.20	101.40	101.50	101.60	101.70
Foundation Level, m	98.50	98.50	98.50	98.50	98.50	98.50
Height of Non-Over Flow H_{NOF}, m	2.50	2.70	2.90	3.00	3.10	3.20
Height H1, mm	2000	2160	2320	2400	2480	2560
Height H2, mm	500	540	580	600	620	640
Top Width of Non-Over Flow TW_{NOF}, mm	1000	1000	1000	1000	1000	1000
Base Width of Non-Over Flow B_{NOF}, mm	2160	2250	2340	2380	2420	2470
Width B1, mm	1160	1250	1340	1380	1420	1470
Pressure at Toe, t/m^2	8.76	9.57	10.39	10.81	11.23	11.61
Pressure at Heel, t/m^2	0.01	0.01	0.01	0.01	0.00	0.02
Factor of Safety against Overturning	2.52	2.59	2.65	2.67	2.70	2.73
Shear Friction Factor F	5.50	4.97	4.53	4.33	4.14	3.99
Total Volume of Masonry, m^3	14.46	16.20	18.03	18.94	19.86	20.89

TABLE 6.8
Design Results for Masonry Check Dam with Vertical D/S Face Height = 1.50 m and Stream Width L = 20.0 m

Catchment Area, ha	50.00	100.00	150.00	200.00	250.00	300.00
A. WEIR						
Span of Weir W, m	16.00	16.00	16.00	16.00	16.00	16.00
Flood Discharge Q, m^3/s	6.54	11.00	14.91	18.50	21.87	25.07
Head Over Crest, m	0.37	0.52	0.64	0.73	0.82	0.90
Top Level, m	100.00	100.00	100.00	100.00	100.00	100.00
Foundation Level, m	98.50	98.50	98.50	98.50	98.50	98.50
Height of Weir H_{WEIR}, m	1.50	1.50	1.50	1.50	1.50	1.50
Top Width of Weir TW_{WEIR}, mm	1710	1780	1820	1860	1890	1910
Base Width of Weir B_{WEIR}, mm	1710	1780	1820	1860	1890	1910
Pressure at Toe, t/m^2	5.89	6.08	6.26	6.35	6.47	6.59
Pressure at Heel, t/m^2	0.00	0.01	0.00	0.02	0.02	0.01
Factor of Safety against Overturning	2.11	2.12	2.12	2.13	2.13	2.13
Shear Friction Factor F	4.58	4.31	4.10	3.98	3.86	3.74
Total Volume of Masonry, m^3	41.04	42.72	43.68	44.64	45.36	45.84
B. NON-OVER FLOW PORTION						
Length, m	4.00	4.00	4.00	4.00	4.00	4.00
Top Level including Free Board, m	100.90	101.10	101.20	101.30	101.40	101.50
Foundation Level, m	98.50	98.50	98.50	98.50	98.50	98.50
Height of Non-Over Flow H_{NOF}, m	2.40	2.60	2.70	2.80	2.90	3.00
Height H1, mm	1920	2080	2160	2240	2320	2400
Height H2, mm	480	520	540	560	580	600
Top Width of Non-Over Flow TW_{NOF}, mm	1000	1000	1000	1000	1000	1000

(Continued)

TABLE 6.8 (Continued)

Catchment Area, ha	50.00	100.00	150.00	200.00	250.00	300.00
Base Width of Non-Over Flow B_{NOF}, mm	2110	2200	2250	2290	2340	2380
Width B1, mm	1110	1200	1250	1290	1340	1380
Pressure at Toe, t/m^2	8.36	9.18	9.57	10.00	10.39	10.81
Pressure at Heel, t/m^2	0.00	0.00	0.01	0.00	0.01	0.01
Factor of Safety against Overturning	2.48	2.55	2.59	2.61	2.65	2.67
Shear Friction Factor F	5.79	5.21	4.97	4.73	4.53	4.33
Total Volume of Masonry, m^3	18.12	20.38	21.60	22.76	24.04	25.25

6.2.3 Section with Slopes on Both Faces

The tables developed after the rigorous design of the different components for 1.50 m high masonry check dams for different combinations are presented from Table 6.9 to Table 6.12.

TABLE 6.9
Design Results for Masonry Check Dam with Slopes on Both Faces Height = 1.50 m and Stream Width L = 5.0 m

Catchment Area, ha	50.00	100.00
A. WEIR		
Span of Weir W, m	4.00	4.00
Flood Discharge Q, m^3/s	6.54	11.00
Head Over Crest, m	0.92	1.31
Top Level, m	100.00	100.00
Foundation Level, m	98.50	98.50
Height of Weir H_{WEIR}, m	1.50	1.50
Top Width of Weir TW_{WEIR}, mm	500	500
Base Width of Weir B_{WEIR}, mm	1720	1800
U/S Slope – H_{US}: 1.0(V)	0.404	0.430
D/S Slope – H_{DS}: 1.0(V)	0.404	0.430
Width B1', mm	610	650
Width B2', mm	610	650
Pressure at Toe, t/m^2	4.98	5.37
Pressure at Heel, t/m^2	0.01	0.04
Factor of Safety against Overturning	1.59	1.61
Shear Friction Factor F	3.14	2.76
Total Volume of Masonry, m^3	6.66	6.90
B. NON-OVER FLOW PORTION		
Length, m	1.00	1.00
Top Level including Free Board, m	101.50	102.00
Foundation Level, m	98.50	98.50
Height of Non-Over Flow H_{NOF}, m	3.00	3.50
Height H1, mm	2500	3000

(Continued)

TABLE 6.9 (Continued)

Catchment Area, ha	50.00	100.00
Height H2, mm	500	500
Top Width of Non-Over Flow TW_{NOF}, mm	500	500
Base Width of Non-Over Flow B_{NOF}, mm	1940	2300
U/S Slope – H_{US}: 1.0(V)	0.286	0.300
D/S Slope – H_{DS}: 1.0(V)	0.286	0.300
Width B1, mm	720	900
Width B2, mm	720	900
Pressure at Toe, t/m^2	7.50	8.81
Pressure at Heel, t/m^2	0.02	0.00
Factor of Safety against Overturning	1.92	1.98
Shear Friction Factor F	3.17	2.81
Total Volume of Masonry, m^3	3.30	4.45

TABLE 6.10
Design Results for Masonry Check Dam with Slopes on Both Faces Height = 1.50 m and Stream Width L = 10.0 m

Catchment Area, ha	50.00	100.00	150.00	200.00
A. WEIR				
Span of Weir W, m	8.00	8.00	8.00	8.00
Flood Discharge Q, m^3/s	6.54	11.00	14.91	18.50
Head Over Crest, m	0.58	0.82	1.01	1.16
Top Level, m	100.00	100.00	100.00	100.00
Foundation Level, m	98.50	98.50	98.50	98.50
Height of Weir H_{WEIR}, m	1.50	1.50	1.50	1.50
Top Width of Weir TW_{WEIR}, mm	500	500	500	500
Base Width of Weir B_{WEIR}, mm	1640	1700	1760	1760
U/S Slope – H_{US}: 1.0(V)	0.378	0.400	0.420	0.420
D/S Slope – H_{DS}: 1.0(V)	0.378	0.400	0.420	0.420
Width B1', mm	570	600	630	630
Width B2', mm	570	600	630	630
Pressure at Toe, t/m^2	4.58	4.84	4.94	5.24
Pressure at Heel, t/m^2	0.04	0.04	0.14	0.02
Factor of Safety against Overturning	1.59	1.59	1.62	1.60
Shear Friction Factor F	3.61	3.26	3.07	2.88
Total Volume of Masonry, m^3	12.84	13.20	13.56	13.56
B. NON-OVER FLOW PORTION				
Length, m	2.00	2.00	2.00	2.00
Top Level including Free Board, m	101.10	101.40	101.60	101.80
Foundation Level, m	98.50	98.50	98.50	98.50
Height of Non-Over Flow H_{NOF}, m	2.60	2.90	3.10	3.30
Height H1, mm	2100	2400	2600	2800

(Continued)

TABLE 6.10 (Continued)

Catchment Area, ha	50.00	100.00	150.00	200.00
Height H2, mm	500	500	500	500
Top Width of Non-Over Flow TW_{NOF}, mm	500	500	500	500
Base Width of Non-Over Flow B_{NOF}, mm	1680	1880	2020	2160
U/S Slope – $H_{US'}$: 1.0(V)	0.280	0.286	0.292	0.295
D/S Slope – $H_{DS'}$: 1.0(V)	0.280	0.286	0.292	0.295
Width B1, mm	590	690	760	830
Width B2, mm	590	690	760	830
Pressure at Toe, t/m^2	6.36	7.18	7.74	8.25
Pressure at Heel, t/m^2	0.12	0.06	0.02	0.03
Factor of Safety against Overturning	1.87	1.92	1.93	1.97
Shear Friction Factor F	3.62	3.29	3.11	2.95
Total Volume of Masonry, m^3	5.07	6.20	7.05	7.94

TABLE 6.11
Design Results for Masonry Check Dam with Slopes on Both Faces Height = 1.50 m and Stream Width L = 15.0 m

Catchment Area, ha	50.00	100.00	150.00	200.00	250.00	300.00
A. WEIR						
Span of Weir W, m	12.00	12.00	12.00	12.00	12.00	12.00
Flood Discharge Q, m^3/s	6.54	11.00	14.91	18.50	21.87	25.07
Head Over Crest, m	0.44	0.63	0.77	0.89	0.99	1.09
Top Level, m	100.00	100.00	100.00	100.00	100.00	100.00
Foundation Level, m	98.50	98.50	98.50	98.50	98.50	98.50
Height of Weir H_{WEIR}, m	1.50	1.50	1.50	1.50	1.50	1.50
Top Width of Weir TW_{WEIR}, mm	500	500	500	500	500	500
Base Width of Weir B_{WEIR}, mm	1580	1660	1680	1720	1740	1760
U/S Slope – $H_{US'}$: 1.0(V)	0.360	0.384	0.390	0.404	0.410	0.420
D/S Slope – $H_{DS'}$: 1.0(V)	0.360	0.384	0.390	0.404	0.410	0.420
Width B1', mm	540	580	590	610	620	630
Width B2', mm	540	580	590	610	620	630
Pressure at Toe, t/m^2	4.46	4.62	4.83	4.93	5.03	5.10
Pressure at Heel, t/m^2	0.02	0.05	0.00	0.03	0.03	0.07
Factor of Safety against Overturning	1.58	1.59	1.58	1.59	1.60	1.61
Shear Friction Factor F	3.80	3.53	3.31	3.18	3.06	2.96
Total Volume of Masonry, m^3	18.72	19.44	19.62	19.98	20.16	20.34
B. NON-OVER FLOW PORTION						
Length, m	3.00	3.00	3.00	3.00	3.00	3.00
Top Level including Free Board, m	101.00	101.20	101.40	101.50	101.60	101.70
Foundation Level, m	98.50	98.50	98.50	98.50	98.50	98.50
Height of Non-Over Flow H_{NOF}, m	2.50	2.70	2.90	3.00	3.10	3.20
Height H1, mm	2000	2200	2400	2500	2600	2700
Height H2, mm	500	500	500	500	500	500

(Continued)

TABLE 6.11 (Continued)

Catchment Area, ha	50.00	100.00	150.00	200.00	250.00	300.00
Top Width of Non-Over Flow TW_{NOF}, mm	500	500	500	500	500	500
Base Width of Non-Over Flow B_{NOF}, mm	1600	1720	1880	1940	2020	2100
U/S Slope – H_{US}: 1.0(V)	0.274	0.278	0.286	0.286	0.292	0.295
D/S Slope – H_{DS}: 1.0(V)	0.274	0.278	0.286	0.286	0.292	0.295
Width B1, mm	550	610	690	720	760	800
Width B2, mm	550	610	690	720	760	800
Pressure at Toe, t/m^2	6.18	6.75	7.18	7.50	7.74	7.95
Pressure at Heel, t/m^2	0.09	0.02	0.06	0.02	0.02	0.06
Factor of Safety against Overturning	1.85	1.87	1.92	1.92	1.93	1.96
Shear Friction Factor F	3.74	3.46	3.29	3.17	3.11	3.03
Total Volume of Masonry, m^3	7.06	8.07	9.31	9.90	10.57	11.27

TABLE 6.12
Design Results for Masonry Check Dam with Slopes on Both Faces Height = 1.50 m and Stream Width L = 20.0 m

Catchment Area, ha	50.00	100.00	150.00	200.00	250.00	300.00
A. WEIR						
Span of Weir W, m	16.00	16.00	16.00	16.00	16.00	16.00
Flood Discharge Q, m^3/s	6.54	11.00	14.91	18.50	21.87	25.07
Head Over Crest, m	0.37	0.52	0.64	0.73	0.82	0.90
Top Level, m	100.00	100.00	100.00	100.00	100.00	100.00
Foundation Level, m	98.50	98.50	98.50	98.50	98.50	98.50
Height of Weir H_{WEIR}, m	1.50	1.50	1.50	1.50	1.50	1.50
Top Width of Weir TW_{WEIR}, mm	500	500	500	500	500	500
Base Width of Weir B_{WEIR}, mm	1580	1620	1660	1680	1700	1720
U/S Slope – H_{US}: 1.0(V)	0.358	0.370	0.385	0.390	0.398	0.405
D/S Slope – H_{DS}: 1.0(V)	0.358	0.370	0.385	0.390	0.398	0.405
Width B1', mm	540	560	580	590	600	610
Width B2', mm	540	560	580	590	600	610
Pressure at Toe, t/m^2	4.30	4.52	4.65	4.73	4.84	4.94
Pressure at Heel, t/m^2	0.09	0.03	0.03	0.05	0.04	0.03
Factor of Safety against Overturning	1.59	1.58	1.58	1.59	1.59	1.59
Shear Friction Factor F	3.98	3.70	3.52	3.38	3.26	3.17
Total Volume of Masonry, m^3	24.96	25.44	25.92	26.16	26.40	26.64
B. NON-OVER FLOW PORTION						
Length, m	4.00	4.00	4.00	4.00	4.00	4.00
Top Level including Free Board, m	100.90	101.10	101.20	101.30	101.40	101.50
Foundation Level, m	98.50	98.50	98.50	98.50	98.50	98.50
Height of Non-Over Flow H_{NOF}, m	2.40	2.60	2.70	2.80	2.90	3.00
Height H1, mm	1900	2100	2200	2300	2400	2500
Height H2, mm	500	500	500	500	500	500
Top Width of Non-Over Flow TW_{NOF}, mm	500	500	500	500	500	500

(Continued)

TABLE 6.12 (Continued)

Catchment Area, ha	50.00	100.00	150.00	200.00	250.00	300.00
Base Width of Non-Over Flow B_{NOF}, mm	1520	1680	1720	1800	1880	1940
U/S Slope – H_{US},: 1.0(V)	0.266	0.280	0.276	0.284	0.286	0.286
D/S Slope – H_{DS},: 1.0(V)	0.266	0.280	0.276	0.284	0.286	0.286
Width B1, mm	510	590	610	650	690	720
Width B2, mm	510	590	610	650	690	720
Pressure at Toe, t/m^2	5.96	6.36	6.76	7.01	7.18	7.50
Pressure at Heel, t/m^2	0.04	0.12	0.00	0.00	0.06	0.02
Factor of Safety against Overturning	1.82	1.87	1.87	1.88	1.92	1.92
Shear Friction Factor F	3.83	3.62	3.46	3.37	3.29	3.17
Total Volume of Masonry, m^3	8.66	10.14	10.76	11.58	12.41	13.20

6.3 DESIGN OF 2.00 M HIGH MASONRY CHECK DAMS

6.3.1 Section with Vertical U/S Face

The tables developed after the rigorous design of the different components for 2.00 m high masonry check dams for different combinations are presented from Table 6.13 to Table 6.16.

TABLE 6.13
Design Results for Masonry Check Dam with Vertical U/S Face Height = 2.00 m and Stream Width L = 5.0 m

Catchment Area, ha	50.00	100.00
A. WEIR		
Span of Weir W, m	4.00	4.00
Flood Discharge Q, m^3/s	6.54	11.00
Head Over Crest, m	0.92	1.31
Top Level, m	100.00	100.00
Foundation Level, m	98.00	98.00
Height of Weir H_{WEIR}, m	2.00	2.00
Top Width of Weir TW_{WEIR}, mm	500	500
Base Width of Weir B_{WEIR}, mm	2360	2580
D/S Slope – H_{DS},: 1.0(V)	0.930	1.040
Width B1', mm	1860	2080
Pressure at Toe, t/m^2	3.83	3.81
Pressure at Heel, t/m^2	0.01	0.02
Factor of Safety against Overturning	1.49	1.49
Shear Friction Factor F	2.65	2.46
Total Volume of Masonry, m^3	11.44	12.32
Length of D/S Apron Floor $L_{D/S\ APRON}$, m	5.70	5.50

(Continued)

TABLE 6.13 (Continued)

Catchment Area, ha	50.00	100.00
B. NON-OVER FLOW PORTION		
Length, m	1.00	1.00
Top Level including Free Board, m	101.50	102.00
Foundation Level, m	98.00	98.00
Height of Non-Over Flow H_{NOF}, m	3.50	4.00
Height H1, mm	2658	3169
Height H2, mm	842	831
Top Width of Non-Over Flow TW_{NOF}, mm	500	500
Base Width of Non-Over Flow B_{NOF}, mm	2080	2410
D/S Slope – H_{DS}: 1.0(V)	0.594	0.602
Width B1, mm	1580	1910
Pressure at Toe, t/m²	6.50	7.54
Pressure at Heel, t/m²	0.02	0.05
Factor of Safety against Overturning	1.61	1.65
Shear Friction Factor F	2.24	2.03
Total Volume of Masonry, m³	3.85	5.02
C. WING WALLS (ABUTMENTS) ON BOTH BANKS		
Length in U/S $L_{U/SABUT}$, m	1.00	1.00
Length in D/S $L_{D/SABUT}$, m	3.08	3.41
Top Level including Free Board, m	101.50	102.00
Foundation Level, m	98.00	98.00
Total Height H_{ABUT}, m	3.50	4.00
D. D/S WING WALLS ON BOTH BANKS		
Length in D/S $L_{D/SWING}$, m	4.92	4.59
Top Level including Free Board, m	99.52	99.95
Foundation Level, m	98.00	98.00
Total Height $H_{D/SWING}$, m	1.52	1.95

TABLE 6.14
Design Results for Masonry Check Dam with Vertical U/S Face Height = 2.00 m and Stream Width L = 10.0 m

Catchment Area, ha	50.00	100.00	150.00	200.00
A. WEIR				
Span of Weir W, m	8.00	8.00	8.00	8.00
Flood Discharge Q, m³/s	6.54	11.00	14.91	18.50
Head Over Crest, m	0.58	0.82	1.01	1.16
Top Level, m	100.00	100.00	100.00	100.00
Foundation Level, m	98.00	98.00	98.00	98.00
Height of Weir H_{WEIR}, m	2.00	2.00	2.00	2.00
Top Width of Weir TW_{WEIR}, mm	500	500	500	500
Base Width of Weir B_{WEIR}, mm	2150	2300	2410	2500
D/S Slope – H_{DS}.: 1.0(V)	0.825	0.900	0.955	1.000

(Continued)

TABLE 6.14 (Continued)

Catchment Area, ha	50.00	100.00	150.00	200.00
Width B1', mm	1650	1800	1910	2000
Pressure at Toe, t/m^2	3.83	3.82	3.83	3.81
Pressure at Heel, t/m^2	0.02	0.01	0.01	0.02
Factor of Safety against Overturning	1.50	1.49	1.49	1.49
Shear Friction Factor F	2.86	2.71	2.60	2.53
Total Volume of Masonry, m^3	21.20	22.40	23.28	24.00
Length of D/S Apron Floor $L_{D/S\ APRON}$, m	5.90	5.70	5.60	5.50
B. NON-OVER FLOW PORTION				
Length, m	2.00	2.00	2.00	2.00
Top Level including Free Board, m	101.10	101.40	101.60	101.80
Foundation Level, m	98.00	98.00	98.00	98.00
Height of Non-Over Flow H_{NOF}, m	3.10	3.40	3.60	3.80
Height H1, mm	2257	2562	2762	2962
Height H2, mm	843	838	838	838
Top Width of Non-Over Flow TW_{NOF}, mm	500	500	500	500
Base Width of Non-Over Flow B_{NOF}, mm	1840	2030	2150	2270
D/S Slope – H_{DS}: 1.0(V)	0.593	0.597	0.597	0.597
Width B1, mm	1340	1530	1650	1770
Pressure at Toe, t/m^2	5.61	6.22	6.63	7.14
Pressure at Heel, t/m^2	0.06	0.08	0.09	0.02
Factor of Safety against Overturning	1.58	1.61	1.63	1.63
Shear Friction Factor F	2.47	2.30	2.20	2.10
Total Volume of Masonry, m^3	6.13	7.32	8.15	9.04
C. WING WALLS (ABUTMENTS) ON BOTH BANKS				
Length in U/S $L_{U/SABUT}$, m	1.00	1.00	1.00	1.00
Length in D/S $L_{D/SABUT}$, m	2.84	3.03	3.15	3.27
Top Level including Free Board, m	101.10	101.40	101.60	101.80
Foundation Level, m	98.00	98.00	98.00	98.00
Total Height H_{ABUT}, m	3.10	3.40	3.60	3.80
D. D/S WING WALLS ON BOTH BANKS				
Length in D/S $L_{D/SWING}$, m	5.16	4.97	4.85	4.73
Top Level including Free Board, m	99.14	99.41	99.62	99.79
Foundation Level, m	98.00	98.00	98.00	98.00
Total Height $H_{D/SWING}$, m	1.14	1.41	1.62	1.79

TABLE 6.15
Design Results for Masonry Check Dam with Vertical U/S Face Height = 2.00 m and Stream Width L = 15.0 m

Catchment Area, ha	50.00	100.00	150.00	200.00	250.00	300.00
A. WEIR						
Span of Weir W, m	12.00	12.00	12.00	12.00	12.00	12.00
Flood Discharge Q, m^3/s	6.54	11.00	14.91	18.50	21.87	25.07

(Continued)

TABLE 6.15 (Continued)

Catchment Area, ha	50.00	100.00	150.00	200.00	250.00	300.00
Head Over Crest, m	0.44	0.63	0.77	0.89	0.99	1.09
Top Level, m	100.00	100.00	100.00	100.00	100.00	100.00
Foundation Level, m	98.00	98.00	98.00	98.00	98.00	98.00
Height of Weir H_{WEIR}, m	2.00	2.00	2.00	2.00	2.00	2.00
Top Width of Weir TW_{WEIR}, mm	500	500	500	500	500	500
Base Width of Weir B_{WEIR}, mm	2050	2180	2270	2340	2400	2460
D/S Slope – $H_{DS'}$: 1.0(V)	0.775	0.840	0.885	0.920	0.950	0.980
Width B1', mm	1550	1680	1770	1840	1900	1960
Pressure at Toe, t/m²	3.85	3.83	3.83	3.83	3.81	3.81
Pressure at Heel, t/m²	0.01	0.01	0.01	0.01	0.02	0.02
Factor of Safety against Overturning	1.51	1.50	1.49	1.49	1.49	1.49
Shear Friction Factor F	2.95	2.82	2.74	2.66	2.61	2.56
Total Volume of Masonry, m³	30.60	32.16	33.24	34.08	34.80	35.52
Length of D/S Apron Floor $L_{D/S\ APRON}$, m	6.00	5.90	5.80	5.70	5.60	5.60
B. NON-OVER FLOW PORTION						
Length, m	3.00	3.00	3.00	3.00	3.00	3.00
Top Level including Free Board, m	101.00	101.20	101.40	101.50	101.60	101.70
Foundation Level, m	98.00	98.00	98.00	98.00	98.00	98.00
Height of Non-Over Flow H_{NOF}, m	3.00	3.20	3.40	3.50	3.60	3.70
Height H1, mm	2153	2353	2562	2658	2758	2862
Height H2, mm	847	847	838	842	842	838
Top Width of Non-Over Flow TW_{NOF}, mm	500	500	500	500	500	500
Base Width of Non-Over Flow B_{NOF}, mm	1770	1890	2030	2080	2140	2210
D/S Slope – H_{DS}: 1.0(V)	0.590	0.590	0.597	0.594	0.594	0.597
Width B1, mm	1270	1390	1530	1580	1640	1710
Pressure at Toe, t/m²	5.44	5.86	6.22	6.50	6.73	6.92
Pressure at Heel, t/m²	0.03	0.03	0.08	0.02	0.01	0.02
Factor of Safety against Overturning	1.57	1.58	1.61	1.61	1.61	1.62
Shear Friction Factor F	2.52	2.39	2.30	2.24	2.19	2.15
Total Volume of Masonry, m³	8.61	9.72	10.98	11.55	12.19	12.89
C. WING WALLS (ABUTMENTS) ON BOTH BANKS						
Length in U/S $L_{U/SABUT}$, m	1.00	1.00	1.00	1.00	1.00	1.00
Length in D/S $L_{D/SABUT}$, m	2.77	2.89	3.03	3.08	3.14	3.21
Top Level including Free Board, m	101.00	101.20	101.40	101.50	101.60	101.70
Foundation Level, m	98.00	98.00	98.00	98.00	98.00	98.00
Total Height H_{ABUT}, m	3.00	3.20	3.40	3.50	3.60	3.70
D. D/S WING WALLS ON BOTH BANKS						
Length in D/S $L_{D/SWING}$, m	5.23	5.11	4.97	4.92	4.86	4.79
Top Level including Free Board, m	98.99	99.20	99.35	99.49	99.60	99.71
Foundation Level, m	98.00	98.00	98.00	98.00	98.00	98.00
Total Height $H_{D/SWING}$, m	0.99	1.20	1.35	1.49	1.60	1.71

TABLE 6.16
Design Results for Masonry Check Dam with Vertical U/S Face Height = 2.00 m and Stream Width L = 20.0 m

Catchment Area, ha	50.00	100.00	150.00	200.00	250.00	300.00
A. WEIR						
Span of Weir W, m	16.00	16.00	16.00	16.00	16.00	16.00
Flood Discharge Q, m^3/s	6.54	11.00	14.91	18.50	21.87	25.07
Head Over Crest, m	0.37	0.52	0.64	0.73	0.82	0.90
Top Level, m	100.00	100.00	100.00	100.00	100.00	100.00
Foundation Level, m	98.00	98.00	98.00	98.00	98.00	98.00
Height of Weir H_{WEIR}, m	2.00	2.00	2.00	2.00	2.00	2.00
Top Width of Weir TW_{WEIR}, mm	500	500	500	500	500	500
Base Width of Weir B_{WEIR}, mm	2000	2110	2190	2250	2300	2350
D/S Slope – H_{DS},: 1.0(V)	0.750	0.805	0.845	0.875	0.900	0.925
Width B1', mm	1500	1610	1690	1750	1800	1850
Pressure at Toe, t/m^2	3.85	3.83	3.83	3.81	3.82	3.81
Pressure at Heel, t/m^2	0.02	0.02	0.02	0.03	0.01	0.02
Factor of Safety against Overturning	1.51	1.50	1.50	1.50	1.49	1.49
Shear Friction Factor F	3.00	2.90	2.82	2.77	2.71	2.66
Total Volume of Masonry, m^3	40.00	41.76	43.04	44.00	44.80	45.60
Length of D/S Apron Floor $L_{D/S\ APRON}$, m	6.00	5.90	5.90	5.80	5.70	5.70
B. NON-OVER FLOW PORTION						
Length, m	4.00	4.00	4.00	4.00	4.00	4.00
Top Level including Free Board, m	100.90	101.10	101.20	101.30	101.40	101.50
Foundation Level, m	98.00	98.00	98.00	98.00	98.00	98.00
Height of Non-Over Flow H_{NOF}, m	2.90	3.10	3.20	3.30	3.40	3.50
Height H1, mm	2051	2251	2353	2453	2562	2658
Height H2, mm	849	849	847	847	838	842
Top Width of Non-Over Flow TW_{NOF}, mm	500	500	500	500	500	500
Base Width of Non-Over Flow B_{NOF}, mm	1710	1830	1890	1950	2030	2080
D/S Slope – H_{DS}: 1.0(V)	0.589	0.589	0.590	0.590	0.597	0.594
Width B1, mm	1210	1330	1390	1450	1530	1580
Pressure at Toe, t/m^2	5.23	5.68	5.86	6.08	6.22	6.50
Pressure at Heel, t/m^2	0.03	0.01	0.03	0.03	0.08	0.02
Factor of Safety against Overturning	1.56	1.57	1.58	1.59	1.61	1.61
Shear Friction Factor F	2.59	2.46	2.39	2.34	2.30	2.24
Total Volume of Masonry, m^3	10.79	12.21	12.96	13.73	14.63	15.41
C. WING WALLS (ABUTMENTS) ON BOTH BANKS						
Length in U/S $L_{U/SABUT}$, m	1.00	1.00	1.00	1.00	1.00	1.00
Length in D/S $L_{D/SABUT}$, m	2.71	2.83	2.89	2.95	3.03	3.08
Top Level including Free Board, m	100.90	101.10	101.20	101.30	101.40	101.50
Foundation Level, m	98.00	98.00	98.00	98.00	98.00	98.00
Total Height H_{ABUT}, m	2.90	3.10	3.20	3.30	3.40	3.50
D. D/S WING WALLS ON BOTH BANKS						
Length in D/S $L_{D/SWING}$, m	5.29	5.17	5.11	5.05	4.97	4.92
Top Level including Free Board, m	98.91	99.08	99.21	99.31	99.41	99.50
Foundation Level, m	98.00	98.00	98.00	98.00	98.00	98.00
Total Height $H_{D/SWING}$, m	0.91	1.08	1.21	1.31	1.41	1.50

6.3.2 Section with Vertical D/S Face

The tables developed after the rigorous design of the different components for 2.00 m high masonry check dams for different combinations are presented from Table 6.17 to Table 6.20.

TABLE 6.17
Design Results for Masonry Check Dam with Vertical D/S Face Height = 2.00 m and Stream Width L = 5.0 m

Catchment Area, ha	50.00	100.00
A. WEIR		
Span of Weir W, m	4.00	4.00
Flood Discharge Q, m^3/s	6.54	11.00
Head Over Crest, m	0.92	1.31
Top Level, m	100.00	100.00
Foundation Level, m	98.00	98.00
Height of Weir H_{WEIR}, m	2.00	2.00
Top Width of Weir TW_{WEIR}, mm	2460	2570
Base Width of Weir B_{WEIR}, mm	2460	2570
Pressure at Toe, t/m^2	8.40	8.95
Pressure at Heel, t/m^2	0.02	0.00
Factor of Safety against Overturning	2.13	2.13
Shear Friction Factor F	3.26	2.94
Total Volume of Masonry, m^3	19.68	20.56
Length of D/S Apron Floor $L_{D/S\ APRON}$, m	5.60	5.50
B. NON-OVER FLOW PORTION		
Length, m	1.00	1.00
Top Level including Free Board, m	101.50	102.00
Foundation Level, m	98.00	98.00
Height of Non-Over Flow H_{NOF}, m	3500	4000
Height H1, mm	2800	3200
Height H2, mm	700	800
Top Width of Non-Over Flow TW, mm	1000	1000
Base Width of Non-Over Flow B_{NOF}, mm	2510	2750
Width B1, mm	1510	1750
Pressure at Toe, t/m^2	12.39	14.42
Pressure at Heel, t/m^2	0.01	0.02
Factor of Safety against Overturning	2.33	2.50
Shear Friction Factor F	3.25	2.84
Total Volume of Masonry, m^3	7.73	9.60
C. WING WALLS (ABUTMENTS) ON BOTH BANKS		
Length in U/S $L_{U/SABUT}$, m	3.51	3.75
Length in D/S $L_{D/SABUT}$, m	2.00	2.00
Top Level including Free Board, m	101.50	102.00
Foundation Level, m	98.00	98.00
Total Height H_{ABUT}, m	3.50	4.00

(Continued)

TABLE 6.17 (Continued)

Catchment Area, ha	50.00	100.00
D. D/S WING WALLS ON BOTH BANKS		
Length in D/S $L_{D/SWING}$, m	3.60	3.50
Top Level including Free Board, m	99.52	99.95
Foundation Level, m	98.00	98.00
Total Height $H_{D/SWING}$, m	1.52	1.95

TABLE 6.18
Design Results for Masonry Check Dam with Vertical D/S Face Height = 2.00 m and Stream Width L = 10.0 m

Catchment Area, ha	50.00	100.00	150.00	200.00
A. WEIR				
Span of Weir W, m	8.00	8.00	8.00	8.00
Flood Discharge Q, m^3/s	6.54	11.00	14.91	18.50
Head Over Crest, m	0.58	0.82	1.01	1.16
Top Level, m	100.00	100.00	100.00	100.00
Foundation Level, m	98.00	98.00	98.00	98.00
Height of Weir H_{WEIR}, m	2.00	2.00	2.00	2.00
Top Width of Weir TW_{WEIR}, mm	2320	2420	2490	2530
Base Width of Weir B_{WEIR}, mm	2320	2420	2490	2530
Pressure at Toe, t/m^2	7.97	8.28	8.52	8.74
Pressure at Heel, t/m^2	0.01	0.01	0.02	0.00
Factor of Safety against Overturning	2.11	2.12	2.13	2.13
Shear Friction Factor F	3.58	3.34	3.18	3.05
Total Volume of Masonry, m^3	37.12	38.72	39.84	40.48
Length of D/S Apron Floor $L_{D/S\ APRON}$, m	5.70	5.60	5.60	5.50
B. NON-OVER FLOW PORTION				
Length, m	2.00	2.00	2.00	2.00
Top Level including Free Board, m	101.10	101.40	101.60	101.80
Foundation Level, m	98.00	98.00	98.00	98.00
Height of Non-Over Flow H_{NOF}, m	3.10	3.40	3.60	3.80
Height H1, mm	2330	2550	2700	2740
Height H2, mm	770	850	900	1060
Top Width of Non-Over Flow TW_{NOF}, mm	1000	1000	1000	1000
Base Width of Non-Over Flow B_{NOF}, mm	2600	2780	2890	2930
Width B1, mm	1600	1780	1890	1930
Pressure at Toe, t/m^2	10.40	11.58	12.37	12.92
Pressure at Heel, t/m^2	0.00	0.01	0.01	0.08
Factor of Safety against Overturning	2.30	2.43	2.51	2.45
Shear Friction Factor F	4.06	3.71	3.49	3.20

(Continued)

TABLE 6.18 (Continued)

Catchment Area, ha	50.00	100.00	150.00	200.00
Total Volume of Masonry, m^3	13.64	15.88	17.41	18.16
C. WING WALLS (ABUTMENTS) ON BOTH BANKS				
Length in U/S $L_{U/SABUT}$, m	3.60	3.78	3.89	3.93
Length in D/S $L_{D/SABUT}$, m	2.00	2.00	2.00	2.00
Top Level including Free Board, m	101.10	101.40	101.60	101.80
Foundation Level, m	98.00	98.00	98.00	98.00
Total Height H_{ABUT}, m	3.10	3.40	3.60	3.80
D. D/S WING WALLS ON BOTH BANKS				
Length in D/S $L_{/SWING}$, m	3.70	3.60	3.60	3.50
Top Level including Free Board, m	99.14	99.41	99.62	99.79
Foundation Level, m	98.00	98.00	98.00	98.00
Total Height $H_{D/SWING}$, m	1.14	1.41	1.62	1.79

TABLE 6.19
Design Results for Masonry Check Dam with Vertical D/S Face Height = 2.00 m and Stream Width L = 15.0 m

Catchment Area, ha	50.00	100.00	150.00	200.00	250.00	300.00
A. WEIR						
Span of Weir W, m	12.00	12.00	12.00	12.00	12.00	12.00
Flood Discharge Q, m^3/s	6.54	11.00	14.91	18.50	21.87	25.07
Head Over Crest, m	0.44	0.63	0.77	0.89	0.99	1.09
Top Level, m	100.00	100.00	100.00	100.00	100.00	100.00
Foundation Level, m	98.00	98.00	98.00	98.00	98.00	98.00
Height of Weir H_{WEIR}, m	2.00	2.00	2.00	2.00	2.00	2.00
Top Width of Weir TW_{WEIR}, mm	2250	2340	2400	2440	2480	2510
Base Width of Weir B_{WEIR}, mm	2250	2340	2400	2440	2480	2510
Pressure at Toe, t/m^2	7.79	8.04	8.22	8.39	8.51	8.65
Pressure at Heel, t/m^2	0.00	0.00	0.01	0.00	0.01	0.01
Factor of Safety against Overturning	2.11	2.11	2.12	2.12	2.13	2.13
Shear Friction Factor F	3.74	3.53	3.39	3.27	3.19	3.11
Total Volume of Masonry, m^3	54.00	56.16	57.60	58.56	59.52	60.24
Length of D/S Apron Floor $L_{D/S\ APRON}$, m	5.80	5.70	5.60	5.60	5.60	5.50
B. NON-OVER FLOW PORTION						
Length, m	3.00	3.00	3.00	3.00	3.00	3.00
Top Level including Free Board, m	101.00	101.20	101.40	101.50	101.60	101.70
Foundation Level, m	98.00	98.00	98.00	98.00	98.00	98.00
Height of Non-Over Flow H_{NOF}, m	3.00	3.20	3.40	3.50	3.60	3.70
Height H1, mm	2250	2400	2550	2630	2700	2630
Height H2, mm	750	800	850	870	900	1070
Top Width of Non-Over Flow TW_{NOF}, mm	1000	1000	1000	1000	1000	1000
Base Width of Non-Over Flow B_{NOF}, mm	2540	2660	2780	2830	2890	2910

(Continued)

TABLE 6.19 (Continued)

Catchment Area, ha	50.00	100.00	150.00	200.00	250.00	300.00
Width B1, mm	1540	1660	1780	1830	1890	1910
Pressure at Toe, t/m^2	10.00	10.80	11.58	11.98	12.37	12.53
Pressure at Heel, t/m^2	0.01	0.00	0.01	0.00	0.01	0.01
Factor of Safety against Overturning	2.26	2.35	2.43	2.47	2.51	2.41
Shear Friction Factor F	4.19	3.93	3.71	3.59	3.49	3.31
Total Volume of Masonry, m^3	19.40	21.55	23.82	24.91	26.11	26.15
C. WING WALLS (ABUTMENTS) ON BOTH BANKS						
Length in U/S $L_{U/SABUT}$, m	3.54	3.66	3.78	3.83	3.89	3.91
Length in D/S $L_{D/SABUT}$, m	2.00	2.00	2.00	2.00	2.00	2.00
Top Level including Free Board, m	101.00	101.20	101.40	101.50	101.60	101.70
Foundation Level, m	98.00	98.00	98.00	98.00	98.00	98.00
Total Height H_{ABUT}, m	3.00	3.20	3.40	3.50	3.60	3.70
D. D/S WING WALLS ON BOTH BANKS						
Length in D/S $L_{D/SWING}$, m	3.80	3.70	3.60	3.60	3.60	3.50
Top Level including Free Board, m	98.99	99.20	99.35	99.49	99.60	99.71
Foundation Level, m	98.00	98.00	98.00	98.00	98.00	98.00
Total Height $H_{D/SWING}$, m	0.99	1.20	1.35	1.49	1.60	1.71

TABLE 6.20
Design Results for Masonry Check Dam with Vertical D/S Face Height = 2.00 m and Stream Width L = 20.0 m

Catchment Area, ha	50.00	100.00	150.00	200.00	250.00	300.00
A. WEIR						
Span of Weir W, m	16.00	16.00	16.00	16.00	16.00	16.00
Flood Discharge Q, m^3/s	6.54	11.00	14.91	18.50	21.87	25.07
Head Over Crest, m	0.37	0.52	0.64	0.73	0.82	0.90
Top Level, m	100.00	100.00	100.00	100.00	100.00	100.00
Foundation Level, m	98.00	98.00	98.00	98.00	98.00	98.00
Height of Weir H_{WEIR}, m	2.00	2.00	2.00	2.00	2.00	2.00
Top Width of Weir TW_{WEIR}, mm	2220	2290	2350	2380	2420	2450
Base Width of Weir B_{WEIR}, mm	2220	2290	2350	2380	2420	2450
Pressure at Toe, t/m^2	7.68	7.89	8.04	8.18	8.28	8.39
Pressure at Heel, t/m^2	0.02	0.00	0.02	0.00	0.01	0.01
Factor of Safety against Overturning	2.11	2.11	2.12	2.11	2.12	2.12
Shear Friction Factor F	3.83	3.65	3.53	3.42	3.34	3.27
Total Volume of Masonry, m^3	71.04	73.28	75.20	76.16	77.44	78.40
Length of D/S Apron Floor $L_{D/S\ APRON}$, m	5.80	5.80	5.70	5.70	5.60	5.60
B. NON-OVER FLOW PORTION						
Length, m	4.00	4.00	4.00	4.00	4.00	4.00
Top Level including Free Board, m	100.90	101.10	101.20	101.30	101.40	101.50
Foundation Level, m	98.00	98.00	98.00	98.00	98.00	98.00

(Continued)

TABLE 6.20 (Continued)

Catchment Area, ha	50.00	100.00	150.00	200.00	250.00	300.00
Height of Non-Over Flow H_{NOF}, m	2900	3100	3200	3300	3400	3500
Height H1, mm	2180	2330	2400	2480	2550	2630
Height H2, mm	720	770	800	820	850	870
Top Width of Non-Over Flow TW_{NOF}, mm	1000	1000	1000	1000	1000	1000
Base Width of Non-Over Flow B_{NOF}, mm	2470	2600	2660	2720	2780	2830
Width B1, mm	1470	1600	1660	1720	1780	1830
Pressure at Toe, t/m^2	9.61	10.40	10.80	11.19	11.58	11.98
Pressure at Heel, t/m^2	0.01	0.00	0.00	0.00	0.01	0.00
Factor of Safety against Overturning	2.20	2.30	2.35	2.39	2.43	2.47
Shear Friction Factor F	4.32	4.06	3.93	3.82	3.71	3.59
Total Volume of Masonry, m^3	24.39	27.28	28.74	30.23	31.76	33.22
C. WING WALLS (ABUTMENTS) ON BOTH BANKS						
Length in U/S $L_{U/SABUT}$, m	3.47	3.60	3.66	3.72	3.78	3.83
Length in D/S $L_{D/SABUT}$, m	2.00	2.00	2.00	2.00	2.00	2.00
Top Level including Free Board, m	100.90	101.10	101.20	101.30	101.40	101.50
Foundation Level, m	98.00	98.00	98.00	98.00	98.00	98.00
Total Height H_{ABUT}, m	2.90	3.10	3.20	3.30	3.40	3.50
D. D/S WING WALLS ON BOTH BANKS						
Length in D/S $L_{D/SWING}$, m	3.80	3.80	3.70	3.70	3.60	3.60
Top Level including Free Board, m	98.91	99.08	99.21	99.31	99.41	99.50
Foundation Level, m	98.00	98.00	98.00	98.00	98.00	98.00
Total Height $H_{D/SWING}$, m	0.91	1.08	1.21	1.31	1.41	1.50

6.3.3 Section with Slopes on Both Faces

The tables developed after the rigorous design of the different components for 2.00 m high masonry check dams for different combinations are presented from Table 6.21 to Table 6.24.

TABLE 6.21
Design Results for Masonry Check Dam with Slopes on Both Faces Height = 2.00 m and Stream Width L = 5.0 m

Catchment Area, ha	50.00	100.00
A. WEIR		
Span of Weir W, m	4.00	4.00
Flood Discharge Q, m^3/s	6.54	11.00
Head Over Crest, m	0.92	1.31
Top Level, m	100.00	100.00

(Continued)

TABLE 6.21 (Continued)

Catchment Area, ha	**50.00**	**100.00**
Foundation Level, m	98.00	98.00
Height of Weir H_{WEIR}, m	2.00	2.00
Top Width of Weir TW_{WEIR}, mm	500	500
Base Width of Weir B_{WEIR}, mm	2220	2320
U/S Slope – $H_{US'}$: 1.0(V)	0.430	0.454
D/S Slope – $H_{DS'}$: 1.0(V)	0.430	0.454
Width B1', mm	860	910
Width B2', mm	860	910
Pressure at Toe, t/m^2	6.07	6.45
Pressure at Heel, t/m^2	0.02	0.07
Factor of Safety against Overturning	1.54	1.56
Shear Friction Factor F	2.71	2.44
Total Volume of Masonry, m^3	10.88	11.28
Length of D/S Apron Floor $L_{D/S\ APRON}$, m	5.80	5.70
B. NON-OVER FLOW PORTION		
Length, m	1.00	1.00
Top Level including Free Board, m	101.50	102.00
Foundation Level, m	98.00	98.00
Height of Non-Over Flow H_{NOF}, m	3.50	4.00
Height H1, mm	3000	3500
Height H2, mm	500	500
Top Width of Non-Over Flow TW_{NOF}, mm	500	500
Base Width of Non-Over Flow B_{NOF}, mm	2420	2780
U/S Slope – $H_{US'}$: 1.0(V)	0.320	0.325
D/S Slope – $H_{DS'}$: 1.0(V)	0.320	0.325
Width B1, mm	960	1140
Width B2, mm	960	1140
Pressure at Toe, t/m^2	8.34	9.66
Pressure at Heel, t/m^2	0.07	0.06
Factor of Safety against Overturning	1.82	1.89
Shear Friction Factor F	2.78	2.52
Total Volume of Masonry, m^3	4.63	5.99
C. WING WALLS (ABUTMENTS) ON BOTH BANKS		
Length in U/S $L_{U/SABUT}$, m	2.46	2.64
Length in D/S $L_{D/SABUT}$, m	1.96	2.14
Top Level including Free Board, m	101.50	102.00
Foundation Level, m	98.00	98.00
Total Height H_{ABUT}, m	3.50	4.00
D. D/S WING WALLS ON BOTH BANKS		
Length in D/S $L_{D/SWING}$, m	3.84	3.56
Top Level including Free Board, m	99.52	99.95
Foundation Level, m	98.00	98.00
Total Height $H_{D/SWING}$, m	1.52	1.95

TABLE 6.22
Design Results for Masonry Check Dam with Slopes on Both Faces Height = 2.00 m and Stream Width L = 10.0 m

Catchment Area, ha	50.00	100.00	150.00	200.00
A. WEIR				
Span of Weir W, m	8.00	8.00	8.00	8.00
Flood Discharge Q, m^3/s	6.54	11.00	14.91	18.50
Head Over Crest, m	0.58	0.82	1.01	1.16
Top Level, m	100.00	100.00	100.00	100.00
Foundation Level, m	98.00	98.00	98.00	98.00
Height of Weir H_{WEIR}, m	2.00	2.00	2.00	2.00
Top Width of Weir TW_{WEIR}, mm	500	500	500	500
Base Width of Weir B_{WEIR}, mm	2120	2180	2240	2280
U/S Slope – H_{US}.: 1.0(V)	0.404	0.420	0.434	0.444
D/S Slope – H_{DS}.: 1.0(V)	0.404	0.420	0.434	0.444
Width B1', mm	810	840	870	890
Width B2', mm	810	840	870	890
Pressure at Toe, t/m^2	5.66	5.97	6.15	6.32
Pressure at Heel, t/m^2	0.07	0.02	0.04	0.04
Factor of Safety against Overturning	1.54	1.54	1.55	1.55
Shear Friction Factor F	3.03	2.78	2.64	2.53
Total Volume of Masonry, m^3	20.96	21.44	21.92	22.24
Length of D/S Apron Floor $L_{D/S\ APRON}$, m	5.90	5.90	5.80	5.80
B. NON-OVER FLOW PORTION				
Length, m	2.00	2.00	2.00	2.00
Top Level including Free Board, m	101.10	101.40	101.60	101.80
Foundation Level, m	98.00	98.00	98.00	98.00
Height of Non-Over Flow H_{NOF}, m	3.10	3.40	3.60	3.80
Height H1, mm	2600	2900	3100	3300
Height H2, mm	500	500	500	500
Top Width of Non-Over Flow TW_{NOF}, mm	500	500	500	500
Base Width of Non-Over Flow B_{NOF}, mm	2140	2340	2480	2640
U/S Slope – H_{US}.: 1.0(V)	0.314	0.318	0.318	0.324
D/S Slope – H_{DS}.: 1.0(V)	0.314	0.318	0.318	0.324
Width B1, mm	820	920	990	1070
Width B2, mm	820	920	990	1070
Pressure at Toe, t/m^2	7.35	8.16	8.65	9.10
Pressure at Heel, t/m^2	0.03	0.00	0.03	0.09
Factor of Safety against Overturning	1.74	1.79	1.83	1.87
Shear Friction Factor F	3.06	2.83	2.71	2.62
Total Volume of Masonry, m^3	7.36	8.74	9.74	10.86
C. WING WALLS (ABUTMENTS) ON BOTH BANKS				
Length in U/S $L_{U/SABUT}$, m	2.32	2.42	2.49	2.57
Length in D/S $L_{D/SABUT}$, m	1.82	1.92	1.99	2.07
Top Level including Free Board, m	101.10	101.40	101.60	101.80
Foundation Level, m	98.00	98.00	98.00	98.00

(Continued)

TABLE 6.22 (Continued)

Catchment Area, ha	50.00	100.00	150.00	200.00
Total Height H_{ABUT}, m	3.10	3.40	3.60	3.80
D. D/S WING WALLS ON BOTH BANKS				
Length in D/S $L_{D/SWING}$, m	4.08	3.98	3.81	3.73
Top Level including Free Board, m	99.14	99.41	99.62	99.79
Foundation Level, m	98.00	98.00	98.00	98.00
Total Height $H_{D/SWING}$, m	1.14	1.41	1.62	1.79

TABLE 6.23
Design Results for Masonry Check Dam with Slopes on Both Faces Height = 2.00 m and Stream Width L = 15.0 m

Catchment Area, ha	50.00	100.00	150.00	200.00	250.00	300.00
A. WEIR						
Span of Weir W, m	12.00	12.00	12.00	12.00	12.00	12.00
Flood Discharge Q, m^3/s	6.54	11.00	14.91	18.50	21.87	25.07
Head Over Crest, m	0.44	0.63	0.77	0.89	0.99	1.09
Top Level, m	100.00	100.00	100.00	100.00	100.00	100.00
Foundation Level, m	98.00	98.00	98.00	98.00	98.00	98.00
Height of Weir H_{WEIR}, m	2.00	2.00	2.00	2.00	2.00	2.00
Top Width of Weir TW_{WEIR}, mm	500	500	500	500	500	500
Base Width of Weir B_{WEIR}, mm	2060	2120	2180	2220	2240	2260
U/S Slope – H_{US}.: 1.0(V)	0.390	0.406	0.418	0.428	0.434	0.440
D/S Slope – H_{DS}.: 1.0(V)	0.390	0.406	0.418	0.428	0.434	0.440
Width B1', mm	780	810	840	860	870	880
Width B2', mm	780	810	840	860	870	880
Pressure at Toe, t/m^2	5.56	5.78	5.86	6.01	6.11	6.26
Pressure at Heel, t/m^2	0.04	0.01	0.07	0.05	0.07	0.03
Factor of Safety against Overturning	1.53	1.53	1.54	1.55	1.55	1.55
Shear Friction Factor F	3.17	2.95	2.84	2.75	2.66	2.58
Total Volume of Masonry, m^3	30.72	31.44	32.16	32.64	32.88	33.12
Length of D/S Apron Floor $L_{D/S\ APRON}$, m	6.00	5.90	5.90	5.80	5.80	5.80
B. NON-OVER FLOW PORTION						
Length, m	3.00	3.00	3.00	3.00	3.00	3.00
Top Level including Free Board, m	101.00	101.20	101.40	101.50	101.60	101.70
Foundation Level, m	98.00	98.00	98.00	98.00	98.00	98.00
Height of Non-Over Flow H_{NOF}, m	3.00	3.20	3.40	3.50	3.60	3.70
Height H1, mm	2500	2700	2900	3000	3100	3200
Height H2, mm	500	500	500	500	500	500
Top Width of Non-Over Flow TW_{NOF}, mm	500	500	500	500	500	500
Base Width of Non-Over Flow B_{NOF}, mm	2060	2200	2340	2420	2480	2560
U/S Slope – H_{US}.: 1.0(V)	0.310	0.315	0.316	0.320	0.320	0.322
D/S Slope – H_{DS}.: 1.0(V)	0.310	0.315	0.316	0.320	0.320	0.322
Width B1, mm	780	850	920	960	990	1030

(Continued)

TABLE 6.23 (Continued)

Catchment Area, ha	50.00	100.00	150.00	200.00	250.00	300.00
Width B2, mm	780	850	920	960	990	1030
Pressure at Toe, t/m^2	7.09	7.62	8.16	8.34	8.64	8.90
Pressure at Heel, t/m^2	0.04	0.03	0.00	0.07	0.04	0.03
Factor of Safety against Overturning	1.72	1.76	1.79	1.82	1.83	1.84
Shear Friction Factor F	3.12	2.97	2.83	2.78	2.71	2.67
Total Volume of Masonry, m^3	10.34	11.69	13.11	13.88	14.61	15.44
C. WING WALLS (ABUTMENTS) ON BOTH BANKS						
Length in U/S $L_{U/SABUT}$, m	2.28	2.35	2.42	2.46	2.49	2.53
Length in D/S $L_{D/SABUT}$, m	1.78	1.85	1.92	1.96	1.99	2.03
Top Level including Free Board, m	101.00	101.20	101.40	101.50	101.60	101.70
Foundation Level, m	98.00	98.00	98.00	98.00	98.00	98.00
Total Height H_{ABUT}, m	3.00	3.20	3.40	3.50	3.60	3.70
D. D/S WING WALLS ON BOTH BANKS						
Length in D/S $L_{D/SWING}$, m	4.22	4.05	3.98	3.84	3.81	3.77
Top Level including Free Board, m	98.99	99.20	99.35	99.49	99.60	99.71
Foundation Level, m	98.00	98.00	98.00	98.00	98.00	98.00
Total Height $H_{D/SWING}$, m	0.99	1.20	1.35	1.49	1.60	1.71

TABLE 6.24

Design Results for Masonry Check Dam with Slopes on Both Faces Height = 2.00 m and Stream Width L = 20.0 m

Catchment Area, ha	50.00	100.00	150.00	200.00	250.00	300.00
A. WEIR						
Span of Weir W, m	16.00	16.00	16.00	16.00	16.00	16.00
Flood Discharge Q, m^3/s	6.54	11.00	14.91	18.50	21.87	25.07
Head Over Crest, m	0.37	0.52	0.64	0.73	0.82	0.90
Top Level, m	100.00	100.00	100.00	100.00	100.00	100.00
Foundation Level, m	98.00	98.00	98.00	98.00	98.00	98.00
Height of Weir H_{WEIR}, m	2.00	2.00	2.00	2.00	2.00	2.00
Top Width of Weir TW_{WEIR}, mm	500	500	500	500	500	500
Base Width of Weir B_{WEIR}, mm	2040	2080	2120	2160	2180	2220
U/S Slope – $H_{US'}$: 1.0(V)	0.384	0.394	0.404	0.414	0.420	0.430
D/S Slope – $H_{DS'}$: 1.0(V)	0.384	0.394	0.404	0.414	0.420	0.430
Width B1', mm	770	790	810	830	840	860
Width B2', mm	770	790	810	830	840	860
Pressure at Toe, t/m^2	5.45	5.69	5.80	5.88	5.97	6.03
Pressure at Heel, t/m^2	0.07	0.00	0.00	0.02	0.02	0.05
Factor of Safety against Overturning	1.53	1.52	1.53	1.53	1.54	1.55

(Continued)

TABLE 6.24 (Continued)

Catchment Area, ha	50.00	100.00	150.00	200.00	250.00	300.00
Shear Friction Factor F	3.26	3.06	2.94	2.87	2.78	2.74
Total Volume of Masonry, m^3	40.64	41.28	41.92	42.56	42.88	43.52
Length of D/S Apron Floor $L_{D/S\ APRON}$, m	6.00	6.00	5.90	5.90	5.90	5.80
B. NON-OVER FLOW PORTION						
Length, m	4.00	4.00	4.00	4.00	4.00	4.00
Top Level including Free Board, m	100.90	101.10	101.20	101.30	101.40	101.50
Foundation Level, m	98.00	98.00	98.00	98.00	98.00	98.00
Height of Non-Over Flow H_{NOF}, m	2.90	3.10	3.20	3.30	3.40	3.50
Height H1, mm	2400	2600	2700	2800	2900	3000
Height H2, mm	500	500	500	500	500	500
Top Width of Non-Over Flow TW_{NOF}, mm	500	500	500	500	500	500
Base Width of Non-Over Flow B_{NOF}, mm	2000	2140	2200	2260	2340	2420
U/S Slope – $H_{US'}$: 1.0(V)	0.312	0.314	0.314	0.314	0.318	0.320
D/S Slope – $H_{DS'}$: 1.0(V)	0.312	0.314	0.314	0.314	0.318	0.320
Width B1, mm	750	820	850	880	920	960
Width B2, mm	750	820	850	880	920	960
Pressure at Toe, t/m^2	6.79	7.35	7.62	7.90	8.16	8.34
Pressure at Heel, t/m^2	0.09	0.03	0.03	0.00	0.00	0.07
Factor of Safety against Overturning	1.71	1.74	1.76	1.77	1.79	1.82
Shear Friction Factor F	3.22	3.06	2.97	2.89	2.83	2.78
Total Volume of Masonry, m^3	13.00	14.72	15.58	16.44	17.48	18.51
C. WING WALLS (ABUTMENTS) ON BOTH BANKS						
Length in U/S $L_{U/SABUT}$, m	2.25	2.32	2.35	2.38	2.42	2.46
Length in D/S $L_{D/SABUT}$, m	1.75	1.82	1.85	1.88	1.92	1.96
Top Level including Free Board, m	100.90	101.10	101.20	101.30	101.40	101.50
Foundation Level, m	98.00	98.00	98.00	98.00	98.00	98.00
Total Height H_{ABUT}, m	2.90	3.10	3.20	3.30	3.40	3.50
D. D/S WING WALLS ON BOTH BANKS						
Length in D/S $L_{D/SWING}$, m	4.25	4.18	4.05	4.02	3.98	3.84
Top Level including Free Board, m	98.91	99.08	99.21	99.31	99.41	99.50
Foundation Level, m	98.00	98.00	98.00	98.00	98.00	98.00
Total Height $H_{D/SWING}$, m	0.91	1.08	1.21	1.31	1.41	1.50

6.4 DESIGN OF 2.50 M HIGH MASONRY CHECK DAMS

6.4.1 Section with Vertical U/S Face

The tables developed after the rigorous design of the different components for 2.50 m high masonry check dams for different combinations are presented from Table 6.25 to Table 6.28.

TABLE 6.25
Design Results for Masonry Check Dam with Vertical U/S Face Height = 2.50 m and Stream Width L = 5.0 m

Catchment Area, ha	50.00	100.00
A. WEIR		
Span of Weir W, m	4.00	4.00
Flood Discharge Q, m^3/s	6.54	11.00
Head Over Crest, m	0.92	1.31
Top Level, m	100.00	100.00
Foundation Level, m	97.50	97.50
Height of Weir H_{WEIR}, m	2.50	2.50
Top Width of Weir TW_{WEIR}, mm	500	500
Base Width of Weir B_{WEIR}, mm	2880	3130
D/S Slope – H_{DS}.: 1.0(V)	0.952	1.052
Width B1', mm	2380	2630
Pressure at Toe, t/m^2	4.46	4.44
Pressure at Heel, t/m^2	0.00	0.00
Factor of Safety against Overturning	1.47	1.47
Shear Friction Factor F	2.31	2.17
Total Volume of Masonry, m^3	16.90	18.15
Length of D/S Apron Floor $L_{D/S\ APRON}$, m	7.20	6.90
B. NON-OVER FLOW PORTION		
Length, m	1.00	1.00
Top Level including Free Board, m	101.50	102.00
Foundation Level, m	97.50	97.50
Height of Non-Over Flow H_{NOF}, m	4.00	4.50
Height H1, mm	3215	3710
Height H2, mm	785	790
Top Width of Non-Over Flow TW_{NOF}, mm	500	500
Base Width of Non-Over Flow B_{NOF}, mm	2550	2850
D/S Slope – H_{DS}: 1.0(V)	0.637	0.633
Width B1, mm	2050	2350
Pressure at Toe, t/m^2	7.04	8.16
Pressure at Heel, t/m^2	0.02	0.03
Factor of Safety against Overturning	1.58	1.61
Shear Friction Factor F	2.02	1.85
Total Volume of Masonry, m^3	5.30	6.62

(Continued)

TABLE 6.25 (Continued)

Catchment Area, ha	50.00	100.00
C. WING WALLS (ABUTMENTS) ON BOTH BANKS		
Length in U/S $L_{U/SABUT}$, m	1.00	1.00
Length in D/S $L_{D/SABUT}$, m	3.55	3.85
Top Level including Free Board, m	101.50	102.00
Foundation Level, m	97.50	97.50
Total Height H_{ABUT}, m	4.00	4.50
D. D/S WING WALLS ON BOTH BANKS		
Length in D/S $L_{D/SWING}$, m	6.45	6.15
Top Level including Free Board, m	99.02	99.45
Foundation Level, m	97.50	97.50
Total Height $H_{D/SWING}$, m	1.52	1.95

TABLE 6.26
Design Results for Masonry Check Dam with Vertical U/S Face Height = 2.50 m and Stream Width L = 10.0 m

Catchment Area, ha	50.00	100.00	150.00	200.00
A. WEIR				
Span of Weir W, m	8.00	8.00	8.00	8.00
Flood Discharge Q, m^3/s	6.54	11.00	14.91	18.50
Head Over Crest, m	0.58	0.82	1.01	1.16
Top Level, m	100.00	100.00	100.00	100.00
Foundation Level, m	97.50	97.50	97.50	97.50
Height of Weir H_{WEIR}, m	2.50	2.50	2.50	2.50
Top Width of Weir TW_{WEIR}, mm	500	500	500	500
Base Width of Weir B_{WEIR}, mm	2640	2820	2950	3040
D/S Slope – $H_{DS'}$: 1.0(V)	0.856	0.928	0.980	1.016
Width B1', mm	2140	2320	2450	2540
Pressure at Toe, t/m^2	4.49	4.44	4.44	4.44
Pressure at Heel, t/m^2	0.00	0.02	0.02	0.01
Factor of Safety against Overturning	1.48	1.48	1.47	1.47
Shear Friction Factor F	2.45	2.36	2.28	2.23
Total Volume of Masonry, m^3	31.40	33.20	34.50	35.40
Length of D/S Apron Floor $L_{D/S\ APRON}$, m	7.40	7.20	7.10	7.00
B. NON-OVER FLOW PORTION				
Length, m	2.00	2.00	2.00	2.00
Top Level including Free Board, m	101.10	101.40	101.60	101.80
Foundation Level, m	97.50	97.50	97.50	97.50
Height of Non-Over Flow H_{NOF}, m	3.60	3.90	4.10	4.30
Height H1, mm	2820	3120	3320	3515
Height H2, mm	780	780	780	785
Top Width of Non-Over Flow TW_{NOF}, mm	500	500	500	500

(Continued)

TABLE 6.26 (Continued)

Catchment Area, ha	50.00	100.00	150.00	200.00
Base Width of Non-Over Flow B_{NOF}, mm	2310	2480	2610	2740
D/S Slope – H_{DS}: 1.0(V)	0.641	0.636	0.636	0.637
Width B1, mm	1810	1980	2110	2240
Pressure at Toe, t/m^2	6.13	6.82	7.24	7.69
Pressure at Heel, t/m^2	0.05	0.02	0.05	0.05
Factor of Safety against Overturning	1.55	1.57	1.59	1.60
Shear Friction Factor F	2.19	2.06	1.99	1.92
Total Volume of Masonry, m^3	8.70	10.08	11.11	12.18
C. WING WALLS (ABUTMENTS) ON BOTH BANKS				
Length in U/S $L_{U/SABUT}$, m	1.00	1.00	1.00	1.00
Length in D/S $L_{D/SABUT}$, m	3.31	3.48	3.61	3.74
Top Level including Free Board, m	101.10	101.40	101.60	101.80
Foundation Level, m	97.50	97.50	97.50	97.50
Total Height H_{ABUT}, m	3.60	3.90	4.10	4.30
D. D/S WING WALLS ON BOTH BANKS				
Length in D/S $L_{D/SWING}$, m	6.69	6.52	6.39	6.26
Top Level including Free Board, m	98.64	98.91	99.12	99.29
Foundation Level, m	97.50	97.50	97.50	97.50
Total Height $H_{D/SWING}$, m	1.14	1.41	1.62	1.79

TABLE 6.27
Design Results for Masonry Check Dam with Vertical U/S Face Height = 2.50 m and Stream Width L = 15.0 m

Catchment Area, ha	50.00	100.00	150.00	200.00	250.00	300.00
A. WEIR						
Span of Weir W, m	12.00	12.00	12.00	12.00	12.00	12.00
Flood Discharge Q, m^3/s	6.54	11.00	14.91	18.50	21.87	25.07
Head Over Crest, m	0.44	0.63	0.77	0.89	0.99	1.09
Top Level, m	100.00	100.00	100.00	100.00	100.00	100.00
Foundation Level, m	97.50	97.50	97.50	97.50	97.50	97.50
Height of Weir H_{WEIR}, m	2.50	2.50	2.50	2.50	2.50	2.50
Top Width of Weir TW_{WEIR}, mm	500	500	500	500	500	500
Base Width of Weir B_{WEIR}, mm	2540	2680	2780	2870	2930	3000
D/S Slope – $H_{DS'}$: 1.0(V)	0.816	0.872	0.912	0.948	0.972	1.000
Width B1', mm	2040	2180	2280	2370	2430	2500
Pressure at Toe, t/m^2	4.47	4.48	4.47	4.43	4.45	4.43
Pressure at Heel, t/m^2	0.02	0.00	0.00	0.03	0.01	0.02
Factor of Safety against Overturning	1.48	1.48	1.47	1.48	1.47	1.47
Shear Friction Factor F	2.53	2.43	2.37	2.33	2.29	2.25
Total Volume of Masonry, m^3	45.60	47.70	49.20	50.55	51.45	52.50

(Continued)

TABLE 6.27 (Continued)

Catchment Area, ha	50.00	100.00	150.00	200.00	250.00	300.00
Length of D/S Apron Floor $L_{D/S\ APRON}$, m	7.50	7.40	7.30	7.20	7.10	7.00
B. NON-OVER FLOW PORTION						
Length, m	3.00	3.00	3.00	3.00	3.00	3.00
Top Level including Free Board, m	101.00	101.20	101.40	101.50	101.60	101.70
Foundation Level, m	97.50	97.50	97.50	97.50	97.50	97.50
Height of Non-Over Flow H_{NOF}, m	3.50	3.70	3.90	4.00	4.10	4.20
Height H1, mm	2721	2919	3114	3215	3314	3413
Height H2, mm	779	781	786	785	786	787
Top Width of Non-Over Flow TW_{NOF}, mm	500	500	500	500	500	500
Base Width of Non-Over Flow B_{NOF}, mm	2250	2370	2480	2550	2610	2670
D/S Slope – H_{DS}: 1.0(V)	0.642	0.640	0.636	0.637	0.636	0.635
Width B1, mm	1750	1870	1980	2050	2110	2170
Pressure at Toe, t/m^2	5.93	6.37	6.82	7.04	7.24	7.46
Pressure at Heel, t/m^2	0.03	0.03	0.02	0.02	0.05	0.06
Factor of Safety against Overturning	1.53	1.55	1.57	1.58	1.59	1.60
Shear Friction Factor F	2.24	2.15	2.06	2.02	1.99	1.95
Total Volume of Masonry, m^3	12.39	13.74	15.12	15.90	16.66	17.43
C. WING WALLS (ABUTMENTS) ON BOTH BANKS						
Length in U/S $L_{U/SABUT}$, m	1.00	1.00	1.00	1.00	1.00	1.00
Length in D/S $L_{D/SABUT}$, m	3.25	3.37	3.48	3.55	3.61	3.67
Top Level including Free Board, m	101.00	101.20	101.40	101.50	101.60	101.70
Foundation Level, m	97.50	97.50	97.50	97.50	97.50	97.50
Total Height H_{ABUT}, m	3.50	3.70	3.90	4.00	4.10	4.20
D. D/S WING WALLS ON BOTH BANKS						
Length in D/S $L_{D/SWING}$, m	6.75	6.63	6.52	6.45	6.39	6.33
Top Level including Free Board, m	98.49	98.70	98.85	98.99	99.10	99.21
Foundation Level, m	97.50	97.50	97.50	97.50	97.50	97.50
Total Height $H_{D/SWING}$, m	0.99	1.20	1.35	1.49	1.60	1.71

TABLE 6.28
Design Results for Masonry Check Dam with Vertical U/S Face Height = 2.50 m and Stream Width L = 20.0 m

Catchment Area, ha	50.00	100.00	150.00	200.00	250.00	300.00
A. WEIR						
Span of Weir W, m	16.00	16.00	16.00	16.00	16.00	16.00
Flood Discharge Q, m^3/s	6.54	11.00	14.91	18.50	21.87	25.07
Head Over Crest, m	0.37	0.52	0.64	0.73	0.82	0.90
Top Level, m	100.00	100.00	100.00	100.00	100.00	100.00
Foundation Level, m	97.50	97.50	97.50	97.50	97.50	97.50

(Continued)

TABLE 6.28 (Continued)

Catchment Area, ha	50.00	100.00	150.00	200.00	250.00	300.00
Height of Weir H_{WEIR}, m	2.50	2.50	2.50	2.50	2.50	2.50
Top Width of Weir TW_{WEIR}, mm	500	500	500	500	500	500
Base Width of Weir B_{WEIR}, mm	2480	2600	2690	2750	2820	2870
D/S Slope – $H_{DS'}$: 1.0(V)	0.792	0.840	0.876	0.900	0.928	0.948
Width B1', mm	1980	2100	2190	2250	2320	2370
Pressure at Toe, t/m^2	4.51	4.47	4.46	4.47	4.44	4.45
Pressure at Heel, t/m^2	0.00	0.02	0.02	0.00	0.02	0.01
Factor of Safety against Overturning	1.48	1.48	1.48	1.47	1.48	1.47
Shear Friction Factor F	2.55	2.49	2.43	2.39	2.36	2.32
Total Volume of Masonry, m^3	59.60	62.00	63.80	65.00	66.40	67.40
Length of D/S Apron Floor $L_{D/S\ APRON}$, m	7.60	7.40	7.40	7.30	7.20	7.20
B. NON-OVER FLOW PORTION						
Length, m	4.00	4.00	4.00	4.00	4.00	4.00
Top Level including Free Board, m	100.90	101.10	101.20	101.30	101.40	101.50
Foundation Level, m	97.50	97.50	97.50	97.50	97.50	97.50
Height of Non-Over Flow H_{NOF}, m	3.40	3.60	3.70	3.80	3.90	4.00
Height H1, mm	2627	2820	2919	3018	3116	3216
Height H2, mm	773	780	781	782	784	784
Top Width of Non-Over Flow TW_{NOF}, mm	500	500	500	500	500	500
Base Width of Non-Over Flow B_{NOF}, mm	2200	2310	2370	2430	2490	2550
D/S Slope – H_{DS}: 1.0(V)	0.647	0.641	0.640	0.639	0.638	0.638
Width B1, mm	1700	1810	1870	1930	1990	2050
Pressure at Toe, t/m^2	5.70	6.13	6.37	6.57	6.77	7.04
Pressure at Heel, t/m^2	0.01	0.05	0.03	0.05	0.07	0.01
Factor of Safety against Overturning	1.52	1.55	1.55	1.56	1.58	1.58
Shear Friction Factor F	2.30	2.19	2.15	2.10	2.06	2.03
Total Volume of Masonry, m^3	15.71	17.41	18.32	19.26	20.22	21.20
C. WING WALLS (ABUTMENTS) ON BOTH BANKS						
Length in U/S $L_{U/SABUT}$, m	1.00	1.00	1.00	1.00	1.00	1.00
Length in D/S $L_{D/SABUT}$, m	3.20	3.31	3.37	3.43	3.49	3.55
Top Level including Free Board, m	100.90	101.10	101.20	101.30	101.40	101.50
Foundation Level, m	97.50	97.50	97.50	97.50	97.50	97.50
Total Height H_{ABUT}, m	3.40	3.60	3.70	3.80	3.90	4.00
D. D/S WING WALLS ON BOTH BANKS						
Length in D/S $L_{D/SWING}$, m	6.80	6.69	6.63	6.57	6.51	6.45
Top Level including Free Board, m	98.41	98.58	98.71	98.81	98.91	99.00
Foundation Level, m	97.50	97.50	97.50	97.50	97.50	97.50
Total Height $H_{D/SWING}$, m	0.91	1.08	1.21	1.31	1.41	1.50

6.4.2 Section with Vertical D/S Face

The tables developed after the rigorous design of the different components for 2.50 m high masonry check dams for different combinations are presented from Table 6.29 to Table 6.32.

TABLE 6.29
Design Results for Masonry Check Dam with Vertical D/S Face Height = 2.50 m and Stream Width L = 5.0 m

Catchment Area, ha	50.00	100.00
A. WEIR		
Span of Weir W, m	4.00	4.00
Flood Discharge Q, m^3/s	6.54	11.00
Head Over Crest, m	0.92	1.31
Top Level, m	100.00	100.00
Foundation Level, m	97.50	97.50
Height of Weir H_{WEIR}, m	2.50	2.50
Top Width of Weir TW_{WEIR}, mm	2980	3120
Base Width of Weir B_{WEIR}, mm	2980	3120
Pressure at Toe, t/m^2	10.23	10.74
Pressure at Heel, t/m^2	0.00	0.01
Factor of Safety against Overturning	2.12	2.13
Shear Friction Factor F	2.92	2.69
Total Volume of Masonry, m^3	29.80	31.20
Length of D/S Apron Floor $L_{D/S\ APRON}$, m	7.10	6.90
B. NON-OVER FLOW PORTION		
Length, m	1.00	1.00
Top Level including Free Board, m	101.50	102.00
Foundation Level, m	97.50	97.50
Height of Non-Over Flow H_{NOF}, m	4.00	4.50
Height H1, mm	3040	3375
Height H2, mm	960	1125
Top Width of Non-Over Flow TW_{NOF}, mm	1000	1000
Base Width of Non-Over Flow B_{NOF}, mm	3040	3110
Width B1, mm	2040	2110
Pressure at Toe, t/m^2	13.54	15.51
Pressure at Heel, t/m^2	0.01	0.02
Factor of Safety against Overturning	2.30	2.32
Shear Friction Factor F	2.91	2.46
Total Volume of Masonry, m^3	10.20	11.62
C. WING WALLS (ABUTMENTS) ON BOTH BANKS		
Length in U/S $L_{U/SABUT}$, m	4.04	4.11
Length in D/S $L_{D/SABUT}$, m	2.00	2.00
Top Level including Free Board, m	101.50	102.00
Foundation Level, m	97.50	97.50
Total Height H_{ABUT}, m	4.00	4.50

(Continued)

TABLE 6.29 (Continued)

Catchment Area, ha	50.00	100.00
D. D/S WING WALLS ON BOTH BANKS		
Length in D/S $L_{D/SWING}$, m	5.10	4.90
Top Level including Free Board, m	99.02	99.45
Foundation Level, m	97.50	97.50
Total Height $H_{D/SWING}$, m	1.52	1.95

TABLE 6.30
Design Results for Masonry Check Dam with Vertical D/S Face Height = 2.50 m and Stream Width L = 10.0 m

Catchment Area, ha	50.00	100.00	150.00	200.00
A. WEIR				
Span of Weir W, m	8.00	8.00	8.00	8.00
Flood Discharge Q, m^3/s	6.54	11.00	14.91	18.50
Head Over Crest, m	0.58	0.82	1.01	1.16
Top Level, m	100.00	100.00	100.00	100.00
Foundation Level, m	97.50	97.50	97.50	97.50
Height of Weir H_{WEIR}, m	2.50	2.50	2.50	2.50
Top Width of Weir TW_{WEIR}, mm	2830	2940	3020	3070
Base Width of Weir B_{WEIR}, mm	2830	2940	3020	3070
Pressure at Toe, t/m^2	9.77	10.09	10.33	10.54
Pressure at Heel, t/m^2	0.00	0.00	0.02	0.00
Factor of Safety against Overturning	2.11	2.11	2.12	2.12
Shear Friction Factor F	3.17	2.99	2.87	2.78
Total Volume of Masonry, m^3	56.60	58.80	60.40	61.40
Length of D/S Apron Floor $L_{D/S\ APRON}$, m	7.20	7.10	7.00	7.00
B. NON-OVER FLOW PORTION				
Length, m	2.00	2.00	2.00	2.00
Top Level including Free Board, m	101.10	101.40	101.60	101.80
Foundation Level, m	97.50	97.50	97.50	97.50
Height of Non-Over Flow H_{NOF}, m	3.60	3.90	4.10	4.30
Height H1, mm	2700	2930	3080	3270
Height H2, mm	900	975	1020	1030
Top Width of Non-Over Flow TW_{NOF}, mm	1000	1000	1000	1000
Base Width of Non-Over Flow B_{NOF}, mm	2920	3080	3180	3190
Width B1, mm	1920	2080	2180	2190
Pressure at Toe, t/m^2	11.83	13.04	13.85	14.74
Pressure at Heel, t/m^2	0.03	0.01	0.00	0.01
Factor of Safety against Overturning	2.18	2.30	2.37	2.41
Shear Friction Factor F	3.29	3.06	2.91	2.72
Total Volume of Masonry, m^3	17.57	19.97	21.61	22.91

(Continued)

TABLE 6.30 (Continued)

Catchment Area, ha	50.00	100.00	150.00	200.00
C. WING WALLS (ABUTMENTS) ON BOTH BANKS				
Length in U/S $L_{U/SABUT}$, m	3.92	4.08	4.18	4.19
Length in D/S $L_{D/SABUT}$, m	2.00	2.00	2.00	2.00
Top Level including Free Board, m	101.10	101.40	101.60	101.80
Foundation Level, m	97.50	97.50	97.50	97.50
Total Height H_{ABUT}, m	3.60	3.90	4.10	4.30
D. D/S WING WALLS ON BOTH BANKS				
Length in D/S $L_{D/SWING}$, m	5.20	5.10	5.00	5.00
Top Level including Free Board, m	98.64	98.91	99.12	99.29
Foundation Level, m	97.50	97.50	97.50	97.50
Total Height $H_{D/SWING}$, m	1.14	1.41	1.62	1.79

TABLE 6.31
Design Results for Masonry Check Dam with Vertical D/S Face Height = 2.50 m and Stream Width L = 15.0 m

Catchment Area, ha	50.00	100.00	150.00	200.00	250.00	300.00
A. WEIR						
Span of Weir W, m	12.00	12.00	12.00	12.00	12.00	12.00
Flood Discharge Q, m^3/s	6.54	11.00	14.91	18.50	21.87	25.07
Head Over Crest, m	0.44	0.63	0.77	0.89	0.99	1.09
Top Level, m	100.00	100.00	100.00	100.00	100.00	100.00
Foundation Level, m	97.50	97.50	97.50	97.50	97.50	97.50
Height of Weir H_{WEIR}, m	2.50	2.50	2.50	2.50	2.50	2.50
Top Width of Weir TW_{WEIR}, mm	2760	2860	2920	2970	3010	3050
Base Width of Weir B_{WEIR}, mm	2760	2860	2920	2970	3010	3050
Pressure at Toe, t/m^2	9.58	9.82	10.02	10.18	10.31	10.44
Pressure at Heel, t/m^2	0.01	0.02	0.01	0.01	0.01	0.02
Factor of Safety against Overturning	2.11	2.11	2.11	2.12	2.12	2.12
Shear Friction Factor F	3.28	3.13	3.03	2.95	2.88	2.82
Total Volume of Masonry, m^3	82.80	85.80	87.60	89.10	90.30	91.50
Length of D/S Apron Floor $L_{D/S\ APRON}$, m	7.30	7.20	7.10	7.10	7.00	7.00
B. NON-OVER FLOW PORTION						
Length, m	3.00	3.00	3.00	3.00	3.00	3.00
Top Level including Free Board, m	101.00	101.20	101.40	101.50	101.60	101.70
Foundation Level, m	97.50	97.50	97.50	97.50	97.50	97.50
Height of Non-Over Flow H_{NOF}, m	3.50	3.70	3.90	4.00	4.10	4.20
Height H1, mm	2590	2740	2890	2960	2920	2860
Height H2, mm	910	960	1010	1040	1180	1340
Top Width of Non-Over Flow TW_{NOF}, mm	1000	1000	1000	1000	1000	1000

(Continued)

TABLE 6.31 (Continued)

Catchment Area, ha	50.00	100.00	150.00	200.00	250.00	300.00
Base Width of Non-Over Flow B_{NOF}, mm	2940	3060	3180	3210	3280	3370
Width B1, mm	1940	2060	2180	2210	2280	2370
Pressure at Toe, t/m^2	11.38	12.17	12.94	12.80	13.51	13.66
Pressure at Heel, t/m^2	0.00	0.00	0.00	0.55	0.01	0.00
Factor of Safety against Overturning	2.16	2.25	2.33	2.35	2.32	2.28
Shear Friction Factor F	3.45	3.29	3.14	3.04	2.95	2.88
Total Volume of Masonry, m^3	25.57	28.02	30.57	31.62	32.21	32.91
C. WING WALLS (ABUTMENTS) ON BOTH BANKS						
Length in U/S $L_{U/SABUT}$, m	3.94	4.06	4.18	4.21	4.28	4.37
Length in D/S $L_{D/SABUT}$, m	2.00	2.00	2.00	2.00	2.00	2.00
Top Level including Free Board, m	101.00	101.20	101.40	101.50	101.60	101.70
Foundation Level, m	97.50	97.50	97.50	97.50	97.50	97.50
Total Height H_{ABUT}, m	3.50	3.70	3.90	4.00	4.10	4.20
D. D/S WING WALLS ON BOTH BANKS						
Length in D/S $L_{D/SWING}$, m	5.30	5.20	5.10	5.10	5.00	5.00
Top Level including Free Board, m	98.49	98.70	98.85	98.99	99.10	99.21
Foundation Level, m	97.50	97.50	97.50	97.50	97.50	97.50
Total Height $H_{D/SWING}$, m	0.99	1.20	1.35	1.49	1.60	1.71

TABLE 6.32
Design Results for Masonry Check Dam with Vertical D/S Face Height = 2.50 m and Stream Width L = 20.0 m

Catchment Area, ha	50.00	100.00	150.00	200.00	250.00	300.00
A. WEIR						
Span of Weir W, m	16.00	16.00	16.00	16.00	16.00	16.00
Flood Discharge Q, m^3/s	6.54	11.00	14.91	18.50	21.87	25.07
Head Over Crest, m	0.37	0.52	0.64	0.73	0.82	0.90
Top Level, m	100.00	100.00	100.00	100.00	100.00	100.00
Foundation Level, m	97.50	97.50	97.50	97.50	97.50	97.50
Height of Weir H_{WEIR}, m	2.50	2.50	2.50	2.50	2.50	2.50
Top Width of Weir TW_{WEIR}, mm	2720	2800	2860	2900	2940	2970
Base Width of Weir B_{WEIR}, mm	2720	2800	2860	2900	2940	2970
Pressure at Toe, t/m^2	0	0	0	0	0	0
Pressure at Heel, t/m^2	9.48	9.69	9.85	9.97	10.09	10.20
Factor of Safety against Overturning	0.01	0.00	0.01	0.00	0.00	0.00
Shear Friction Factor F	2.10	2.11	2.11	2.11	2.11	2.11

(Continued)

TABLE 6.32 (Continued)

Catchment Area, ha	50.00	100.00	150.00	200.00	250.00	300.00
Total Volume of Masonry, m^3	3.34	3.21	3.12	3.05	2.99	2.93
Length of D/S Apron Floor $L_{D/S\ APRON}$, m	108.80	112.00	114.40	116.00	117.60	118.80
Length of D/S Apron Floor, m	7.30	7.20	7.20	7.10	7.10	7.10
B. NON-OVER FLOW PORTION						
Length, m	4.00	4.00	4.00	4.00	4.00	4.00
Top Level including Free Board, m	100.90	101.10	101.20	101.30	101.40	101.50
Foundation Level, m	97.50	97.50	97.50	97.50	97.50	97.50
Height of Non-Over Flow H_{NOF}, m	3.40	3.60	3.70	3.80	3.90	4.00
Height H1, mm	2520	2670	2740	2820	2890	2840
Height H2, mm	880	930	960	980	1010	1160
Top Width of Non-Over Flow TW_{NOF}, mm	1000	1000	1000	1000	1000	1000
Base Width of Non-Over Flow B_{NOF}, mm	2970	3000	3060	3120	3180	3210
Width B1, mm	1970	2000	2060	2120	2180	2210
Pressure at Toe, t/m^2	10.82	11.77	12.17	12.56	12.94	13.12
Pressure at Heel, t/m^2	0.13	0.00	0.00	0.00	0.00	0.02
Factor of Safety against Overturning	2.17	2.20	2.25	2.29	2.33	2.27
Shear Friction Factor F	3.64	3.37	3.29	3.21	3.14	3.01
Total Volume of Masonry, m^3	33.43	35.71	37.36	39.05	40.77	41.11
C. WING WALLS (ABUTMENTS) ON BOTH BANKS						
Length in U/S $L_{U/SABUT}$, m	3.97	4.00	4.06	4.12	4.18	4.21
Length in D/S $L_{D/SABUT}$, m	2.00	2.00	2.00	2.00	2.00	2.00
Top Level including Free Board, m	100.90	101.10	101.20	101.30	101.40	101.50
Foundation Level, m	97.50	97.50	97.50	97.50	97.50	97.50
Total Height H_{ABUT}, m	3.40	3.60	3.70	3.80	3.90	4.00
D. D/S WING WALLS ON BOTH BANKS						
Length in D/S $L_{D/SWING}$, m	5.30	5.20	5.20	5.10	5.10	5.10
Top Level including Free Board, m	98.41	98.58	98.71	98.81	98.91	99.00
Foundation Level, m	97.50	97.50	97.50	97.50	97.50	97.50
Total Height $H_{D/SWING}$, m	0.91	1.08	1.21	1.31	1.41	1.50

6.4.3 Section with Slopes on Both Faces

The tables developed after the rigorous design of the different components for 2.50 m high masonry check dams for different combinations are presented from Table 6.33 to Table 6.36.

TABLE 6.33
Design Results for Masonry Check Dam with Slopes on Both Faces Height = 2.50 m and Stream Width L = 5.0 m

Catchment Area, ha	50.00	100.00
A. WEIR		
Span of Weir W, m	4.00	4.00
Flood Discharge Q, m^3/s	6.54	11.00
Head Over Crest, m	0.92	1.31
Top Level, m	100.00	100.00
Foundation Level, m	97.50	97.50
Height of Weir H_{WEIR}, m	2.50	2.50
Top Width of Weir TW_{WEIR}, mm	500	500
Base Width of Weir B_{WEIR}, mm	2700	2820
U/S Slope – H_{US},: 1.0(V)	0.440	0.462
D/S Slope – H_{DS},: 1.0(V)	0.440	0.462
Width B1', mm	1100	1160
Width B2', mm	1100	1160
Pressure at Toe, t/m^2	7.17	7.55
Pressure at Heel, t/m^2	0.04	0.07
Factor of Safety against Overturning	1.51	1.53
Shear Friction Factor F	2.40	2.19
Total Volume of Masonry, m^3	16.00	16.60
Length of D/S Apron Floor $L_{D/S\ APRON}$, m	7.30	7.20
B. NON-OVER FLOW PORTION		
Length, m	1.00	1.00
Top Level including Free Board, m	101.50	102.00
Foundation Level, m	97.50	97.50
Height of Non-Over Flow H_{NOF}, m	4.00	4.50
Height H1, mm	3500	4000
Height H2, mm	500	500
Top Width of Non-Over Flow TW_{NOF}, mm	500	500
Base Width of Non-Over Flow B_{NOF}, mm	2880	3220
U/S Slope – H_{US},: 1.0(V)	0.340	0.340
D/S Slope – H_{DS},: 1.0(V)	0.340	0.340
Width B1, mm	1190	1360
Width B2, mm	1190	1360
Pressure at Toe, t/m^2	9.34	10.69
Pressure at Heel, t/m^2	0.04	0.02
Factor of Safety against Overturning	1.73	1.80
Shear Friction Factor F	2.47	2.27
Total Volume of Masonry, m^3	6.17	7.69
C. WING WALLS (ABUTMENTS) ON BOTH BANKS		
Length in U/S $L_{U/SABUT}$, m	2.69	2.86
Length in D/S $L_{D/SABUT}$, m	2.19	2.36
Top Level including Free Board, m	101.50	102.00
Foundation Level, m	97.50	97.50
Total Height H_{ABUT}, m	4.00	4.50

(Continued)

TABLE 6.33 (Continued)

Catchment Area, ha	50.00	100.00
D. D/S WING WALLS ON BOTH BANKS		
Length in D/S $L_{D/SWING}$, m	5.11	4.84
Top Level including Free Board, m	99.02	99.45
Foundation Level, m	97.50	97.50
Total Height $H_{D/SWING}$, m	1.52	1.95

TABLE 6.34
Design Results for Masonry Check Dam with Slopes on Both Faces Height = 2.50 m and Stream Width L = 10.0 m

Catchment Area, ha	50.00	100.00	150.00	200.00
A. WEIR				
Span of Weir W, m	8.00	8.00	8.00	8.00
Flood Discharge Q, m^3/s	6.54	11.00	14.91	18.50
Head Over Crest, m	0.58	0.82	1.01	1.16
Top Level, m	100.00	100.00	100.00	100.00
Foundation Level, m	97.50	97.50	97.50	97.50
Height of Weir H_{WEIR}, m	2.50	2.50	2.50	2.50
Top Width of Weir TW_{WEIR}, mm	500	500	500	500
Base Width of Weir B_{WEIR}, mm	2580	2660	2720	2760
U/S Slope – H_{US}.: 1.0(V)	0.414	0.430	0.445	0.450
D/S Slope – H_{DS}.: 1.0(V)	0.414	0.430	0.445	0.450
Width B1', mm	1040	1080	1110	1130
Width B2', mm	1040	1080	1110	1130
Pressure at Toe, t/m^2	6.82	7.09	7.29	7.47
Pressure at Heel, t/m^2	0.02	0.01	0.01	0.00
Factor of Safety against Overturning	1.50	1.50	1.51	1.51
Shear Friction Factor F	2.62	2.45	2.34	2.26
Total Volume of Masonry, m^3	30.80	31.60	32.20	32.60
Length of D/S Apron Floor $L_{D/S\ APRON}$, m	7.50	7.40	7.30	7.30
B. NON-OVER FLOW PORTION				
Length, m	2.00	2.00	2.00	2.00
Top Level including Free Board, m	101.10	101.40	101.60	101.80
Foundation Level, m	97.50	97.50	97.50	97.50
Height of Non-Over Flow H_{NOF}, m	3.60	3.90	4.10	4.30
Height H1, mm	3100	3400	3600	3800
Height H2, mm	500	500	500	500
Top Width of Non-Over Flow TW_{NOF}, mm	500	500	500	500
Base Width of Non-Over Flow B_{NOF}, mm	2600	2800	2940	3080
U/S Slope – H_{US}.: 1.0(V)	0.338	0.338	0.338	0.340
D/S Slope – H_{DS}.: 1.0(V)	0.338	0.338	0.338	0.340
Width B1, mm	1050	1150	1220	1290

(Continued)

TABLE 6.34 (Continued)

Catchment Area, ha	50.00	100.00	150.00	200.00
Width B2, mm	1050	1150	1220	1290
Pressure at Toe, t/m^2	8.28	9.11	9.63	10.16
Pressure at Heel, t/m^2	0.06	0.02	0.02	0.02
Factor of Safety against Overturning	1.67	1.71	1.74	1.77
Shear Friction Factor F	2.66	2.51	2.42	2.34
Total Volume of Masonry, m^3	10.10	11.73	12.88	14.11
C. WING WALLS (ABUTMENTS) ON BOTH BANKS				
Length in U/S $L_{U/SABUT}$, m	2.55	2.65	2.72	2.79
Length in D/S $L_{D/SABUT}$, m	2.05	2.15	2.22	2.29
Top Level including Free Board, m	101.10	101.40	101.60	101.80
Foundation Level, m	97.50	97.50	97.50	97.50
Total Height H_{ABUT}, m	3.60	3.90	4.10	4.30
D. D/S WING WALLS ON BOTH BANKS				
Length in D/S $L_{D/SWING}$, m	5.45	5.25	5.08	5.01
Top Level including Free Board, m	98.64	98.91	99.12	99.29
Foundation Level, m	97.50	97.50	97.50	97.50
Total Height $H_{D/SWING}$, m	1.14	1.41	1.62	1.79

TABLE 6.35
Design Results for Masonry Check Dam with Slopes on Both Faces Height = 2.50 m and Stream Width L = 15.0 m

Catchment Area, ha	50.00	100.00	150.00	200.00	250.00	300.00
A. WEIR						
Span of Weir W, m	12.00	12.00	12.00	12.00	12.00	12.00
Flood Discharge Q, m^3/s	6.54	11.00	14.91	18.50	21.87	25.07
Head Over Crest, m	0.44	0.63	0.77	0.89	0.99	1.09
Top Level, m	100.00	100.00	100.00	100.00	100.00	100.00
Foundation Level, m	97.50	97.50	97.50	97.50	97.50	97.50
Height of Weir H_{WEIR}, m	2.50	2.50	2.50	2.50	2.50	2.50
Top Width of Weir TW_{WEIR}, mm	500	500	500	500	500	500
Base Width of Weir B_{WEIR}, mm	2520	2600	2660	2700	2720	2760
U/S Slope – H_{US}: 1.0(V)	0.402	0.418	0.430	0.440	0.442	0.450
D/S Slope – H_{DS}: 1.0(V)	0.402	0.418	0.430	0.440	0.442	0.450
Width B1', mm	1010	1050	1080	1100	1110	1130
Width B2', mm	1010	1050	1080	1100	1110	1130
Pressure at Toe, t/m^2	6.65	6.85	6.97	7.10	7.25	7.32
Pressure at Heel, t/m^2	0.05	0.05	0.07	0.07	0.04	0.07
Factor of Safety against Overturning	1.50	1.50	1.51	1.51	1.51	1.52
Shear Friction Factor F	2.72	2.58	2.49	2.42	2.35	2.31
Total Volume of Masonry, m^3	45.30	46.50	47.40	48.00	48.30	48.90
Length of D/S Apron Floor $L_{D/S\ APRON}$, m	7.50	7.40	7.40	7.30	7.30	7.30

(Continued)

TABLE 6.35 (Continued)

Catchment Area, ha	50.00	100.00	150.00	200.00	250.00	300.00
B. NON-OVER FLOW PORTION						
Length, m	3.00	3.00	3.00	3.00	3.00	3.00
Top Level including Free Board, m	101.00	101.20	101.40	101.50	101.60	101.70
Foundation Level, m	97.50	97.50	97.50	97.50	97.50	97.50
Height of Non-Over Flow H_{NOF}, m	3.50	3.70	3.90	4.00	4.10	4.20
Height H1, mm	3000	3200	3400	3500	3600	3700
Height H2, mm	500	500	500	500	500	500
Top Width of Non-Over Flow TW_{NOF}, mm	500	500	500	500	500	500
Base Width of Non-Over Flow B_{NOF}, mm	2520	2660	2800	2880	2940	3020
U/S Slope – H_{US}: 1.0(V)	0.338	0.338	0.338	0.340	0.340	0.340
D/S Slope – H_{DS}: 1.0(V)	0.338	0.338	0.338	0.340	0.340	0.340
Width B1, mm	1010	1080	1150	1190	1220	1260
Width B2, mm	1010	1080	1150	1190	1220	1260
Pressure at Toe, t/m^2	8.09	8.60	9.11	9.34	9.63	9.84
Pressure at Heel, t/m^2	0.00	0.00	0.02	0.04	0.02	0.06
Factor of Safety against Overturning	1.64	1.67	1.71	1.73	1.74	1.77
Shear Friction Factor F	2.71	2.60	2.51	2.47	2.42	2.39
Total Volume of Masonry, m^3	14.33	15.91	17.59	18.50	19.32	20.28
C. WING WALLS (ABUTMENTS) ON BOTH BANKS						
Length in U/S $L_{U/SABUT}$, m	2.51	2.58	2.65	2.69	2.72	2.76
Length in D/S $L_{D/SABUT}$, m	2.01	2.08	2.15	2.19	2.22	2.26
Top Level including Free Board, m	101.00	101.20	101.40	101.50	101.60	101.70
Foundation Level, m	97.50	97.50	97.50	97.50	97.50	97.50
Total Height H_{ABUT}, m	3.50	3.70	3.90	4.00	4.10	4.20
D. D/S WING WALLS ON BOTH BANKS						
Length in D/S $L_{D/SWING}$, m	5.49	5.32	5.25	5.11	5.08	5.04
Top Level including Free Board, m	98.49	98.70	98.85	98.99	99.10	99.21
Foundation Level, m	97.50	97.50	97.50	97.50	97.50	97.50
Total Height $H_{D/SWING}$, m	0.99	1.20	1.35	1.49	1.60	1.71

TABLE 6.36
Design Results for Masonry Check Dam with Slopes on Both Faces Height = 2.50 m and Stream Width L = 20.0 m

Catchment Area, ha	50.00	100.00	150.00	200.00	250.00	300.00
A. WEIR						
Span of Weir W, m	16.00	16.00	16.00	16.00	16.00	16.00
Flood Discharge Q, m^3/s	6.54	11.00	14.91	18.50	21.87	25.07
Head Over Crest, m	0.37	0.52	0.64	0.73	0.82	0.90
Top Level, m	100.00	100.00	100.00	100.00	100.00	100.00

(Continued)

TABLE 6.36 (Continued)

Catchment Area, ha	50.00	100.00	150.00	200.00	250.00	300.00
Foundation Level, m	97.50	97.50	97.50	97.50	97.50	97.50
Height of Weir H_{WEIR}, m	2.50	2.50	2.50	2.50	2.50	2.50
Top Width of Weir TW_{WEIR}, mm	500	500	500	500	500	500
Base Width of Weir B_{WEIR}, mm	2480	2560	2600	2640	2660	2700
U/S Slope – H_{US}: 1.0(V)	0.394	0.410	0.420	0.426	0.430	0.440
D/S Slope – H_{DS}: 1.0(V)	0.394	0.410	0.420	0.426	0.430	0.440
Width B1', mm	990	1030	1050	1070	1080	1100
Width B2', mm	990	1030	1050	1070	1080	1100
Pressure at Toe, t/m^2	6.61	6.74	6.87	6.97	7.09	7.11
Pressure at Heel, t/m^2	0.02	0.04	0.04	0.04	0.01	0.07
Factor of Safety against Overturning	1.49	1.50	1.50	1.50	1.50	1.51
Shear Friction Factor F	2.76	2.66	2.57	2.51	2.45	2.41
Total Volume of Masonry, m^3	59.60	61.20	62.00	62.80	63.20	64.00
Length of D/S Apron Floor $L_{D/S\ APRON}$, m	7.60	7.50	7.40	7.40	7.40	7.30
B. NON-OVER FLOW PORTION						
Length, m	4.00	4.00	4.00	4.00	4.00	4.00
Top Level including Free Board, m	100.90	101.10	101.20	101.30	101.40	101.50
Foundation Level, m	97.50	97.50	97.50	97.50	97.50	97.50
Height of Non-Over Flow H_{NOF}, m	3.40	3.60	3.70	3.80	3.90	4.00
Height H1, mm	2900	3100	3200	3300	3400	3500
Height H2, mm	500	500	500	500	500	500
Top Width of Non-Over Flow TW_{NOF}, mm	500	500	500	500	500	500
Base Width of Non-Over Flow B_{NOF}, mm	2460	2600	2660	2740	2800	2880
U/S Slope – H_{US}: 1.0(V)	0.338	0.338	0.338	0.338	0.338	0.340
D/S Slope – H_{DS}: 1.0(V)	0.338	0.338	0.338	0.338	0.338	0.340
Width B1, mm	980	1050	1080	1120	1150	1190
Width B2, mm	980	1050	1080	1120	1150	1190
Pressure at Toe, t/m^2	7.83	8.28	8.60	8.83	9.11	9.34
Pressure at Heel, t/m^2	0.01	0.06	0.00	0.03	0.02	0.04
Factor of Safety against Overturning	1.62	1.67	1.67	1.70	1.71	1.73
Shear Friction Factor F	2.77	2.66	2.60	2.56	2.51	2.47
Total Volume of Masonry, m^3	18.17	20.21	21.21	22.38	23.45	24.66
C. WING WALLS (ABUTMENTS) ON BOTH BANKS						
Length in U/S $L_{U/SABUT}$, m	2.48	2.55	2.58	2.62	2.65	2.69
Length in D/S $L_{D/SABUT}$, m	1.98	2.05	2.08	2.12	2.15	2.19
Top Level including Free Board, m	100.90	101.10	101.20	101.30	101.40	101.50
Foundation Level, m	97.50	97.50	97.50	97.50	97.50	97.50
Total Height H_{ABUT}, m	3.40	3.60	3.70	3.80	3.90	4.00
D. D/S WING WALLS ON BOTH BANKS						
Length in D/S $L_{D/SWING}$, m	5.62	5.45	5.32	5.28	5.25	5.11
Top Level including Free Board, m	98.41	98.58	98.71	98.81	98.91	99.00
Foundation Level, m	97.50	97.50	97.50	97.50	97.50	97.50
Total Height $H_{D/SWING}$, m	0.91	1.08	1.21	1.31	1.41	1.50

6.5 DESIGN OF 3.00 M HIGH MASONRY CHECK DAMS

6.5.1 Section with Vertical U/S Face

The tables developed after the rigorous design of the different components for 3.00 m high masonry check dams for different combinations are presented from Table 6.37 to Table 6.40.

TABLE 6.37
Design Results for Masonry Check Dam with Vertical U/S Face Height = 3.00 m and Stream Width L = 5.0 m

Catchment Area, ha	50.00	100.00
A. WEIR		
Span of Weir W, m	4.00	4.00
Flood Discharge Q, m^3/s	6.54	11.00
Head Over Crest, m	0.92	1.31
Top Level, m	100.00	100.00
Foundation Level, m	97.00	97.00
Height of Weir H_{WEIR}, m	3.00	3.00
Top Width of Weir TW_{WEIR}, mm	500	500
Base Width of Weir B_{WEIR}, mm	3410	3680
D/S Slope – $H_{DS'}$: 1.0(V)	0.970	1.060
Width B1', mm	2910	3180
Pressure at Toe, t/m^2	5.08	5.07
Pressure at Heel, t/m^2	0.01	0.01
Factor of Safety against Overturning	1.46	1.46
Shear Friction Factor F	2.07	1.97
Total Volume of Masonry, m^3	23.46	25.08
Length of D/S Apron Floor $L_{D/S\ APRON}$, m	8.60	8.40
B. NON-OVER FLOW PORTION		
Length, m	1.00	1.00
Top Level including Free Board, m	101.50	102.00
Foundation Level, m	97.00	97.00
Height of Non-Over Flow H_{NOF}, m	4.50	5.00
Height H1, mm	3754	4247
Height H2, mm	746	753
Top Width of Non-Over Flow TW_{NOF}, mm	500	500
Base Width of Non-Over Flow B_{NOF}, mm	3020	3320
D/S Slope – H_{DS}: 1.0(V)	0.670	0.664
Width B1, mm	2520	2820
Pressure at Toe, t/m^2	7.60	8.73
Pressure at Heel, t/m^2	0.07	0.01
Factor of Safety against Overturning	1.56	1.59
Shear Friction Factor F	1.85	1.71
Total Volume of Masonry, m^3	6.99	8.48
C. WING WALLS (ABUTMENTS) ON BOTH BANKS		
Length in U/S $L_{U/SABUT}$, m	1.00	1.00

(Continued)

TABLE 6.37 (Continued)

Catchment Area, ha	50.00	100.00
Length in D/S $L_{D/SABUT}$, m	4.02	4.32
Top Level including Free Board, m	101.50	102.00
Foundation Level, m	97.00	97.00
Total Height H_{ABUT}, m	4.50	5.00
D. D/S WING WALLS ON BOTH BANKS		
Length in D/S $L_{D/SWING}$, m	7.99	7.68
Top Level including Free Board, m	98.52	98.95
Foundation Level, m	97.00	97.00
Total Height $H_{D/SWING}$, m	1.52	1.95

TABLE 6.38
Design Results for Masonry Check Dam with Vertical U/S Face Height = 3.00 m and Stream Width L = 10.0 m

Catchment Area, ha	50.00	100.00	150.00	200.00
A. WEIR				
Span of Weir W, m	8.00	8.00	8.00	8.00
Flood Discharge Q, m^3/s	6.54	11.00	14.91	18.50
Head Over Crest, m	0.58	0.82	1.01	1.16
Top Level, m	100.00	100.00	100.00	100.00
Foundation Level, m	97.00	97.00	97.00	97.00
Height of Weir H_{WEIR}, m	3.00	3.00	3.00	3.00
Top Width of Weir TW_{WEIR}, mm	500	500	500	500
Base Width of Weir B_{WEIR}, mm	3150	3340	3480	3580
D/S Slope – H_{DS}: 1.0(V)	0.883	0.946	0.993	1.026
Width B1', mm	2650	2840	2980	3080
Pressure at Toe, t/m^2	5.10	5.07	5.06	5.06
Pressure at Heel, t/m^2	0.02	0.03	0.03	0.02
Factor of Safety against Overturning	1.47	1.46	1.46	1.46
Shear Friction Factor F	2.18	2.11	2.05	2.01
Total Volume of Masonry, m^3	43.80	46.08	47.76	48.96
Length of D/S Apron Floor $L_{D/S\ APRON}$, m	8.90	8.70	8.60	8.50
B. NON-OVER FLOW PORTION				
Length, m	2.00	2.00	2.00	2.00
Top Level including Free Board, m	101.10	101.40	101.60	101.80
Foundation Level, m	97.00	97.00	97.00	97.00
Height of Non-Over Flow H_{NOF}, m	4.10	4.40	4.60	4.80
Height H1, mm	3365	3657	3853	4049
Height H2, mm	735	743	747	751
Top Width of Non-Over Flow TW_{NOF}, mm	500	500	500	500
Base Width of Non-Over Flow B_{NOF}, mm	2790	2960	3080	3200
D/S Slope – H_{DS}: 1.0(V)	0.680	0.673	0.669	0.666

(Continued)

TABLE 6.38 (Continued)

Catchment Area, ha	50.00	100.00	150.00	200.00
Width B1, mm	2290	2460	2580	2700
Pressure at Toe, t/m^2	6.71	7.40	7.85	8.25
Pressure at Heel, t/m^2	0.02	0.01	0.03	0.05
Factor of Safety against Overturning	1.52	1.55	1.56	1.58
Shear Friction Factor F	1.98	1.88	1.82	1.76
Total Volume of Masonry, m^3	11.79	13.40	14.56	15.71
C. WING WALLS (ABUTMENTS) ON BOTH BANKS				
Length in U/S $L_{U/SABUT}$, m	1.00	1.00	1.00	1.00
Length in D/S $L_{D/SABUT}$, m	3.79	3.96	4.08	4.20
Top Level including Free Board, m	101.10	101.40	101.60	101.80
Foundation Level, m	97.00	97.00	97.00	97.00
Total Height H_{ABUT}, m	4.10	4.40	4.60	4.80
D. D/S WING WALLS ON BOTH BANKS				
Length in D/S $L_{D/SWING}$, m	8.21	8.04	7.92	7.80
Top Level including Free Board, m	98.14	98.41	98.62	98.79
Foundation Level, m	97.00	97.00	97.00	97.00
Total Height $H_{D/SWING}$, m	1.14	1.41	1.62	1.79

TABLE 6.39
Design Results for Masonry Check Dam with Vertical U/S Face Height = 3.00 m and Stream Width L = 15.0 m

Catchment Area, ha	50.00	100.00	150.00	200.00	250.00	300.00
A. WEIR						
Span of Weir W, m	12.00	12.00	12.00	12.00	12.00	12.00
Flood Discharge Q, m^3/s	6.54	11.00	14.91	18.50	21.87	25.07
Head Over Crest, m	0.44	0.63	0.77	0.89	0.99	1.09
Top Level, m	100.00	100.00	100.00	100.00	100.00	100.00
Foundation Level, m	97.00	97.00	97.00	97.00	97.00	97.00
Height of Weir H_{WEIR}, m	3.00	3.00	3.00	3.00	3.00	3.00
Top Width of Weir TW_{WEIR}, mm	500	500	500	500	500	500
Base Width of Weir B_{WEIR}, mm	3030	3190	3300	3380	3460	3530
D/S Slope – $H_{DS'}$: 1.0(V)	0.842	0.896	0.933	0.960	0.986	1.010
Width B1', mm	2530	2690	2800	2880	2960	3030
Pressure at Toe, t/m^2	5.13	5.09	5.08	5.10	5.08	5.07
Pressure at Heel, t/m^2	0.02	0.02	0.02	0.00	0.01	0.01
Factor of Safety against Overturning	1.47	1.47	1.46	1.46	1.46	1.46
Shear Friction Factor F	2.22	2.16	2.12	2.08	2.05	2.02
Total Volume of Masonry, m^3	63.54	66.42	68.40	69.84	71.28	72.54
Length of D/S Apron Floor $L_{D/S\ APRON}$, m	9.00	8.90	8.80	8.70	8.60	8.50

(Continued)

TABLE 6.39 (Continued)

Catchment Area, ha	50.00	100.00	150.00	200.00	250.00	300.00
B. NON-OVER FLOW PORTION						
Length, m	3.00	3.00	3.00	3.00	3.00	3.00
Top Level including Free Board, m	101.00	101.20	101.40	101.50	101.60	101.70
Foundation Level, m	97.00	97.00	97.00	97.00	97.00	97.00
Height of Non-Over Flow H_{NOF}, m	4.00	4.20	4.40	4.50	4.60	4.70
Height H1, mm	3270	3463	3656	3756	3853	3953
Height H2, mm	730	737	744	744	747	747
Top Width of Non-Over Flow TW_{NOF}, mm	500	500	500	500	500	500
Base Width of Non-Over Flow B_{NOF}, mm	2740	2850	2960	3020	3080	3140
D/S Slope – H_{DS}: 1.0(V)	0.685	0.678	0.672	0.672	0.669	0.669
Width B1, mm	2240	2350	2460	2520	2580	2640
Pressure at Toe, t/m^2	6.51	6.91	7.40	7.62	7.85	8.08
Pressure at Heel, t/m^2	0.01	0.05	0.02	0.00	0.03	0.00
Factor of Safety against Overturning	1.51	1.53	1.55	1.55	1.56	1.57
Shear Friction Factor F	2.03	1.95	1.88	1.85	1.82	1.80
Total Volume of Masonry, m^3	17.02	18.50	20.11	20.96	21.84	22.73
C. WING WALLS (ABUTMENTS) ON BOTH BANKS						
Length in U/S $L_{U/SABUT}$, m	1.00	1.00	1.00	1.00	1.00	1.00
Length in D/S $L_{D/SABUT}$, m	3.74	3.85	3.96	4.02	4.08	4.14
Top Level including Free Board, m	101.00	101.20	101.40	101.50	101.60	101.70
Foundation Level, m	97.00	97.00	97.00	97.00	97.00	97.00
Total Height H_{ABUT}, m	4.00	4.20	4.40	4.50	4.60	4.70
D. D/S WING WALLS ON BOTH BANKS						
Length in D/S $L_{D/SWING}$, m	8.26	8.15	8.04	7.98	7.92	7.86
Top Level including Free Board, m	97.99	98.20	98.35	98.49	98.60	98.71
Foundation Level, m	97.00	97.00	97.00	97.00	97.00	97.00
Total Height $H_{D/SWING}$, m	0.99	1.20	1.35	1.49	1.60	1.71

TABLE 6.40
Design Results for Masonry Check Dam with Vertical U/S Face Height = 3.00 m and Stream Width L = 20.0 m

Catchment Area, ha	50.00	100.00	150.00	200.00	250.00	300.00
A. WEIR						
Span of Weir W, m	16.00	16.00	16.00	16.00	16.00	16.00
Flood Discharge Q, m^3/s	6.54	11.00	14.91	18.50	21.87	25.07
Head Over Crest, m	0.37	0.52	0.64	0.73	0.82	0.90
Top Level, m	100.00	100.00	100.00	100.00	100.00	100.00
Foundation Level, m	97.00	97.00	97.00	97.00	97.00	97.00
Height of Weir H_{WEIR}, m	3.00	3.00	3.00	3.00	3.00	3.00
Top Width of Weir TW_{WEIR}, mm	500	500	500	500	500	500

(Continued)

TABLE 6.40 (Continued)

Catchment Area, ha	**50.00**	**100.00**	**150.00**	**200.00**	**250.00**	**300.00**
Base Width of Weir B_{WEIR}, mm	2970	3100	3190	3270	3330	3400
D/S Slope – $H_{DS'}$: 1.0(V)	0.823	0.866	0.896	0.923	0.943	0.965
Width B1', mm	2470	2600	2690	2770	2830	2900
Pressure at Toe, t/m^2	5.14	5.11	5.12	5.08	5.10	5.06
Pressure at Heel, t/m^2	0.01	0.02	0.00	0.03	0.00	0.04
Factor of Safety against Overturning	1.47	1.47	1.46	1.46	1.46	1.46
Shear Friction Factor F	2.25	2.20	2.16	2.13	2.10	2.08
Total Volume of Masonry, m^3	83.28	86.40	88.56	90.48	91.92	93.60
Length of D/S Apron Floor $L_{D/S\ APRON}$, m	9.10	9.00	8.90	8.80	8.70	8.70
B. NON-OVER FLOW PORTION						
Length, m	4.00	4.00	4.00	4.00	4.00	4.00
Top Level including Free Board, m	100.90	101.10	101.20	101.30	101.40	101.50
Foundation Level, m	97.00	97.00	97.00	97.00	97.00	97.00
Height of Non-Over Flow H_{NOF}, m	3.90	4.10	4.20	4.30	4.40	4.50
Height H1, mm	3172	3365	3460	3560	3660	3755
Height H2, mm	728	735	740	740	740	745
Top Width of Non-Over Flow TW_{NOF}, mm	500	500	500	500	500	500
Base Width of Non-Over Flow B_{NOF}, mm	2680	2790	2840	2910	2970	3020
D/S Slope – H_{DS}: 1.0(V)	0.687	0.680	0.676	0.676	0.676	0.671
Width B1, mm	2180	2290	2340	2410	2470	2520
Pressure at Toe, t/m^2	6.28	6.71	6.96	7.16	7.36	7.60
Pressure at Heel, t/m^2	0.02	0.02	0.00	0.03	0.04	0.05
Factor of Safety against Overturning	1.51	1.52	1.53	1.54	1.55	1.56
Shear Friction Factor F	2.06	1.98	1.94	1.92	1.89	1.85
Total Volume of Masonry, m^3	21.66	23.59	24.59	25.76	26.88	27.95
C. WING WALLS (ABUTMENTS) ON BOTH BANKS						
Length in U/S $L_{U/SABUT}$, m	1.00	1.00	1.00	1.00	1.00	1.00
Length in D/S $L_{D/SABUT}$, m	3.68	3.79	3.84	3.91	3.97	4.02
Top Level including Free Board, m	100.90	101.10	101.20	101.30	101.40	101.50
Foundation Level, m	97.00	97.00	97.00	97.00	97.00	97.00
Total Height H_{ABUT}, m	3.90	4.10	4.20	4.30	4.40	4.50
D. D/S WING WALLS ON BOTH BANKS						
Length in D/S $L_{D/SWING}$, m	8.32	8.21	8.16	8.09	8.03	7.98
Top Level including Free Board, m	97.91	98.08	98.21	98.31	98.41	98.50
Foundation Level, m	97.00	97.00	97.00	97.00	97.00	97.00
Total Height $H_{D/SWING}$, m	0.91	1.08	1.21	1.31	1.41	1.50

6.5.2 Section with Vertical D/S Face

The tables developed after the rigorous design of the different components for 3.00 m high masonry check dams for different combinations are presented from Table 6.41 to Table 6.44.

TABLE 6.41
Design Results for Masonry Check Dam with Vertical D/S Face Height = 3.00 m and Stream Width L = 5.0 m

Catchment Area, ha	50.00	100.00
A. WEIR		
Span of Weir W, m	4.00	4.00
Flood Discharge Q, m^3/s	6.54	11.00
Head Over Crest, m	0.92	1.31
Top Level, m	100.00	100.00
Foundation Level, m	97.00	97.00
Height of Weir H_{WEIR}, m	3.00	3.00
Top Width of Weir TW_{WEIR}, mm	3500	3660
Base Width of Weir B_{WEIR}, mm	3500	3660
Pressure at Toe, t/m^2	12.03	12.53
Pressure at Heel, t/m^2	0.00	0.02
Factor of Safety against Overturning	2.11	2.12
Shear Friction Factor F	2.68	2.50
Total Volume of Masonry, m^3	42.00	43.92
Length of D/S Apron Floor $L_{D/S\ APRON}$, m	8.50	8.40
B. NON-OVER FLOW PORTION		
Length, m	1.00	1.00
Top Level including Free Board, m	101.50	102.00
Foundation Level, m	97.00	97.00
Height of Non-Over Flow H_{NOF}, m	4.50	5.00
Height H1, mm	3600	4000
Height H2, mm	900	1000
Top Width of Non-Over Flow TW_{NOF}, mm	1000	1000
Base Width of Non-Over Flow B_{NOF}, mm	3280	4560
Width B1, mm	2280	3560
Pressure at Toe, t/m^2	15.35	17.19
Pressure at Heel, t/m^2	0.02	0.00
Factor of Safety against Overturning	2.28	2.29
Shear Friction Factor F	2.50	2.95
Total Volume of Masonry, m^3	12.71	19.24
C. WING WALLS (ABUTMENTS) ON BOTH BANKS		
Length in U/S $L_{U/SABUT}$, m	4.28	5.56
Length in D/S $L_{D/SABUT}$, m	2.00	2.00
Top Level including Free Board, m	101.50	102.00
Foundation Level, m	97.00	97.00
Total Height H_{ABUT}, m	4.50	5.00
D. D/S WING WALLS ON BOTH BANKS		
Length in D/S $L_{D/SWING}$, m	6.50	6.40
Top Level including Free Board, m	98.52	98.95
Foundation Level, m	97.00	97.00
Total Height $H_{D/SWING}$, m	1.52	1.95

TABLE 6.42
Design Results for Masonry Check Dam with Vertical D/S Face Height = 3.00 m and Stream Width L = 10.0 m

Catchment Area, ha	50.00	100.00	150.00	200.00
A. WEIR				
Span of Weir W, m	8.00	8.00	8.00	8.00
Flood Discharge Q, m^3/s	6.54	11.00	14.91	18.50
Head Over Crest, m	0.58	0.82	1.01	1.16
Top Level, m	100.00	100.00	100.00	100.00
Foundation Level, m	97.00	97.00	97.00	97.00
Height of Weir H_{WEIR}, m	3.00	3.00	3.00	3.00
Top Width of Weir TW_{WEIR}, mm	3340	3460	3540	3600
Base Width of Weir B_{WEIR}, mm	3340	3460	3540	3600
Pressure at Toe, t/m^2	11.56	11.88	12.14	12.34
Pressure at Heel, t/m^2	0.01	0.01	0.00	0.00
Factor of Safety against Overturning	**2.11**	2.11	2.11	2.12
Shear Friction Factor F	2.87	2.74	2.64	2.57
Total Volume of Masonry, m^3	80.16	83.04	84.96	86.40
Length of D/S Apron Floor $L_{D/S\ APRON}$, m	8.70	8.60	8.50	8.40
B. NON-OVER FLOW PORTION				
Length, m	2.00	2.00	2.00	2.00
Top Level including Free Board, m	101.10	101.40	101.60	101.80
Foundation Level, m	97.00	97.00	97.00	97.00
Height of Non-Over Flow H_{NOF}, m	4.10	4.40	4.60	4.80
Height H1, mm	3040	3190	3360	3410
Height H2, mm	1060	1210	1240	1390
Top Width of Non-Over Flow TW_{NOF}, mm	1000	1000	1000	1000
Base Width of Non-Over Flow B_{NOF}, mm	3440	3460	3510	3610
Width B1, mm	2440	2460	2510	2610
Pressure at Toe, t/m^2	13.19	14.28	15.14	15.74
Pressure at Heel, t/m^2	0.00	0.01	0.00	0.00
Factor of Safety against Overturning	2.18	2.18	2.24	2.24
Shear Friction Factor F	2.93	2.64	2.51	2.40
Total Volume of Masonry, m^3	23.01	24.49	26.06	27.39
C. WING WALLS (ABUTMENTS) ON BOTH BANKS				
Length in U/S $L_{U/SABUT}$, m	4.44	4.46	4.51	4.61
Length in D/S $L_{D/SABUT}$, m	2.00	2.00	2.00	2.00
Top Level including Free Board, m	101.10	101.40	101.60	101.80
Foundation Level, m	97.00	97.00	97.00	97.00
Total Height H_{ABUT}, m	4.10	4.40	4.60	4.80
D. D/S WING WALLS ON BOTH BANKS				
Length in D/S $L_{D/SWING}$, m	6.70	6.60	6.50	6.40
Top Level including Free Board, m	98.14	98.41	98.62	98.79
Foundation Level, m	97.00	97.00	97.00	97.00
Total Height $H_{D/SWING}$, m	1.14	1.41	1.62	1.79

TABLE 6.43
Design Results for Masonry Check Dam with Vertical D/S Face Height = 3.00 m and Stream Width L = 15.0 m

Catchment Area, ha	50.00	100.00	150.00	200.00	250.00	300.00
A. WEIR						
Span of Weir W, m	12.00	12.00	12.00	12.00	12.00	12.00
Flood Discharge Q, m^3/s	6.54	11.00	14.91	18.50	21.87	25.07
Head Over Crest, m	0.44	0.63	0.77	0.89	0.99	1.09
Top Level, m	100.00	100.00	100.00	100.00	100.00	100.00
Foundation Level, m	97.00	97.00	97.00	97.00	97.00	97.00
Height of Weir H_{WEIR}, m	3.00	3.00	3.00	3.00	3.00	3.00
Top Width of Weir TW_{WEIR}, mm	3260	3370	3430	3490	3530	3570
Base Width of Weir B_{WEIR}, mm	3260	3370	3430	3490	3530	3570
Pressure at Toe, t/m^2	11.38	11.62	11.83	11.98	12.12	12.26
Pressure at Heel, t/m^2	0.00	0.02	0.00	0.01	0.00	0.00
Factor of Safety against Overturning	2.10	2.11	2.11	2.11	2.11	2.11
Shear Friction Factor F	2.95	2.85	2.76	2.70	2.65	2.60
Total Volume of Masonry, m^3	117.36	121.32	123.48	125.64	127.08	128.52
Length of D/S Apron Floor $L_{D/S\ APRON}$, m	8.80	8.70	8.60	8.60	8.50	8.50
B. NON-OVER FLOW PORTION						
Length, m	3.00	3.00	3.00	3.00	3.00	3.00
Top Level including Free Board, m	101.00	101.20	101.40	101.50	101.60	101.70
Foundation Level, m	97.00	97.00	97.00	97.00	97.00	97.00
Height of Non-Over Flow H_{NOF}, m	4.00	4.20	4.40	4.50	4.60	4.70
Height H1, mm	2980	3130	3190	3270	3240	3210
Height H2, mm	1020	1070	1210	1230	1360	1490
Top Width of Non-Over Flow TW_{NOF}, mm	1000	1000	1000	1000	1000	1000
Base Width of Non-Over Flow B_{NOF}, mm	3320	3430	3460	3520	3580	3630
Width B1, mm	2320	2430	2460	2520	2580	2630
Pressure at Toe, t/m^2	12.85	13.63	14.28	14.68	14.87	15.08
Pressure at Heel, t/m^2	0.00	0.01	0.01	0.00	0.01	0.01
Factor of Safety against Overturning	2.12	2.20	2.18	2.22	2.20	2.17
Shear Friction Factor F	2.95	2.83	2.64	2.60	2.53	2.46
Total Volume of Masonry, m^3	32.74	35.41	36.74	38.16	38.83	39.43
C. WING WALLS (ABUTMENTS) ON BOTH BANKS						
Length in U/S $L_{U/SABUT}$, m	4.32	4.43	4.46	4.52	4.58	4.63
Length in D/S $L_{D/SABUT}$, m	2.00	2.00	2.00	2.00	2.00	2.00
Top Level including Free Board, m	101.00	101.20	101.40	101.50	101.60	101.70
Foundation Level, m	97.00	97.00	97.00	97.00	97.00	97.00
Total Height H_{ABUT}, m	4.00	4.20	4.40	4.50	4.60	4.70
D. D/S WING WALLS ON BOTH BANKS						
Length in D/S $L_{D/SWING}$, m	6.80	6.70	6.60	6.60	6.50	6.50
Top Level including Free Board, m	97.99	98.20	98.35	98.49	98.60	98.71
Foundation Level, m	97.00	97.00	97.00	97.00	97.00	97.00
Total Height $H_{D/SWING}$, m	0.99	1.20	1.35	1.49	1.60	1.71

TABLE 6.44
Design Results for Masonry Check Dam with Vertical D/S Face Height = 3.00 m and Stream Width L = 20.0 m

Catchment Area, ha	50.00	100.00	150.00	200.00	250.00	300.00
A. WEIR						
Span of Weir W, m	16.00	16.00	16.00	16.00	16.00	16.00
Flood Discharge Q, m^3/s	6.54	11.00	14.91	18.50	21.87	25.07
Head Over Crest, m	0.37	0.52	0.64	0.73	0.82	0.90
Top Level, m	100.00	100.00	100.00	100.00	100.00	100.00
Foundation Level, m	97.00	97.00	97.00	97.00	97.00	97.00
Height of Weir H_{WEIR}, m	3.00	3.00	3.00	3.00	3.00	3.00
Top Width of Weir TW_{WEIR}, mm	3220	3310	3370	3420	3460	3490
Base Width of Weir B_{WEIR}, mm	3220	3310	3370	3420	3460	3490
Pressure at Toe, t/m^2	11.29	11.48	11.65	11.75	11.88	12.00
Pressure at Heel, t/m^2	0.01	0.02	0.01	0.02	0.01	0.00
Factor of Safety against Overturning	2.10	2.11	2.11	2.11	2.11	2.11
Shear Friction Factor F	3.00	2.91	2.84	2.79	2.74	2.69
Total Volume of Masonry, m^3	154.56	158.88	161.76	164.16	166.08	167.52
Length of D/S Apron Floor $L_{D/S\ APRON}$, m	8.80	8.70	8.70	8.60	8.60	8.60
B. NON-OVER FLOW PORTION						
Length, m	4.00	4.00	4.00	4.00	4.00	4.00
Top Level including Free Board, m	100.90	101.10	101.20	101.30	101.40	101.50
Foundation Level, m	97.00	97.00	97.00	97.00	97.00	97.00
Height of Non-Over Flow H_{NOF}, m	3.90	4.10	4.20	4.30	4.40	4.50
Height H1, mm	3050	3200	3280	3230	3210	3290
Height H2, mm	850	900	920	1070	1190	1210
Top Width of Non-Over Flow TW_{NOF}, mm	1000	1000	1000	1000	1000	1000
Base Width of Non-Over Flow B_{NOF}, mm	3240	3310	3350	3400	3420	3440
Width B1, mm	2240	2310	2350	2400	2420	2440
Pressure at Toe, t/m^2	12.71	13.53	13.93	14.11	14.32	14.75
Pressure at Heel, t/m^2	0.00	0.00	0.01	0.00	0.01	0.01
Factor of Safety against Overturning	2.16	2.22	2.25	2.21	2.17	2.19
Shear Friction Factor F	3.03	2.87	2.80	2.72	2.62	2.55
Total Volume of Masonry, m^3	42.86	45.95	47.59	48.20	48.61	50.11
C. WING WALLS (ABUTMENTS) ON BOTH BANKS						
Length in U/S $L_{U/SABUT}$, m	4.24	4.31	4.35	4.40	4.42	4.44
Length in D/S $L_{D/SABUT}$, m	2.00	2.00	2.00	2.00	2.00	2.00
Top Level including Free Board, m	100.90	101.10	101.20	101.30	101.40	101.50
Foundation Level, m	97.00	97.00	97.00	97.00	97.00	97.00
Total Height H_{ABUT}, m	3.90	4.10	4.20	4.30	4.40	4.50
D. D/S WING WALLS ON BOTH BANKS						
Length in D/S $L_{D/SWING}$, m	6.80	6.70	6.70	6.60	6.60	6.60
Top Level including Free Board, m	97.91	98.08	98.21	98.31	98.41	98.50
Foundation Level, m	97.00	97.00	97.00	97.00	97.00	97.00
Total Height $H_{D/SWING}$, m	0.91	1.08	1.21	1.31	1.41	1.50

6.5.3 Section with Slopes on Both Faces

The tables developed after the rigorous design of the different components for 3.00 m high masonry check dam for different combinations are presented from Table 6.45 to Table 6.48.

TABLE 6.45
Design Results for Masonry Check Dam with Slopes on Both Faces Height = 3.00 m and Stream Width L = 5.0 m

Catchment Area, ha	50.00	100.00
A. WEIR		
Span of Weir W, m	4.00	4.00
Flood Discharge Q, m^3/s	6.54	11.00
Head Over Crest, m	0.92	1.31
Top Level, m	100.00	100.00
Foundation Level, m	97.00	97.00
Height of Weir H_{WEIR}, m	3.00	3.00
Top Width of Weir TW_{WEIR}, mm	500	500
Base Width of Weir B_{WEIR}, mm	3180	3300
U/S Slope – H_{US}: 1.0(V)	0.445	0.465
D/S Slope – H_{DS}: 1.0(V)	0.445	0.465
Width B1', mm	1340	1400
Width B2', mm	1340	1400
Pressure at Toe, t/m^2	8.28	8.72
Pressure at Heel, t/m^2	0.03	0.01
Factor of Safety against Overturning	1.48	1.49
Shear Friction Factor F	2.17	2.00
Total Volume of Masonry, m^3	22.08	22.80
Length of D/S Apron Floor $L_{D/S\ APRON}$, m	8.90	8.70
B. NON-OVER FLOW PORTION		
Length, m	1.00	1.00
Top Level including Free Board, m	101.50	102.00
Foundation Level, m	97.00	97.00
Height of Non-Over Flow H_{NOF}, m	4.50	5.00
Height H1, mm	4000	4500
Height H2, mm	500	500
Top Width of Non-Over Flow TW_{NOF}, mm	500	500
Base Width of Non-Over Flow B_{NOF}, mm	3320	3680
U/S Slope – H_{US}: 1.0(V)	0.353	0.353
D/S Slope – H_{DS}: 1.0(V)	0.353	0.353
Width B1, mm	1410	1590
Width B2, mm	1410	1590
Pressure at Toe, t/m^2	10.41	11.69
Pressure at Heel, t/m^2	0.00	0.01
Factor of Safety against Overturning	1.67	1.74
Shear Friction Factor F	2.23	2.09

(Continued)

TABLE 6.45 (Continued)

Catchment Area, ha	50.00	100.00
Total Volume of Masonry, m^3	7.89	9.66
C. WING WALLS (ABUTMENTS) ON BOTH BANKS		
Length in U/S $L_{U/SABUT}$, m	2.91	3.09
Length in D/S $L_{D/SABUT}$, m	2.41	2.59
Top Level including Free Board, m	101.50	102.00
Foundation Level, m	97.00	97.00
Total Height H_{ABUT}, m	4.50	5.00
D. D/S WING WALLS ON BOTH BANKS		
Length in D/S $L_{D/SWING}$, m	6.49	6.11
Top Level including Free Board, m	98.52	98.95
Foundation Level, m	97.00	97.00
Total Height $H_{D/SWING}$, m	1.52	1.95

TABLE 6.46
Design Results for Masonry Check Dam with Slopes on Both Faces Height = 3.00 m and Stream Width L = 10.0 m

Catchment Area, ha	50.00	100.00	150.00	200.00
A. WEIR				
Span of Weir W, m	8.00	8.00	8.00	8.00
Flood Discharge Q, m^3/s	6.54	11.00	14.91	18.50
Head Over Crest, m	0.58	0.82	1.01	1.16
Top Level, m	100.00	100.00	100.00	100.00
Foundation Level, m	97.00	97.00	97.00	97.00
Height of Weir H_{WEIR}, m	3.00	3.00	3.00	3.00
Top Width of Weir TW_{WEIR}, mm	500	500	500	500
Base Width of Weir B_{WEIR}, mm	3060	3140	3200	3260
U/S Slope – $H_{US'}$: 1.0(V)	0.425	0.440	0.450	0.460
D/S Slope – $H_{DS'}$: 1.0(V)	0.425	0.440	0.450	0.460
Width B1', mm	1280	1320	1350	1380
Width B2', mm	1280	1320	1350	1380
Pressure at Toe, t/m^2	7.88	8.18	8.41	8.52
Pressure at Heel, t/m^2	0.06	0.03	0.00	0.05
Factor of Safety against Overturning	1.48	1.48	1.48	1.49
Shear Friction Factor F	2.34	2.21	2.12	2.07
Total Volume of Masonry, m^3	42.72	43.68	44.40	45.12
Length of D/S Apron Floor $L_{D/S\ APRON}$, m	9.00	8.90	8.80	8.80
B. NON-OVER FLOW PORTION				
Length, m	2.00	2.00	2.00	2.00
Top Level including Free Board, m	101.10	101.40	101.60	101.80
Foundation Level, m	97.00	97.00	97.00	97.00
Height of Non-Over Flow H_{NOF}, m	4.10	4.40	4.60	4.80

(Continued)

TABLE 6.46 (Continued)

Catchment Area, ha	50.00	100.00	150.00	200.00
Height H1, mm	3600	3900	4100	4300
Height H2, mm	500	500	500	500
Top Width of Non-Over Flow TW_{NOF}, mm	500	500	500	500
Base Width of Non-Over Flow B_{NOF}, mm	3060	3260	3420	3540
U/S Slope – H_{US}: 1.0(V)	0.355	0.355	0.355	0.353
D/S Slope – H_{DS}: 1.0(V)	0.355	0.355	0.355	0.353
Width B1, mm	1280	1380	1460	1520
Width B2, mm	1280	1380	1460	1520
Pressure at Toe, t/m^2	9.33	10.13	10.58	11.16
Pressure at Heel, t/m^2	0.02	0.00	0.06	0.02
Factor of Safety against Overturning	1.61	1.66	1.69	1.72
Shear Friction Factor F	2.38	2.26	2.21	2.14
Total Volume of Masonry, m^3	13.32	15.17	16.57	17.88
C. WING WALLS (ABUTMENTS) ON BOTH BANKS				
Length in U/S $L_{U/SABUT}$, m	2.78	2.88	2.96	3.02
Length in D/S $L_{D/SABUT}$, m	2.28	2.38	2.46	2.52
Top Level including Free Board, m	101.10	101.40	101.60	101.80
Foundation Level, m	97.00	97.00	97.00	97.00
Total Height H_{ABUT}, m	4.10	4.40	4.60	4.80
D. D/S WING WALLS ON BOTH BANKS				
Length in D/S $L_{D/SWING}$, m	6.72	6.52	6.34	6.28
Top Level including Free Board, m	98.14	98.41	98.62	98.79
Foundation Level, m	97.00	97.00	97.00	97.00
Total Height $H_{D/SWING}$, m	1.14	1.41	1.62	1.79

TABLE 6.47
Design Results for Masonry Check Dam with Slopes on Both Faces Height = 3.00 m and Stream Width L = 15.0 m

Catchment Area, ha	50.00	100.00	150.00	200.00	250.00	300.00
A. WEIR						
Span of Weir W, m	12.00	12.00	12.00	12.00	12.00	12.00
Flood Discharge Q, m^3/s	6.54	11.00	14.91	18.50	21.87	25.07
Head Over Crest, m	0.44	0.63	0.77	0.89	0.99	1.09
Top Level, m	100.00	100.00	100.00	100.00	100.00	100.00
Foundation Level, m	97.00	97.00	97.00	97.00	97.00	97.00
Height of Weir H_{WEIR}, m	3.00	3.00	3.00	3.00	3.00	3.00
Top Width of Weir TW_{WEIR}, mm	500	500	500	500	500	500
Base Width of Weir B_{WEIR}, mm	2980	3060	3120	3180	3200	3240

(Continued)

TABLE 6.47 (Continued)

Catchment Area, ha	50.00	100.00	150.00	200.00	250.00	300.00
U/S Slope – $H_{US'}$: 1.0(V)	0.414	0.428	0.438	0.445	0.450	0.455
D/S Slope – $H_{DS'}$: 1.0(V)	0.414	0.428	0.438	0.445	0.450	0.455
Width B1', mm	1240	1280	1310	1340	1350	1370
Width B2', mm	1240	1280	1310	1340	1350	1370
Pressure at Toe, t/m^2	7.81	8.01	8.15	8.21	8.36	8.45
Pressure at Heel, t/m^2	0.00	0.00	0.00	0.07	0.03	0.05
Factor of Safety against Overturning	1.47	1.47	1.48	1.49	1.49	1.49
Shear Friction Factor F	2.41	2.30	2.23	2.19	2.13	2.10
Total Volume of Masonry, m^3	62.64	64.08	65.16	66.24	66.60	67.32
Length of D/S Apron Floor $L_{D/S\ APRON}$, m	9.10	9.00	8.90	8.90	8.80	8.80
B. NON-OVER FLOW PORTION						
Length, m	3.00	3.00	3.00	3.00	3.00	3.00
Top Level including Free Board, m	101.00	101.20	101.40	101.50	101.60	101.70
Foundation Level, m	97.00	97.00	97.00	97.00	97.00	97.00
Height of Non-Over Flow H_{NOF}, m	4.00	4.20	4.40	4.50	4.60	4.70
Height H1, mm	3500	3700	3900	4000	4100	4200
Height H2, mm	500	500	500	500	500	500
Top Width of Non-Over Flow TW_{NOF}, mm	500	500	500	500	500	500
Base Width of Non-Over Flow B_{NOF}, mm	3000	3140	3260	3320	3420	3480
U/S Slope – $H_{US'}$: 1.0(V)	0.356	0.356	0.355	0.353	0.355	0.355
D/S Slope – $H_{DS'}$: 1.0(V)	0.356	0.356	0.355	0.353	0.355	0.355
Width B1, mm	1250	1320	1380	1410	1460	1490
Width B2, mm	1250	1320	1380	1410	1460	1490
Pressure at Toe, t/m^2	9.07	9.52	10.13	10.41	10.58	10.87
Pressure at Heel, t/m^2	0.02	0.08	0.00	0.00	0.06	0.04
Factor of Safety against Overturning	1.59	1.63	1.66	1.67	1.69	1.71
Shear Friction Factor F	2.42	2.35	2.26	2.23	2.21	2.18
Total Volume of Masonry, m^3	19.12	20.96	22.75	23.68	24.85	25.83
C. WING WALLS (ABUTMENTS) ON BOTH BANKS						
Length in U/S $L_{U/SABUT}$, m	2.75	2.82	2.88	2.91	2.96	2.99
Length in D/S $L_{D/SABUT}$, m	2.25	2.32	2.38	2.41	2.46	2.49
Top Level including Free Board, m	101.00	101.20	101.40	101.50	101.60	101.70
Foundation Level, m	97.00	97.00	97.00	97.00	97.00	97.00
Total Height H_{ABUT}, m	4.00	4.20	4.40	4.50	4.60	4.70
D. D/S WING WALLS ON BOTH BANKS						
Length in D/S $L_{D/SWING}$, m	6.85	6.68	6.52	6.49	6.34	6.31
Top Level including Free Board, m	97.99	98.20	98.35	98.49	98.60	98.71
Foundation Level, m	97.00	97.00	97.00	97.00	97.00	97.00
Total Height $H_{D/SWING}$, m	0.99	1.20	1.35	1.49	1.60	1.71

TABLE 6.48
Design Results for Masonry Check Dam with Slopes on Both Faces Height = 3.00 m and Stream Width L = 20.0 m

Catchment Area, ha	50.00	100.00	150.00	200.00	250.00	300.00
A. WEIR						
Span of Weir W, m	16.00	16.00	16.00	16.00	16.00	16.00
Flood Discharge Q, m^3/s	6.54	11.00	14.91	18.50	21.87	25.07
Head Over Crest, m	0.37	0.52	0.64	0.73	0.82	0.90
Top Level, m	100.00	100.00	100.00	100.00	100.00	100.00
Foundation Level, m	97.00	97.00	97.00	97.00	97.00	97.00
Height of Weir H_{WEIR}, m	3.00	3.00	3.00	3.00	3.00	3.00
Top Width of Weir TW_{WEIR}, mm	500	500	500	500	500	500
Base Width of Weir B_{WEIR}, mm	2940	3020	3080	3120	3140	3180
U/S Slope – $H_{US'}$: 1.0(V)	0.406	0.420	0.430	0.435	0.440	0.445
D/S Slope – $H_{DS'}$: 1.0(V)	0.406	0.420	0.430	0.435	0.440	0.445
Width B1', mm	1220	1260	1290	1310	1320	1340
Width B2', mm	1220	1260	1290	1310	1320	1340
Pressure at Toe, t/m^2	7.74	7.86	7.95	8.04	8.18	8.24
Pressure at Heel, t/m^2	0.00	0.03	0.06	0.07	0.03	0.06
Factor of Safety against Overturning	1.46	1.47	1.48	1.48	1.48	1.49
Shear Friction Factor F	2.44	2.37	2.31	2.27	2.21	2.18
Total Volume of Masonry, m^3	82.56	84.48	85.92	86.88	87.36	88.32
Length of D/S Apron Floor $L_{D/S\ APRON}$, m	9.10	9.00	9.00	8.90	8.90	8.90
B. NON-OVER FLOW PORTION						
Length, m	4.00	4.00	4.00	4.00	4.00	4.00
Top Level including Free Board, m	100.90	101.10	101.20	101.30	101.40	101.50
Foundation Level, m	97.00	97.00	97.00	97.00	97.00	97.00
Height of Non-Over Flow H_{NOF}, m	3.90	4.10	4.20	4.30	4.40	4.50
Height H1, mm	3400	3600	3700	3800	3900	4000
Height H2, mm	500	500	500	500	500	500
Top Width of Non-Over Flow TW_{NOF}, mm	500	500	500	500	500	500
Base Width of Non-Over Flow B_{NOF}, mm	2920	3060	3140	3200	3260	3340
U/S Slope – $H_{US'}$: 1.0(V)	0.355	0.355	0.356	0.356	0.355	0.355
D/S Slope – $H_{DS'}$: 1.0(V)	0.355	0.355	0.356	0.356	0.355	0.355
Width B1, mm	1210	1280	1320	1350	1380	1420
Width B2, mm	1210	1280	1320	1350	1380	1420
Pressure at Toe, t/m^2	8.83	9.33	9.52	9.84	10.13	10.34
Pressure at Heel, t/m^2	0.00	0.02	0.08	0.03	0.00	0.04
Factor of Safety against Overturning	1.57	1.61	1.63	1.64	1.66	1.68
Shear Friction Factor F	2.45	2.38	2.35	2.31	2.26	2.24
Total Volume of Masonry, m^3	24.25	26.63	27.95	29.12	30.33	31.71
C. WING WALLS (ABUTMENTS) ON BOTH BANKS						
Length in U/S $L_{U/SABUT}$, m	2.71	2.78	2.82	2.85	2.88	2.92
Length in D/S $L_{D/SABUT}$, m	2.21	2.28	2.32	2.35	2.38	2.42
Top Level including Free Board, m	100.90	101.10	101.20	101.30	101.40	101.50

(Continued)

TABLE 6.48 (Continued)

Catchment Area, ha	50.00	100.00	150.00	200.00	250.00	300.00
Foundation Level, m	97.00	97.00	97.00	97.00	97.00	97.00
Total Height H_{ABUT}, m	3.90	4.10	4.20	4.30	4.40	4.50
D. D/S WING WALLS ON BOTH BANKS						
Length in D/S $L_{D/SWING}$, m	6.89	6.72	6.68	6.55	6.52	6.48
Top Level including Free Board, m	97.91	98.08	98.21	98.31	98.41	98.50
Foundation Level, m	97.00	97.00	97.00	97.00	97.00	97.00
Total Height $H_{D/SWING}$, m	0.91	1.08	1.21	1.31	1.41	1.50

6.6 DESIGN OF MASONRY WING WALLS

The following forces are considered, in general, in the design of masonry check dams:

- Self weight of wall;
- Water pressure;
- Uplift pressures; and
- Active earth pressures in design of abutment and wing walls.

The following parameters were varied for developing the design tables for masonry walls:

Height of Masonry Wing Walls	:	1.00 m to 6.00 m
Unit Weight of Water, γ_{water}	:	1.00 t/m³ or 9.81 kN/m³
Unit Weight of Masonry, $\gamma_{Masonry}$	:	2.30 t/m³ or 22.56 kN/m³
Effect of Water for Uplift Pressure	:	30% of Total Height
Unit Weight of Saturated Earthfill, γ_{sat}	:	1.80 t/m³ or 17.66 kN/m³
Angle of Repose of Backfill, ϕ	:	30^0
Angle of Friction between the Masonry Wall and Backfill, ϕ	:	⅔ ϕ
Angle of Surcharge of Backfill, ι	:	0°
Coefficient of Friction between the Masonry and Soil/rock mass μ	:	0.55

The base width of the masonry wing wall/abutment was worked out on the principles discussed in Chapter 3 on design philosophy.

The generalised cross section of the masonry wing wall (abutment) and downstream wing wall with vertical water face and vertical earth face (Figure 6.4) for heights varying from 1.00 m to 6.00 m have been adopted for carrying out the analysis.

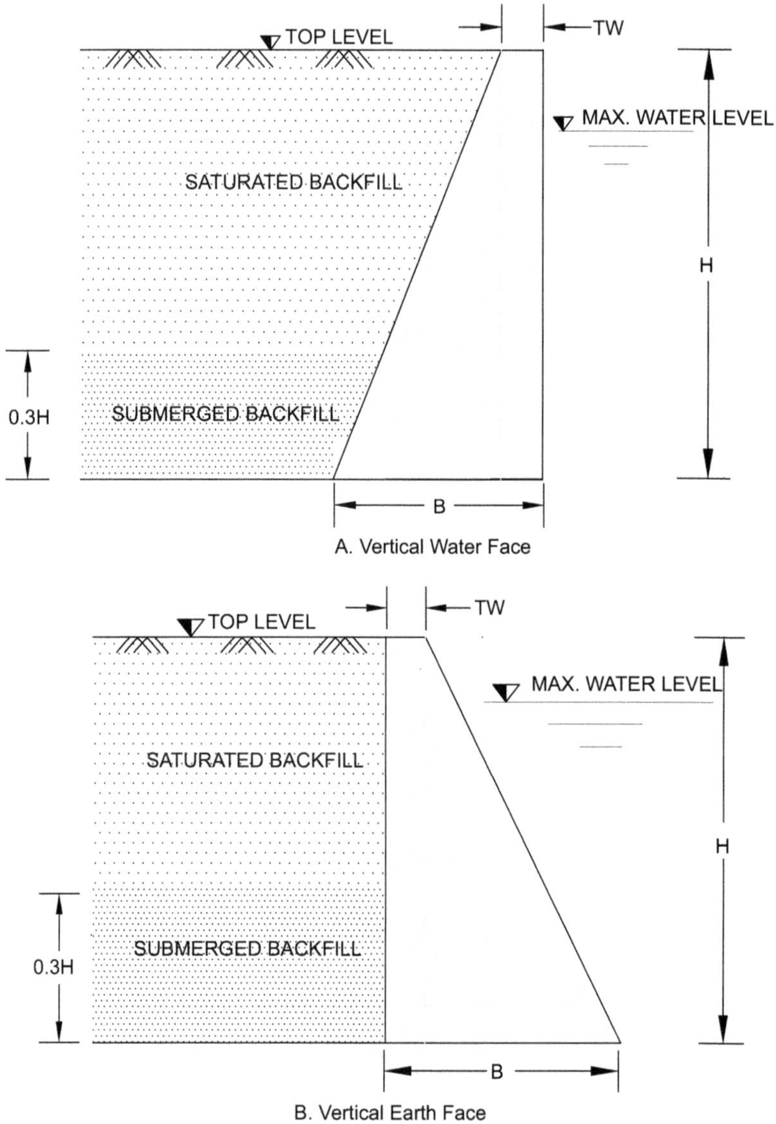

FIGURE 6.4A, 6.4B Typical Cross Sections of Masonry Wing Wall (Abutment) and Downstream Wing Wall.

6.6.1 Section with Vertical Water Face

The design details of masonry wing walls with vertical water face, computed for heights from 1.00 m to 6.00 m, are presented in Table 6.49, and the variation of the base width and the volume of masonry with the height of the wall have been depicted in Figure 6.5.

TABLE 6.49
Design Details for Masonry Wing Wall with Vertical Water Face

Height m	Top Width TW mm	Base Width B mm	Pressure at Toe t/m²	Pressure at Heel t/m²	F_{OVT}	F_{SLD}	Volume of Masonry m³	B/H Ratio
1.00	300	710	3.82	0.00	2.97	1.71	0.505	0.71
1.25	300	930	4.65	0.01	3.00	1.69	0.769	0.74
1.50	500	1020	5.85	0.00	2.93	1.71	1.140	0.68
1.75	500	1260	6.64	0.01	2.98	1.70	1.540	0.72
2.00	500	1480	7.47	0.02	3.00	1.69	1.980	0.74
2.25	500	1680	8.35	0.00	2.99	1.67	2.453	0.75
2.50	500	1890	9.19	0.02	3.00	1.67	2.988	0.76
2.75	500	2090	10.06	0.02	2.99	1.66	3.561	0.76
3.00	500	2290	10.92	0.02	2.99	1.65	4.185	0.76
3.25	500	2490	11.78	0.03	2.99	1.65	4.859	0.77
3.50	500	2680	12.67	0.01	2.97	1.64	5.565	0.77
3.75	500	2880	13.53	0.02	2.98	1.63	6.338	0.77
4.00	500	3070	14.42	0.01	2.97	1.63	7.140	0.77
4.25	500	3270	15.27	0.02	2.97	1.63	8.011	0.77
4.50	500	3460	16.16	0.01	2.96	1.62	8.910	0.77
4.75	500	3650	17.04	0.00	2.95	1.62	9.856	0.77
5.00	500	3850	17.89	0.02	2.96	1.61	10.875	0.77
5.25	500	4040	18.78	0.01	2.95	1.61	11.918	0.77
5.50	500	4230	19.66	0.00	2.94	1.61	13.008	0.77
5.75	500	4430	20.51	0.02	2.95	1.61	14.174	0.77
6.00	500	4620	21.39	0.01	2.94	1.60	15.360	0.77

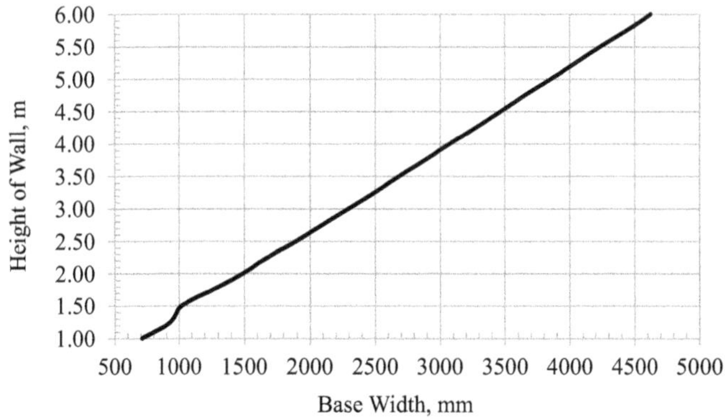

A. Variation of Base Width with Height

FIGURE 6.5A, 6.5B Variation of Base Width and Volume for Vertical Water Face Masonry Wing Wall.

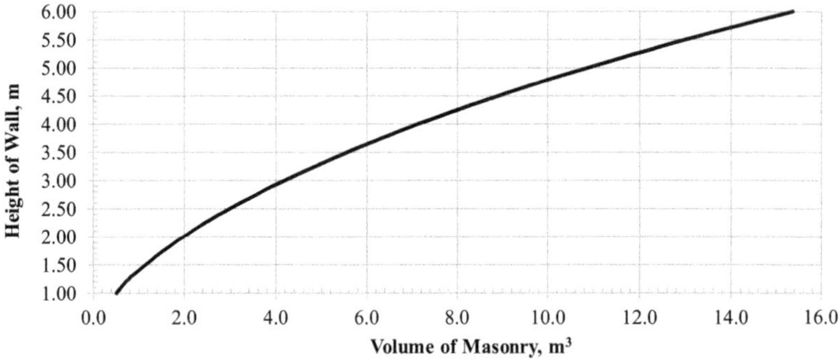

FIGURE 6.5A, 6.5B (Continued)

6.6.2 Section with Vertical Earth Face

The design details of masonry wing walls with vertical earth face, computed for heights from 1.00 m to 6.00 m, are presented in Table 6.50, and the variation of the base width and the volume of masonry with the height of the wall have been depicted in Figure 6.6.

TABLE 6.50
Design Details for Masonry Wing Wall with Vertical Earth Face

Height m	Top Width TW mm	Base Width B mm	Pressure at Toe t/m²	Pressure at Heel t/m²	FOVT	FSLD	Depth Shear Key for μ = 0.55 mm	Volume of Masonry m³	B/H Ratio
1.00	300	400	4.06	0.00	2.12	1.48	60	0.350	0.40
1.25	300	490	4.69	0.00	2.00	1.34	110	0.494	0.39
1.50	500	610	6.33	0.00	2.19	1.56	40	0.833	0.41
1.75	500	690	7.11	0.00	2.06	1.44	110	1.041	0.39
2.00	500	780	7.74	0.00	2.00	1.36	160	1.280	0.39
2.25	500	870	8.41	0.00	1.94	1.29	210	1.541	0.39
2.50	500	970	8.92	0.00	1.92	1.25	250	1.838	0.39
2.75	500	1060	9.65	0.00	1.88	1.21	290	2.145	0.39
3.00	500	1160	10.22	0.00	1.86	1.18	340	2.490	0.39
3.25	500	1260	10.80	0.00	1.84	1.16	370	2.860	0.39
3.50	500	1360	11.40	0.00	1.83	1.13	410	3.255	0.39
3.75	500	1460	12.02	0.00	1.82	1.12	450	3.675	0.39
4.00	500	1560	12.64	0.00	1.80	1.10	490	4.120	0.39
4.25	500	1660	13.27	0.00	1.79	1.09	520	4.590	0.39
4.50	500	1770	13.75	0.00	1.80	1.08	560	5.108	0.39

(Continued)

TABLE 6.50 (Continued)

Height m	Top Width TW mm	Base Width B mm	Pressure at Toe t/m²	Pressure at Heel t/m²	FOVT	FSLD	Depth Shear Key for μ = 0.55 mm	Volume of Masonry m³	B/H Ratio
4.75	500	1870	14.39	0.00	1.79	1.07	600	5.629	0.39
5.00	500	1970	15.04	0.00	1.78	1.06	640	6.175	0.39
5.25	500	2080	15.54	0.00	1.78	1.05	660	6.773	0.40
5.50	500	2180	16.20	0.00	1.77	1.04	700	7.370	0.40
5.75	500	2280	16.86	0.00	1.76	1.03	750	7.993	0.40
6.00	500	2390	17.37	0.00	1.76	1.03	780	8.670	0.40

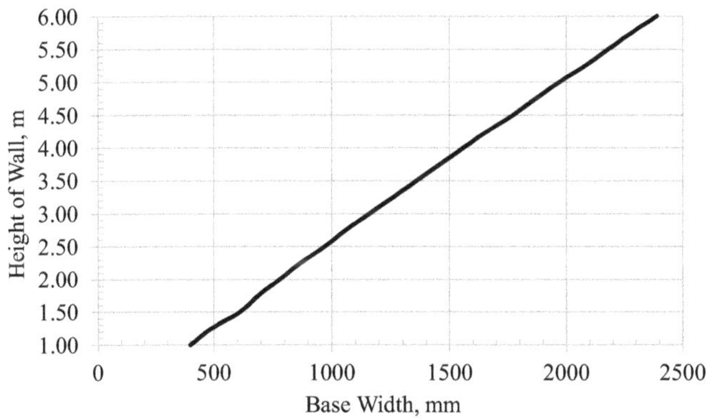

A. Variation of Base Width with Height

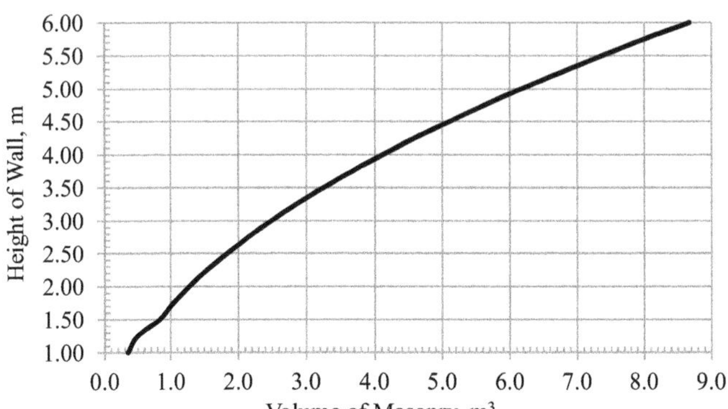

B. Variation of Volume of Masonry with Height

FIGURE 6.6A, 6.6B Variation of Base Width and Volume for Vertical Earth Face Masonry Wing Wall.

6.7 APPLICATION OF DESIGN TABLES

Steps to carry out the design of masonry check dams include the following:

Step 1 : For the chosen shape of the cross section, given the value of the height of a check dam H and the width of the stream, select the appropriate table.

Step 2 : From the appropriate table, for the known catchment area, select the appropriate column of the table and read from section A, top width, base width for the weir section, top level, foundation level and factor of safety against overturning, base pressures at toe and heel, length of downstream apron for check dams greater than 1.50 m high, total volume of masonry, etc.

Step 3 : From section B, read the top width, base width for non-over flow section, top level, foundation level and factor of safety against overturning, base pressures at toe and pressure at heel, total volume of masonry, etc.

Step 4 : Read the design details for masonry wing walls (abutments) from Table 6.49 or Table 6.50 for the height of the wing wall. For intermediate values, the data can be interpolated and partly read from Figure 6.5 or Figure 6.6.

Step 5 : Read the design details for downstream masonry wing walls from Table 6.49 or Table 6.50 for the height of the D/S wing wall. For intermediate values, the data can be interpolated and partly read from Figure 6.5 or Figure 6.6.

Step 6 : Engineering drawing for the masonry check dam can be prepared by selecting the appropriate figure (from 9.7 to 9.12) on the basis of the design data obtained from Step 2 to Step 5.

Step 7 : Specifications and general guidelines to be followed during the construction of masonry check dams can be read from Figure 9.13.

6.8 DESIGN EXAMPLES

To illustrate the use of tables, the following examples have been chosen:

6.8.1 Example 1

Design a masonry check dam for following parameters:

i. Shape of the masonry section: vertical U/S face
ii. Height of a check dam H = 2.50m
iii. Width of stream/gadhera = 10.0 m
iv. Catchment area = 150 ha

Solution:

Step 1: For a chosen shape of the cross section as vertical U/S face, the given value of the height of a check dam as 2.50 m and width of stream as 10.0 m, select Table 6.26.

Step 2: From Table 6.26, for the known catchment area, select the appropriate column of the table. The values given in section A for the design of masonry weir are as follows:

Span of Weir W, m	8.00
Flood Discharge Q, m^3/s	14.91
Head Over Crest, m	1.01
Top Level, m	100.00
Foundation Level, m	97.50
Height of Weir H_{WEIR}, m	2.50
Top Width of Weir TW_{WEIR}, mm	500
Base Width of Weir B_{WEIR}, mm	2950
D/S Slope – H_{DS}: 1.0(V)	0.980
Width B1', mm	2450
Pressure at Toe, t/m^2	4.44
Pressure at Heel, t/m^2	0.02
Factor of Safety against Overturning	1.47
Shear Friction Factor F	2.28
Total Volume of Masonry, m^3	34.50
Length of D/S Apron Floor $L_{D/S\ APRON}$, m	7.10

Step 3: From Table 6.26, read the values given in section B for the design of the non-over flow section:

Length, m	2.00
Top Level including Free Board, m	101.60
Foundation Level, m	97.50
Height of Non-Over Flow H_{NOF}, m	4.10
Height H1, mm	3320
Height H2, mm	780
Top Width of Non-Over Flow TW_{NOF}, mm	500
Base Width of Non-Over Flow B_{NOF}, mm	2610

D/S Slope – H_{DS}: 1.0(V)	0.636
Width B1, mm	2110
Pressure at Toe, t/m^2	7.24
Pressure at Heel, t/m^2	0.05
Factor of Safety against Overturning	1.59
Shear Friction Factor F	1.99
Total Volume of Masonry, m^3	11.11

Step 4: From Table 6.26, read the values given in section C for the design of wing walls (abutments):

Length in U/S $L_{U/SABUT}$, m	1.00
Length in D/S $L_{D/SABUT}$, m	3.61
Top Level including Free Board, m	101.60
Foundation Level, m	97.50
Total Height H_{ABUT}, m	4.10

From Figure 6.5, for the masonry wing wall with vertical water face, read the values for a height of 4.10 m or interpolate the values from Table 6.49:

Top Width TW, mm	500
Base Width B, mm	3150
F_{OVT}	2.97
F_{SLD}	1.63
Volume of Masonry per m run, m^3	7.50
Total Volume of Masonry, m^3	34.53

Step 5: From Table 6.26, read the values given in section D for the design of D/S wing walls:

Length in D/S $L_{D/SWING}$, m	6.39
Top Level including Free Board, m	99.12
Foundation Level, m	97.50
Total Height $H_{D/SWING}$, m	1.62

From Figure 6.5, for the masonry wing wall with vertical water face, read the values for a height of 1.62 m or interpolate the values from Table 6.49:

Top Width TW, mm	500
Base Width B, mm	1135
Pressure at Toe, t/m^2	6.23
Pressure at Heel, t/m^2	0.00
F_{OVT}	2.95
F_{SLD}	1.71

Volume of Masonry per m run, m^3	1.33
Total Volume of Masonry, m^3	8.50

Step 6: Based on the design data obtained from Step 2 to Step 5, an engineering drawing for the 2.50 m high masonry check dam with vertical U/S face can be prepared with the help of Figure 9.10.

Step 7: Specifications and general guidelines for the construction of masonry check dams can be read from Figure 9.13.

6.8.2 Example 2

Design a masonry check dam for following parameters:

i. Shape of the masonry section: vertical D/S face
ii. All the other data are the same, as given in example 1.

Solution:

Step 1: For a chosen shape of the cross section as vertical D/S face, the given value of the height of a check dam as 2.50 m and width of stream as 10.0 m, select Table 6.30.

Step 2: From Table 6.30, for the known catchment area, select the appropriate column of the table. The values given in section A for the design of masonry weir are as follows:

Span of Weir W, m	8.00
Flood Discharge Q, m^3/s	14.91
Head Over Crest, m	1.01
Top Level, m	100.00
Foundation Level, m	97.50
Height of Weir H_{WEIR}, m	2.50
Top Width of Weir TW_{WEIR}, mm	3020
Base Width of Weir B_{WEIR}, mm	3020
Pressure at Toe, t/m^2	10.33
Pressure at Heel, t/m^2	0.02
Factor of Safety against Overturning	2.12
Shear Friction Factor F	2.87
Total Volume of Masonry, m^3	60.40
Length of D/S Apron Floor $L_{D/S\ APRON}$, m	7.00

Step 3: From Table 6.30, read the values given in section B for the design of the non-over flow section:

Length, m	2.00
Top Level including Free Board, m	101.60

Foundation Level, m	97.50
Height of Non-Over Flow H_{NOF}, m	4.10
Height H1, mm	3080
Height H2, mm	1020
Top Width of Non-Over Flow TW_{NOF}, mm	1000
Base Width of Non-Over Flow B_{NOF}, mm	3180
Width B1, mm	2180
Pressure at Toe, t/m^2	13.85
Pressure at Heel, t/m^2	0.00
Factor of Safety against Overturning	2.37
Shear Friction Factor F	2.91
Total Volume of Masonry, m^3	21.61

Step 4: From Table 6.30, read the values given in section C for the design of wing walls (abutments):

Length in U/S $L_{U/SABUT}$, m	4.18
Length in D/S $L_{D/SABUT}$, m	2.00
Top Level including Free Board, m	101.60
Foundation Level, m	97.50
Total Height H_{ABUT}, m	4.10

From Figure 6.5, for the masonry wall with vertical water face, read the values for a height of 4.10 m or interpolate the values from Table 6.49:

Top Width TW, mm	500
Base Width B, mm	3150
Pressure at Toe, t/m^2	14.76
Pressure at Heel, t/m^2	0.01
F_{OVT}	2.97
F_{SLD}	1.63
Volume of Masonry per m run, m^3	7.49
Total Volume of Masonry, m^3	46.29

Step 5: From Table 6.30, read the values given in section D for the design of D/S wing walls:

Length in D/S $L_{D/SWING}$, m	5.00
Top Level including Free Board, m	99.12
Foundation Level, m	97.50
Total Height $H_{D/SWING}$, m	1.62

From Figure 6.5, for the masonry wall with vertical water face, read the values for a height of 1.62 m or interpolate the values from Table 6.49:

Top Width TW, mm	500
Base Width B, mm	1135
Pressure at Toe, t/m^2	6.23
Pressure at Heel, t/m^2	0.00
F_{OVT}	2.95
F_{SLD}	1.71
Volume of Masonry per m run, m^3	1.33
Total Volume of Masonry, m^3	6.65

Step 6: Based on the design data obtained from Step 2 to Step 5, an engineering drawing for the 2.50 m high masonry check dam with vertical D/S face can be prepared with the help of Figure 9.11.

Step 7: Specifications and general guidelines for the construction of masonry check dams can be read from Figure 9.13.

6.8.3 Example 3

Design a masonry check dam for following parameters:

i. Shape of masonry section: slopes on both faces
ii. All the other data are the same, as given in example 1.

Solution:

Step 1: For a chosen shape of the cross section as slopes on both faces, the given value of the height of a check dam as 2.50 m and width of stream as 10.0 m, select Table 6.34.

Step 2: From Table 6.34, for the known catchment area, select the appropriate column of the table. The values given in section A for the design of the masonry weir are as follows:

Span of Weir W, m	8.00
Flood Discharge Q, m^3/s	14.91
Head Over Crest, m	1.01
Top Level, m	100.00
Foundation Level, m	97.50
Height of Weir H_{WEIR}, m	2.50
Top Width of Weir TW_{WEIR}, mm	500
Base Width of Weir B_{WEIR}, mm	2720

U/S Slope – H_{US}: 1.0(V)	0.45
D/S Slope – H_{DS}: 1.0(V)	0.45
Width B1', mm	1110
Width B2', mm	1110
Pressure at Toe, t/m^2	7.29
Pressure at Heel, t/m^2	0.01
Factor of Safety against Overturning	1.51
Shear Friction Factor F	2.34
Total Volume of Masonry, m^3	32.20
Length of D/S Apron Floor $L_{D/S\ APRON}$, m	7.30

Step 3: From Table 6.34, read the values given in section B for the design of the non-over flow section:

Length, m	2.00
Top Level including Free Board, m	101.60
Foundation Level, m	97.50
Height of Non-Over Flow H_{NOF}, m	4.10
Height H1, mm	3600
Height H2, mm	500
Top Width of Non-Over Flow TW_{NOF}, mm	500
Base Width of Non-Over Flow B_{NOF}, mm	2940
U/S Slope – H_{US}: 1.0(V)	0.34
D/S Slope – H_{DS}: 1.0(V)	0.34
Width B1, mm	1220
Width B2, mm	1220
Pressure at Toe, t/m^2	9.63
Pressure at Heel, t/m^2	0.02
Factor of Safety against Overturning	1.74
Shear Friction Factor F	2.42
Total Volume of Masonry, m^3	12.88

Step 4: From Table 6.34, read the values given in section C for the design of wing walls (abutments):

Length in U/S $L_{U/SABUT}$, m	2.72
Length in D/S $L_{D/SABUT}$, m	2.22
Top Level including Free Board, m	101.60
Foundation Level, m	97.50
Total Height H_{ABUT}, m	4.10

From Figure 6.5, for the masonry wall with vertical water face, read the values for a height of 4.10 m or interpolate the values from Table 6.49:

Top Width TW, mm	500
Base Width B, mm	3150
Pressure at Toe, t/m^2	14.76
Pressure at Heel, t/m^2	0.01
F_{OVT}	2.97
F_{SLD}	1.63
Volume of Masonry per m run, m^3	7.49
Total Volume of Masonry, m^3	37.00

Step 5: From Table 6.34, read the values given in section D for the design of D/S wing walls:

Length in D/S $L_{D/SWING}$, m	5.08
Top Level including Free Board, m	99.12
Foundation Level, m	97.50
Total Height $H_{D/SWING}$, m	1.62

From Figure 6.5, for the masonry wall with vertical water face, read the values for a height of 1.62 m or interpolate the values from Table 6.49:

Top Width TW, mm	500
Base Width B, mm	1135
Pressure at Toe, t/m^2	6.23
Pressure at Heel, t/m^2	0.00
F_{OVT}	2.95
F_{SLD}	1.71
Volume of Masonry per m run, m^3	1.33
Total Volume of Masonry, m^3	6.76

Step 6: Based on the design data obtained from Step 2 to Step 5, an engineering drawing for the 2.50 m high masonry check dam with slopes on both faces can be prepared with the help of Figure 9.12.

Step 7: Specifications and general guidelines for the construction of masonry check dams can be read from Figure 9.13.

REFERENCE

IS:6512 (2019) *Indian Standard Criteria for Design of Solid Gravity Dams*, IS:6512, Bureau of Indian Standards, New Delhi.

7 Design Aids for Concrete Check Dams

7.1 PARAMETERS CONSIDERED FOR THE DEVELOPMENT OF TABLES

Concrete check dams can be constructed with different shapes having vertical U/S face or vertical D/S face or steps on both faces. The following forces are considered, in general, in the design of concrete check dams:

- Self weight of wall;
- Water pressure;
- Uplift pressures; and
- Active earth pressures in the design of abutment and wing walls.

The following parameters were varied for developing the design tables for concrete check dams:

Height of a Check Dam Weir	:	1.50 m, 2.00 m, 2.50 m, and 3.00 m
Catchment Area[1]	:	50 ha, 100 ha, 150 ha, 200 ha, 250 ha, and 300 ha
Width of Stream/Gadhera	:	5.0 m, 10.0 m, 15.0 m, and 20.0 m
Shape of a Check Dam Section	:	• Vertical U/S Face • Vertical D/S Face • Stepped U/S and D/S Face
Unit Weight of Water, γ_{water}	:	1.00 t/m^3 or 9.81 kN/m^3
Unit Weight of Concrete, $\gamma_{Concrete}$	:	2.50 t/m^3 or 24.52 kN/m^3
Effect of Water for Uplift Pressure	:	100% of Total Height
Coefficient of Friction between the Concrete and Soil/Rock Mass μ	:	0.55

[1]For small width of streams, the design of check dams has been carried out by limiting the head over crest to about 1.30 m.

7.1.1 Typical Cross Sections Adopted

The generalised cross sections of concrete check dam weir and non-over flow for heights varying from 1.50 m to 3.00 m for carrying out the analysis for different shapes are depicted in Figure 7.1 to Figure 7.3.

DOI: 10.1201/9781003605621-7

7.1.1.1 Section with Vertical U/S Face

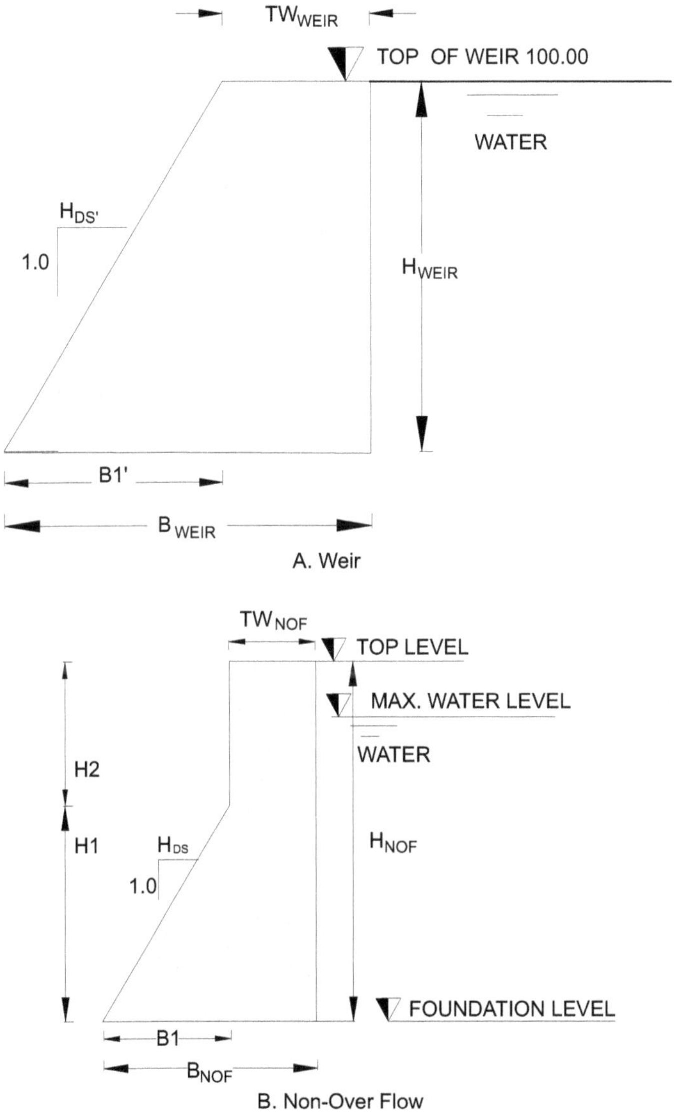

FIGURE 7.1A, 7.1B Typical Cross Sections of Concrete Weir and Non-Over Flow with Vertical U/S Face.

7.1.1.2 Section with Vertical D/S Face

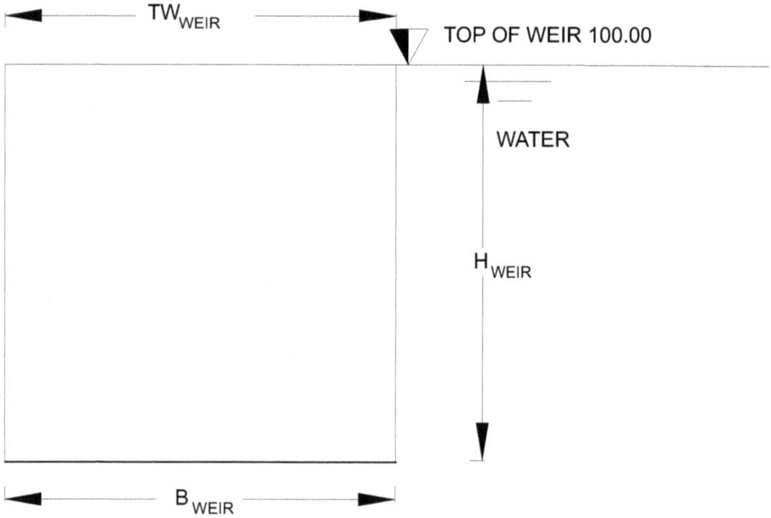

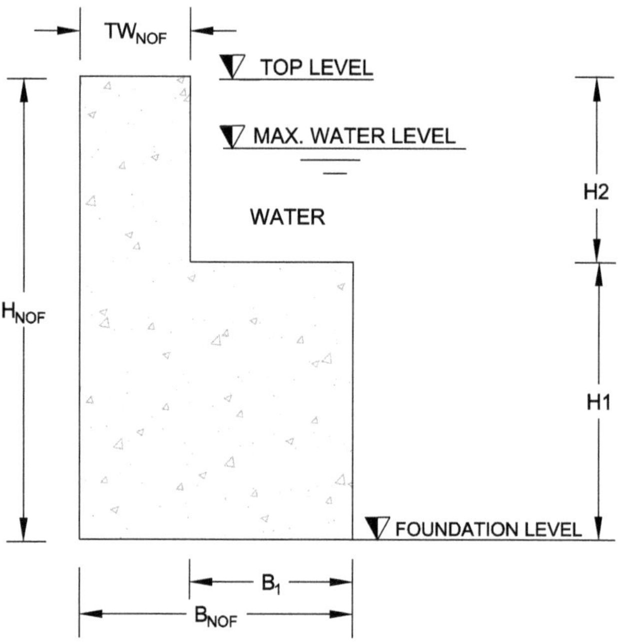

FIGURE 7.2A, 7.2B Typical Cross Sections of Concrete Weir and Non-Over Flow with Vertical D/S Face.

7.1.1.3 Section with Steps on Both Faces

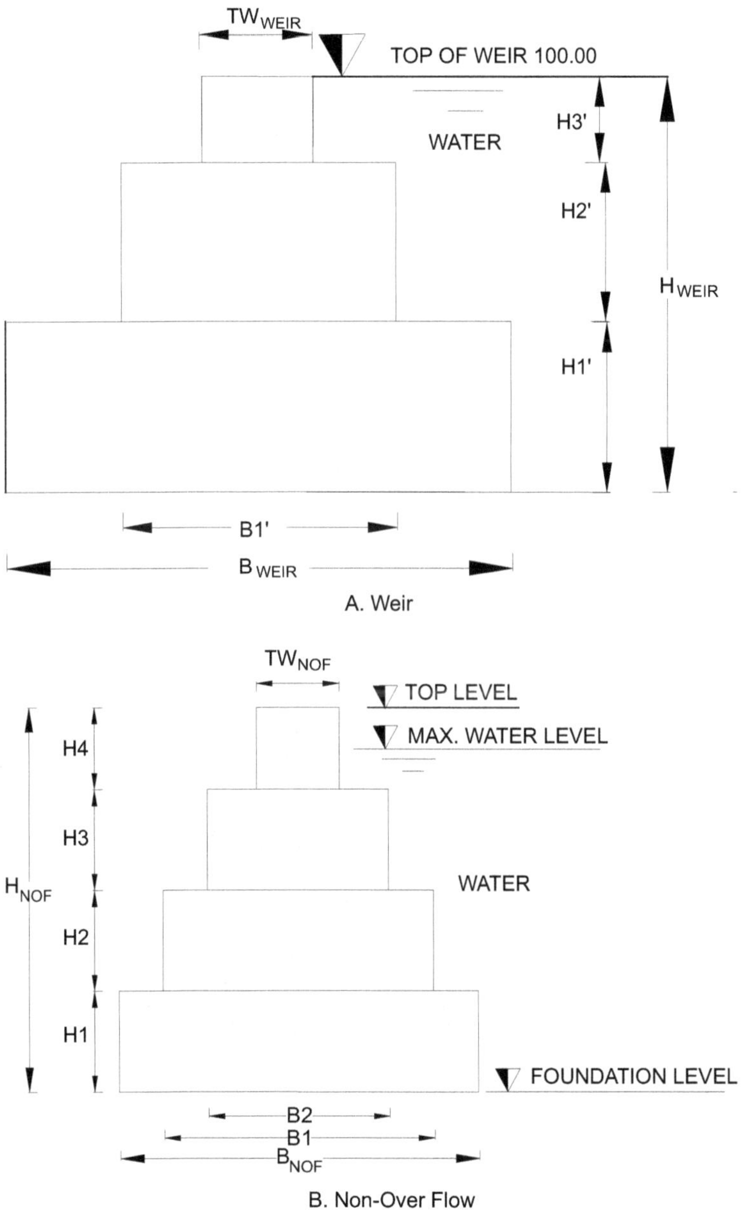

FIGURE 7.3A, 7.3B Typical Cross Sections of Concrete Weir and Non-Over Flow with Steps on Both Faces

7.1.2 Design Approach

For each concrete check dam of specified height, the stability analysis was carried out by the conventional gravity approach (IS:6512, 2019), discussed in Chapter 3 on design philosophy, for six sets of catchment areas and four sets of stream widths for each shape to develop the design tables. For a small width of streams, the design of concrete check dams has been limited to the head over crest to about 1.30 m.

The base width of the concrete check dam (non-over flow and overflow sections) has been so proportioned in the analysis that there is nearly zero stress near the heel under static loading; the factor of safety against overturning is equal to or more than 1.50.

7.2 DESIGN OF 1.50 M HIGH CONCRETE CHECK DAMS

7.2.1 Section with Vertical U/S Face

The tables developed after the rigorous design of the different components for 1.50 m high concrete check dams for different combinations are presented from Table 7.1 to Table 7.4.

TABLE 7.1
Design Results for Concrete Check Dam with Vertical U/S Face Height = 1.50 m and Stream Width = 5.0 m

Catchment Area, ha	50.00	100.00
A. WEIR		
Span of Weir W, m	4.00	4.00
Flood Discharge Q, m^3/s	6.54	11.00
Head Over Crest, m	0.92	1.31
Top Level, m	100.00	100.00
Foundation Level, m	98.50	98.50
Height of Weir H_{WEIR}, m	1.50	1.50
Top Width of Weir TW_{WEIR}, mm	500	500
Base Width of Weir B_{WEIR}, mm	1710	1870
Width B1', mm	1210	1370
D/S Slope – H_{DS}.: 1.0(V)	0.805	0.910
Pressure at Toe, t/m^2	3.68	3.72
Pressure at Heel, t/m^2	0.03	0.01
Factor of Safety against Overturning	1.57	1.55
Shear Friction Factor F	2.97	2.69
Total Volume of Concrete, m^3	6.63	7.11
B. NON-OVER FLOW PORTION		
Length, m	1.00	1.00
Top Level including Free Board, m	101.50	102.00
Foundation Level, m	98.50	98.50
Height of Non-Over Flow H_{NOF}, m	3.00	3.50
Height H1, mm	1500	1400
Height H2, mm	1500	2100
Top Width of Non-Over Flow TW_{NOF}, mm	500	590
Base Width of Non-Over Flow B_{NOF}, mm	1710	1870

(Continued)

TABLE 7.1 (Continued)

Catchment Area, ha	50.00	100.00
Width B1, mm	1210	1370
D/S Slope – H_{DS}: 1.0(V)	0.805	0.910
Pressure at Toe, t/m²	4.72	6.36
Pressure at Heel, t/m²	0.82	0.06
Factor of Safety against Overturning	1.74	1.59
Shear Friction Factor F	2.61	2.11
Total Volume of Concrete, m³	2.41	2.96

TABLE 7.2
Design Results for Concrete Check Dam with Vertical U/S Face Height = 1.50 m and Stream Width = 10.0 m

Catchment Area, ha	50.00	100.00	150.00	200.00
A. WEIR				
Span of Weir W, m	8.00	8.00	8.00	8.00
Flood Discharge Q, m³/s	6.54	11.00	14.91	18.50
Head Over Crest, m	0.58	0.82	1.01	1.16
Top Level, m	100.00	100.00	100.00	100.00
Foundation Level, m	98.50	98.50	98.50	98.50
Height of Weir H_{WEIR}, m	1.50	1.50	1.50	1.50
Top Width of Weir TW_{WEIR}, mm	500	500	500	500
Base Width of Weir B_{WEIR}, mm	1540	1670	1750	1820
Width B1', mm	1040	1170	1250	1320
D/S Slope – H_{DS}.: 1.0(V)	0.690	0.780	0.830	0.880
Pressure at Toe, t/m²	3.71	3.66	3.71	3.66
Pressure at Heel, t/m²	0.01	0.04	0.00	0.04
Factor of Safety against Overturning	1.57	1.58	1.56	1.56
Shear Friction Factor F	3.27	3.06	2.90	2.80
Total Volume of Concrete, m³	12.24	13.02	13.50	13.92
B. NON-OVER FLOW PORTION				
Length, m	2.00	2.00	2.00	2.00
Top Level including Free Board, m	101.10	101.40	101.60	101.80
Foundation Level, m	98.50	98.50	98.50	98.50
Height of Non-Over Flow H_{NOF}, m	2.60	2.90	3.10	3.30
Height H1, mm	1500	1500	1500	1500
Height H2, mm	1100	1400	1600	1800
Top Width of Non-Over Flow TW_{NOF}, mm	500	500	500	500
Base Width of Non-Over Flow B_{NOF}, mm	1540	1670	1750	1820
Width B1, mm	1040	1170	1250	1320
D/S Slope – H_{DS}: 1.0(V)	0.690	0.780	0.830	0.880
Pressure at Toe, t/m²	3.78	4.45	5.07	5.71
Pressure at Heel, t/m²	1.47	1.02	0.54	0.04
Factor of Safety against Overturning	1.92	1.78	1.68	1.59
Shear Friction Factor F	3.18	2.74	2.50	2.27
Total Volume of Concrete, m³	4.16	4.66	4.98	5.28

TABLE 7.3
Design Results for Concrete Check Dam with Vertical U/S Face Height = 1.50 m and Stream Width = 15.0 m

Catchment Area, ha	50.00	100.00	150.00	200.00	250.00	300.00
A. WEIR						
Span of Weir W, m	12.00	12.00	12.00	12.00	12.00	12.00
Flood Discharge Q, m^3/s	6.54	11.00	14.91	18.50	21.87	25.07
Head Over Crest, m	0.44	0.63	0.77	0.89	0.99	1.09
Top Level, m	100.00	100.00	100.00	100.00	100.00	100.00
Foundation Level, m	98.50	98.50	98.50	98.50	98.50	98.50
Height of Weir H_{WEIR}, m	1.50	1.50	1.50	1.50	1.50	1.50
Top Width of Weir TW_{WEIR}, mm	500	500	500	500	500	500
Base Width of Weir B_{WEIR}, mm	1470	1570	1640	1700	1750	1790
Width B1', mm	970	1070	1140	1200	1250	1290
D/S Slope – H_{DS}: 1.0(V)	0.645	0.710	0.760	0.800	0.830	0.860
Pressure at Toe, t/m^2	3.67	3.70	3.68	3.67	3.67	3.65
Pressure at Heel, t/m^2	0.06	0.01	0.02	0.03	0.03	0.06
Factor of Safety against Overturning	1.60	1.57	1.57	1.57	1.57	1.57
Shear Friction Factor F	3.44	3.22	3.09	2.99	2.92	2.85
Total Volume of Concrete, m^3	17.73	18.63	19.26	19.80	20.25	20.61
B. NON-OVER FLOW PORTION						
Length, m	3.00	3.00	3.00	3.00	3.00	3.00
Top Level including Free Board, m	101.00	101.20	101.40	101.50	101.60	101.70
Foundation Level, m	98.50	98.50	98.50	98.50	98.50	98.50
Height of Non-Over Flow H_{NOF}, m	2.50	2.70	2.90	3.00	3.10	3.20
Height H1, mm	1500	1500	1500	1500	1500	1500
Height H2, mm	1000	1200	1400	1500	1600	1700
Top Width of Non-Over Flow TW_{NOF}, mm	500	500	500	500	500	500
Base Width of Non-Over Flow B_{NOF}, mm	1470	1570	1640	1700	1750	1790
Width B1, mm	970	1070	1140	1200	1250	1290
D/S Slope – H_{DS}: 1.0(V)	0.645	0.710	0.760	0.800	0.830	0.860
Pressure at Toe, t/m^2	3.71	4.05	4.66	4.79	5.07	5.33
Pressure at Heel, t/m^2	1.52	1.30	0.87	0.76	0.54	0.35
Factor of Safety against Overturning	1.95	1.87	1.76	1.72	1.68	1.64
Shear Friction Factor F	3.31	3.00	2.70	2.60	2.50	2.38
Total Volume of Concrete, m^3	5.93	6.46	6.92	7.20	7.46	7.70

TABLE 7.4
Design Results for Concrete Check Dam with Vertical U/S Face Height = 1.50 m and Stream Width = 20.0 m

Catchment Area, ha	50.00	100.00	150.00	200.00	250.00	300.00
A. WEIR						
Span of Weir W, m	16.00	16.00	16.00	16.00	16.00	16.00
Flood Discharge Q, m^3/s	6.54	11.00	14.91	18.50	21.87	25.07

(Continued)

TABLE 7.4 (Continued)

Catchment Area, ha	50.00	100.00	150.00	200.00	250.00	300.00
Head Over Crest, m	0.37	0.52	0.64	0.73	0.82	0.90
Top Level, m	100.00	100.00	100.00	100.00	100.00	100.00
Foundation Level, m	98.50	98.50	98.50	98.50	98.50	98.50
Height of Weir H_{WEIR}, m	1.50	1.50	1.50	1.50	1.50	1.50
Top Width of Weir TW_{WEIR}, mm	500	500	500	500	500	500
Base Width of Weir B_{WEIR}, mm	1430	1510	1570	1620	1670	1700
Width B1', mm	930	1010	1070	1120	1170	1200
D/S Slope – H_{DS}: 1.0(V)	0.620	0.670	0.710	0.745	0.780	0.800
Pressure at Toe, t/m²	3.70	3.70	3.71	3.69	3.66	3.68
Pressure at Heel, t/m²	0.04	0.03	0.00	0.01	0.04	0.02
Factor of Safety against Overturning	1.60	1.58	1.57	1.57	1.58	1.56
Shear Friction Factor F	3.51	3.34	3.21	3.12	3.06	2.98
Total Volume of Concrete, m³	23.16	24.12	24.84	25.44	26.04	26.40
B. NON-OVER FLOW PORTION						
Length, m	4.00	4.00	4.00	4.00	4.00	4.00
Top Level including Free Board, m	100.90	101.10	101.20	101.30	101.40	101.50
Foundation Level, m	98.50	98.50	98.50	98.50	98.50	98.50
Height of Non-Over Flow H_{NOF}, m	2.40	2.60	2.70	2.80	2.90	3.00
Height H1, mm	1500	1500	1500	1500	1500	1500
Height H2, mm	900	1100	1200	1300	1400	1500
Top Width of Non-Over Flow TW_{NOF}, mm	500	500	500	500	500	500
Base Width of Non-Over Flow B_{NOF}, mm	1430	1510	1570	1620	1670	1700
Width B1, mm	930	1010	1070	1120	1170	1200
D/S Slope – H_{DS}: 1.0(V)	0.620	0.670	0.710	0.745	0.780	0.800
Pressure at Toe, t/m²	3.50	3.97	4.05	4.28	4.45	4.79
Pressure at Heel, t/m²	1.64	1.35	1.30	1.13	1.02	0.76
Factor of Safety against Overturning	2.00	1.89	1.87	1.81	1.78	1.72
Shear Friction Factor F	3.49	3.12	3.00	2.86	2.74	2.60
Total Volume of Concrete, m³	7.59	8.23	8.61	8.96	9.31	9.60

7.2.2 Section with Vertical D/S Face

The tables developed after the rigorous design of the different components for 1.50 m high concrete check dams for different combinations are presented from Table 7.5 to Table 7.8.

TABLE 7.5
Design Results for Concrete Check Dam with Vertical D/S Face Height = 1.50 m and Stream Width = 5.0 m

Catchment Area, ha	50.00	100.00
A. WEIR		
Span of Weir W, m	4.00	4.00
Flood Discharge Q, m^3/s	6.54	11.00
Head Over Crest, m	0.92	1.31
Top Level, m	100.00	100.00
Foundation Level, m	98.50	98.50
Height of Weir H_{WEIR}, m	1.50	1.50
Top Width of Weir TW_{WEIR}, mm	1820	1920
Base Width of Weir B_{WEIR}, mm	1820	1920
Pressure at Toe, t/m^2	7.22	7.75
Pressure at Heel, t/m^2	0.00	0.00
Factor of Safety against Overturning	2.17	2.18
Shear Friction Factor F	3.61	3.20
Total Volume of Concrete, m^3	10.92	11.52
B. NON-OVER FLOW PORTION		
Length, m	1.00	1.00
Top Level including Free Board, m	101.50	102.00
Foundation Level, m	98.50	98.50
Height of Non-Over Flow H_{NOF}, m	3000	3500
Height H1, mm	2400	2700
Height H2, mm	600	800
Top Width of Non-Over Flow TW_{NOF}, mm	1000	1000
Base Width of Non-Over Flow B_{NOF}, mm	2230	2260
Pressure at Toe, t/m^2	11.96	14.09
Pressure at Heel, t/m^2	0.00	0.01
Factor of Safety against Overturning	2.72	2.53
Shear Friction Factor F	4.20	3.21
Total Volume of Concrete, m^3	5.95	6.90

TABLE 7.6
Design Results for Concrete Check Dam with Vertical D/S Face Height = 1.50 m and Stream Width = 10.0 m

Catchment Area, ha	50.00	100.00	150.00	200.00
A. WEIR				
Span of Weir W, m	8.00	8.00	8.00	8.00
Flood Discharge Q, m^3/s	6.54	11.00	14.91	18.50
Head Over Crest, m	0.58	0.82	1.01	1.16
Top Level, m	100.00	100.00	100.00	100.00
Foundation Level, m	98.50	98.50	98.50	98.50
Height of Weir H_{WEIR}, m	1.50	1.50	1.50	1.50

(Continued)

TABLE 7.6 (Continued)

Catchment Area, ha	50.00	100.00	150.00	200.00
Top Width of Weir TW_{WEIR}, mm	1700	1790	1850	1890
Base Width of Weir B_{WEIR}, mm	1700	1790	1850	1890
Pressure at Toe, t/m^2	6.78	7.09	7.33	7.53
Pressure at Heel, t/m^2	0.00	0.01	0.02	0.02
Factor of Safety against Overturning	2.16	2.17	2.18	2.18
Shear Friction Factor F	4.05	3.73	3.51	3.36
Total Volume of Concrete, m^3	20.40	21.48	22.20	22.68
B. NON-OVER FLOW PORTION				
Length, m	2.00	2.00	2.00	2.00
Top Level including Free Board, m	101.10	101.40	101.60	101.80
Foundation Level, m	98.50	98.50	98.50	98.50
Height of Non-Over Flow H_{NOF}, m	2600	2900	3100	3300
Height H1, mm	2080	2320	2480	2640
Height H2, mm	520	580	620	660
Top Width of Non-Over Flow TW_{NOF}, mm	1000	1000	1000	1000
Base Width of Non-Over Flow B_{NOF}, mm	2060	2190	2280	2360
Pressure at Toe, t/m^2	10.18	11.51	12.37	13.26
Pressure at Heel, t/m^2	0.01	0.01	0.03	0.02
Factor of Safety against Overturning	2.61	2.70	2.76	2.80
Shear Friction Factor F	5.04	4.38	4.04	3.74
Total Volume of Concrete, m^3	9.61	11.32	12.55	13.78

TABLE 7.7
Design Results for Concrete Check Dam with Vertical D/S Face Height = 1.50 m and Stream Width = 15.0 m

Catchment Area, ha	50.00	100.00	150.00	200.00	250.00	300.00
A. WEIR						
Span of Weir W, m	12.00	12.00	12.00	12.00	12.00	12.00
Flood Discharge Q, m^3/s	6.54	11.00	14.91	18.50	21.87	25.07
Head Over Crest, m	0.44	0.63	0.77	0.89	0.99	1.09
Top Level, m	100.00	100.00	100.00	100.00	100.00	100.00
Foundation Level, m	98.50	98.50	98.50	98.50	98.50	98.50
Height of Weir H_{WEIR}, m	1.50	1.50	1.50	1.50	1.50	1.50
Top Width of Weir TW_{WEIR}, mm	1640	1720	1770	1810	1840	1870
Base Width of Weir B_{WEIR}, mm	1640	1720	1770	1810	1840	1870
Pressure at Toe, t/m^2	6.59	6.84	7.03	7.19	7.32	7.44
Pressure at Heel, t/m^2	0.00	0.00	0.00	0.00	0.00	0.01
Factor of Safety against Overturning	2.16	2.16	2.17	2.17	2.17	2.18
Shear Friction Factor F	4.27	3.98	3.79	3.64	3.53	3.42
Total Volume of Concrete, m^3	29.52	30.96	31.86	32.58	33.12	33.66

(Continued)

TABLE 7.7 (Continued)

Catchment Area, ha	50.00	100.00	150.00	200.00	250.00	300.00
B. NON-OVER FLOW PORTION						
Length, m	3.00	3.00	3.00	3.00	3.00	3.00
Top Level including Free Board, m	101.00	101.20	101.40	101.50	101.60	101.70
Foundation Level, m	98.50	98.50	98.50	98.50	98.50	98.50
Height of Non-Over Flow H_{NOF}, m	2500	2700	2900	3000	3100	3200
Height H1, mm	2000	2160	2320	2400	2480	2560
Height H2, mm	500	540	580	600	620	640
Top Width of Non-Over Flow TW_{NOF}, mm	1000	1000	1000	1000	1000	1000
Base Width of Non-Over Flow B_{NOF}, mm	2020	2110	2190	2230	2280	2320
Pressure at Toe, t/m^2	9.72	10.60	11.51	11.96	12.37	12.82
Pressure at Heel, t/m^2	0.02	0.02	0.01	0.00	0.03	0.02
Factor of Safety against Overturning	2.58	2.65	2.70	2.72	2.76	2.78
Shear Friction Factor F	5.31	4.81	4.38	4.20	4.04	3.89
Total Volume of Concrete, m^3	13.62	15.29	16.98	17.86	18.82	19.74

TABLE 7.8
Design Results for Concrete Check Dam with Vertical D/S Face Height = 1.50 m and Stream Width = 20.0 m

Catchment Area, ha	50.00	100.00	150.00	200.00	250.00	300.00
A. WEIR						
Span of Weir W, m	16.00	16.00	16.00	16.00	16.00	16.00
Flood Discharge Q, m^3/s	6.54	11.00	14.91	18.50	21.87	25.07
Head Over Crest, m	0.37	0.52	0.64	0.73	0.82	0.90
Top Level, m	100.00	100.00	100.00	100.00	100.00	100.00
Foundation Level, m	98.50	98.50	98.50	98.50	98.50	98.50
Height of Weir H_{WEIR}, m	1.50	1.50	1.50	1.50	1.50	1.50
Top Width of Weir TW_{WEIR}, mm	1610	1680	1730	1760	1790	1820
Base Width of Weir B_{WEIR}, mm	1610	1680	1730	1760	1790	1820
Pressure at Toe, t/m^2	6.49	6.68	6.84	6.96	7.09	7.17
Pressure at Heel, t/m^2	0.01	0.01	0.02	0.01	0.01	0.03
Factor of Safety against Overturning	2.16	2.16	2.17	2.17	2.17	2.18
Shear Friction Factor F	4.40	4.16	3.98	3.85	3.73	3.64
Total Volume of Concrete, m^3	38.64	40.32	41.52	42.24	42.96	43.68
B. NON-OVER FLOW PORTION						
Length, m	4.00	4.00	4.00	4.00	4.00	4.00
Top Level including Free Board, m	100.90	101.10	101.20	101.30	101.40	101.50
Foundation Level, m	98.50	98.50	98.50	98.50	98.50	98.50
Height of Non-Over Flow H_{NOF}, m	2400	2600	2700	2800	2900	3000
Height H1, mm	1920	2080	2160	2240	2320	2400
Height H2, mm	480	520	540	560	580	600
Top Width of Non-Over Flow TW_{NOF}, mm	1000	1000	1000	1000	1000	1000
Base Width of Non-Over Flow B_{NOF}, mm	1970	2060	2110	2150	2190	2230
Pressure at Toe, t/m^2	9.29	10.18	10.60	11.05	11.51	11.96
Pressure at Heel, t/m^2	0.01	0.01	0.02	0.01	0.01	0.00

(Continued)

TABLE 7.8 (Continued)

Catchment Area, ha	50.00	100.00	150.00	200.00	250.00	300.00
Factor of Safety against Overturning	2.55	2.61	2.65	2.67	2.70	2.72
Shear Friction Factor F	5.58	5.04	4.81	4.59	4.38	4.20
Total Volume of Concrete, m^3	17.05	19.22	20.39	21.50	22.64	23.81

7.2.3 Section with Steps on Both Faces

The tables developed after the rigorous design of the different components for 1.50 m high concrete check dams for different combinations are presented from Table 7.9 to Table 7.12.

TABLE 7.9
Design Results for Concrete Check Dam with Steps on Both Faces Height = 1.50 m and Stream Width = 5.0 m

Catchment Area, ha	50.00	100.00
A. WEIR		
Span of Weir W, m	4.00	4.00
Flood Discharge Q, m^3/s	6.54	11.00
Head Over Crest, m	0.92	1.31
Top Level, m	100.00	100.00
Foundation Level, m	98.50	98.50
Height of Weir H_{WEIR}, m	1.50	1.50
Size of Concrete Steps (W, H), m		
Base Width B_{WEIR}, Thickness H1'	2.14, 1.00	2.36, 1.00
Top Width TW_{WEIR}, Thickness H2'	1.14, 0.50	1.36, 0.50
Pressure at Toe, t/m^2	5.90	6.31
Pressure at Heel, t/m^2	0.00	0.02
Factor of Safety against Overturning	2.11	2.18
Shear Friction Factor F	3.01	2.74
Total Volume of Concrete, m^3	10.84	12.16
B. NON-OVER FLOW PORTION		
Length, m	1.00	1.00
Top Level including Free Board, m	101.50	102.00
Foundation Level, m	98.50	98.50
Height of Non-Over Flow H_{NOF}, m	3.00	3.50
Size of Concrete Steps (W, H), m		
Base Width B_{NOF}, Thickness H1	1.76, 1.00	2.06, 1.00
Width B1, Thickness H2	0.76, 1.00	1.06, 1.00
Top Width TW_{NOF}, Thickness H3	0.50, 1.00	0.50, 1.50
Pressure at Toe, t/m^2	8.72	10.12
Pressure at Heel, t/m^2	0.03	0.00
Factor of Safety against Overturning	2.06	2.06
Shear Friction Factor F	4.01	3.59
Total Volume of Concrete, m^3	3.02	3.87

TABLE 7.10
Design Results for Concrete Check Dam with Steps on Both Faces Height = 1.50 m and Stream Width = 10.0 m

Catchment Area, ha	50.00	100.00	150.00	200.00
A. WEIR				
Span of Weir W, m	8.00	8.00	8.00	8.00
Flood Discharge Q, m^3/s	6.54	11.00	14.91	18.50
Head Over Crest, m	0.58	0.82	1.01	1.16
Top Level, m	100.00	100.00	100.00	100.00
Foundation Level, m	98.50	98.50	98.50	98.50
Height of Weir H_{WEIR}, m	1.50	1.50	1.50	1.50
Size of Concrete Steps (W, H), m				
Base Width B_{WEIR}, Thickness H1'	1.89, 1.00	2.07, 1.00	2.20, 1.00	2.28, 1.00
Top Width TW_{WEIR}, Thickness H2'	0.89, 0.50	1.07, 0.50	1.20, 0.50	1.28, 0.50
Pressure at Toe, t/m^2	5.52	5.78	5.99	6.15
Pressure at Heel, t/m^2	0.00	0.01	0.02	0.02
Factor of Safety against Overturning	2.04	2.10	2.13	2.16
Shear Friction Factor F	3.26	3.08	2.95	2.84
Total Volume of Concrete, m^3	18.68	20.84	22.40	23.36
B. NON-OVER FLOW PORTION				
Length, m	2.00	2.00	2.00	2.00
Top Level including Free Board, m	101.10	101.40	101.60	101.80
Foundation Level, m	98.50	98.50	98.50	98.50
Height of Non-Over Flow H_{NOF}, m	2.60	2.90	3.10	3.30
Size of Concrete Steps (W, H), m				
Base Width B_{NOF}, Thickness H1	1.55, 1.00	1.70, 1.00	1.82, 1.00	1.94, 1.00
Width B1, Thickness H2	0.50, 1.60	0.70, 1.00	0.82, 1.00	0.94, 1.00
Top Width TW_{NOF}, Thickness H3		0.50, 0.90	0.5, 1.10	0.50, 1.30
Pressure at Toe, t/m^2	7.35	8.45	9.03	9.54
Pressure at Heel, t/m^2	0.07	0.00	0.00	0.05
Factor of Safety against Overturning	2.04	2.04	2.05	2.06
Shear Friction Factor F	4.52	4.12	3.92	3.75
Total Volume of Concrete, m^3	4.70	5.70	6.38	7.06

TABLE 7.11
Design Results for Concrete Check Dam with Steps on Both Faces Height = 1.50 m and Stream Width = 15.0 m

Catchment Area, ha	50.00	100.00	150.00	200.00	250.00	300.00
A. WEIR						
Span of Weir W, m	12.00	12.00	12.00	12.00	12.00	12.00
Flood Discharge Q, m^3/s	6.54	11.00	14.91	18.50	21.87	25.07
Head Over Crest, m	0.44	0.63	0.77	0.89	0.99	1.09
Top Level, m	100.00	100.00	100.00	100.00	100.00	100.00
Foundation Level, m	98.50	98.50	98.50	98.50	98.50	98.50

(Continued)

TABLE 7.11 (Continued)

Catchment Area, ha	50.00	100.00	150.00	200.00	250.00	300.00
Height of Weir H_{WEIR}, m	1.50	1.50	1.50	1.50	1.50	1.50
Size of Concrete Steps (W, H), m						
Base Width B_{WEIR}, Thickness H1'	1.78, 1.00	1.94, 1.00	2.03, 1.00	2.12, 1.00	2.18, 1.00	2.24, 1.00
Top Width TW_{WEIR}, Thickness H2'	0.78, 0.50	0.94, 0.50	1.03, 0.50	1.12, 0.50	1.18, 0.50	1.24, 0.50
Pressure at Toe, t/m²	5.36	5.56	5.72	5.85	5.96	6.06
Pressure at Heel, t/m²	0.00	0.02	0.02	0.02	0.01	0.03
Factor of Safety against Overturning	2.01	2.06	2.09	2.11	2.13	2.15
Shear Friction Factor F	3.38	3.23	3.11	3.03	2.95	2.89
Total Volume of Concrete, m³	26.04	28.92	30.54	32.16	33.24	34.32
B. NON-OVER FLOW PORTION						
Length, m	3.00	3.00	3.00	3.00	3.00	3.00
Top Level including Free Board, m	101.00	101.20	101.40	101.50	101.60	101.70
Foundation Level, m	98.50	98.50	98.50	98.50	98.50	98.50
Height of Non-Over Flow H_{NOF}, m	2.50	2.70	2.90	3.00	3.10	3.20
Size of Concrete Steps (W, H), m						
Base Width B_{NOF}, Thickness H1	1.49, 1.00	1.59, 1.00	1.70, 1.00	1.76, 1.00	1.82, 1.00	1.88, 1.00
Width B1, Thickness H2	0.50, 1.50	0.59, 1.00	0.70, 1.00	0.76, 1.00	0.82, 1.00	0.88, 1.00
Top Width TW_{NOF}, Thickness H3		0.50, 0.70	0.50, 0.90	0.50, 1.00	0.50, 1.10	0.50, 1.20
Pressure at Toe, t/m²	7.16	7.80	8.45	8.72	9.03	9.29
Pressure at Heel, t/m²	0.05	0.04	0.00	0.03	0.00	0.03
Factor of Safety against Overturning	2.03	2.04	2.04	2.06	2.05	2.06
Shear Friction Factor F	4.69	4.37	4.12	4.01	3.92	3.83
Total Volume of Concrete, m³	6.72	7.59	8.55	9.06	9.57	10.08

TABLE 7.12
Design Results for Concrete Check Dam with Steps on Both Faces Height = 1.50 m and Stream Width = 20.0 m

Catchment Area, ha	50.00	100.00	150.00	200.00	250.00	300.00
A. WEIR						
Span of Weir W, m	16.00	16.00	16.00	16.00	16.00	16.00
Flood Discharge Q, m³/s	6.54	11.00	14.91	18.50	21.87	25.07
Head Over Crest, m	0.37	0.52	0.64	0.73	0.82	0.90
Top Level, m	100.00	100.00	100.00	100.00	100.00	100.00
Foundation Level, m	98.50	98.50	98.50	98.50	98.50	98.50
Height of Weir H_{WEIR}, m	1.50	1.50	1.50	1.50	1.50	1.50
Size of Concrete Steps (W, H), m						
Base Width B_{WEIR}, Thickness H1'	1.73, 1.00	1.85, 1.00	1.94, 1.00	2.01, 1.00	2.07, 1.00	2.12, 1.00
Top Width TW_{WEIR}, Thickness H2'	0.73, 0.50	0.85, 0.50	0.94, 0.50	1.01, 0.50	1.07, 0.50	1.12, 0.50
Pressure at Toe, t/m²	5.26	5.44	5.58	5.68	5.78	5.87
Pressure at Heel, t/m²	0.02	0.01	0.00	0.01	0.01	0.01

(Continued)

TABLE 7.12 (Continued)

Catchment Area, ha	50.00	100.00	150.00	200.00	250.00	300.00
Factor of Safety against Overturning	2.00	2.04	2.06	2.08	2.10	2.11
Shear Friction Factor F	3.46	3.32	3.22	3.15	3.08	3.02
Total Volume of Concrete, m^3	33.52	36.40	38.56	40.24	41.68	42.88
B. NON-OVER FLOW PORTION						
Length, m	4.00	4.00	4.00	4.00	4.00	4.00
Top Level including Free Board, m	100.90	101.10	101.20	101.30	101.40	101.50
Foundation Level, m	98.50	98.50	98.50	98.50	98.50	98.50
Height of Non-Over Flow H_{NOF}, m	2.40	2.60	2.70	2.80	2.90	3.00
Size of Concrete Steps (W, H), m						
Base Width B_{NOF}, Thickness H1	1.43, 1.00	1.54, 1.00	1.58, 1.00	1.64, 1.00	1.70, 1.00	1.76, 1.00
Width B1, Thickness H2	0.50, 1.40	0.54, 1.60	0.58, 1.70	0.64, 1.80	0.70, 1.00	0.76, 1.00
Top Width TW_{NOF}, Thickness H3					0.50, 0.90	0.50, 1.00
Pressure at Toe, t/m^2	6.96	7.55	7.93	8.37	8.45	8.72
Pressure at Heel, t/m^2	0.04	0.03	0.05	0.07	0.00	0.03
Factor of Safety against Overturning	2.04	2.04	2.05	2.08	2.04	2.06
Shear Friction Factor F	4.85	4.53	4.37	4.28	4.12	4.01
Total Volume of Concrete, m^3	8.52	9.62	10.26	11.17	11.40	12.08

7.3 DESIGN OF 2.00 M HIGH CONCRETE CHECK DAMS

7.3.1 Section with Vertical U/S Face

The tables developed after the rigorous design of the different components for 2.00 m high concrete check dams for different combinations are presented from Table 7.13 to Table 7.16.

TABLE 7.13
Design Results for Concrete Check Dam with Vertical U/S Face
Height = 2.00 m and Stream Width = 5.0 m

Catchment Area, ha	50.00	100.00
A. WEIR		
Span of Weir W, m	4.00	4.00
Flood Discharge Q, m^3/s	6.54	11.00
Head Over Crest, m	0.92	1.31
Top Level, m	100.00	100.00
Foundation Level, m	98.00	98.00
Height of Weir H_{WEIR}, m	2.00	2.00
Top Width of Weir TW_{WEIR}, mm	500	500
Base Width of Weir B_{WEIR}, mm	2210	2420
Width B1', mm	1710	1920
D/S Slope – H_{DS}: 1.0(V)	0.855	0.960

(Continued)

TABLE 7.13 (Continued)

Catchment Area, ha	50.00	100.00
Pressure at Toe, t/m^2	4.40	4.36
Pressure at Heel, t/m^2	0.01	0.03
Factor of Safety against Overturning	1.53	1.53
Shear Friction Factor F	2.54	2.36
Total Volume of Concrete, m^3	10.84	11.68
Length of D/S Apron Floor $L_{D/S\ APRON}$, m	5.80	5.60
B. NON-OVER FLOW PORTION		
Length, m	1.00	1.00
Top Level including Free Board, m	101.50	102.00
Foundation Level, m	98.00	98.00
Height of Non-Over Flow H_{NOF}, m	3.50	4.00
Height H1, mm	2000	1938
Height H2, mm	1500	2063
Top Width of Non-Over Flow TW_{NOF}, mm	500	560
Base Width of Non-Over Flow B_{NOF}, mm	2210	2420
Width B1, mm	1710	1920
D/S Slope – H_{DS}: 1.0(V)	0.855	0.960
Pressure at Toe, t/m^2	5.09	6.32
Pressure at Heel, t/m^2	0.74	0.03
Factor of Safety against Overturning	1.66	1.55
Shear Friction Factor F	2.32	1.96
Total Volume of Concrete, m^3	3.46	4.04
C. WING WALLS (ABUTMENTS) ON BOTH BANKS		
Length in U/S $L_{U/SABUT}$, m	1.50	1.50
Length in D/S $L_{D/SABUT}$, m	2.71	2.92
Top Level including Free Board, m	101.50	102.00
Foundation Level, m	98.00	98.00
Total Height H_{ABUT}, m	3.50	4.00
D. D/S WING WALLS ON BOTH BANKS		
Length in D/S $L_{D/SWING}$, m	3.09	2.68
Top Level including Free Board, m	99.52	99.95
Foundation Level, m	98.00	98.00
Total Height $H_{D/SWING}$, m	1.52	1.95

TABLE 7.14
Design Results for Concrete Check Dam with Vertical U/S Face Height = 2.00 m and Stream Width = 10.0 m

Catchment Area, ha	50.00	100.00	150.00	200.00
A. WEIR				
Span of Weir W, m	8.00	8.00	8.00	8.00
Flood Discharge Q, m^3/s	6.54	11.00	14.91	18.50
Head Over Crest, m	0.58	0.82	1.01	1.16

(Continued)

TABLE 7.14 (Continued)

Catchment Area, ha	50.00	100.00	150.00	200.00
Top Level, m	100.00	100.00	100.00	100.00
Foundation Level, m	98.00	98.00	98.00	98.00
Height of Weir H_{WEIR}, m	2.00	2.00	2.00	2.00
Top Width of Weir TW_{WEIR}, mm	500	500	500	500
Base Width of Weir B_{WEIR}, mm	2010	2150	2260	2340
Width B1', mm	1510	1650	1760	1840
D/S Slope – $H_{DS'}$: 1.0(V)	0.755	0.825	0.880	0.920
Pressure at Toe, t/m^2	4.41	4.40	4.40	4.38
Pressure at Heel, t/m^2	0.02	0.02	0.01	0.02
Factor of Safety against Overturning	1.54	1.53	1.53	1.53
Shear Friction Factor F	2.73	2.59	2.49	2.42
Total Volume of Concrete, m^3	20.08	21.20	22.08	22.72
Length of D/S Apron Floor $L_{D/S\ APRON}$, m	6.00	5.90	5.80	5.70
B. NON-OVER FLOW PORTION				
Length, m	2.00	2.00	2.00	2.00
Top Level including Free Board, m	101.10	101.40	101.60	101.80
Foundation Level, m	98.00	98.00	98.00	98.00
Height of Non-Over Flow H_{NOF}, m	3.10	3.40	3.60	3.80
Height H1, mm	2000	2000	2000	2000
Height H2, mm	1100	1400	1600	1800
Top Width of Non-Over Flow TW_{NOF}, mm	500	500	500	500
Base Width of Non-Over Flow B_{NOF}, mm	2010	2150	2260	2340
Width B1, mm	1510	1650	1760	1840
D/S Slope – H_{DS}: 1.0(V)	0.755	0.825	0.880	0.920
Pressure at Toe, t/m^2	4.34	4.95	5.30	5.88
Pressure at Heel, t/m^2	1.27	0.84	0.57	0.11
Factor of Safety against Overturning	1.77	1.68	1.63	1.56
Shear Friction Factor F	2.70	2.40	2.24	2.08
Total Volume of Concrete, m^3	6.12	6.70	7.12	7.48
C. WING WALLS (ABUTMENTS) ON BOTH BANKS				
Length in U/S $L_{U/SABUT}$, m	1.50	1.50	1.50	1.50
Length in D/S $L_{D/SABUT}$, m	2.51	2.65	2.76	2.84
Top Level including Free Board, m	101.10	101.40	101.60	101.80
Foundation Level, m	98.00	98.00	98.00	98.00
Total Height H_{ABUT}, m	3.10	3.40	3.60	3.80
D. D/S WING WALLS ON BOTH BANKS				
Length in D/S $L_{D/SWING}$, m	3.49	3.25	3.04	2.86
Top Level including Free Board, m	99.14	99.41	99.62	99.79
Foundation Level, m	98.00	98.00	98.00	98.00
Total Height $H_{D/SWING}$, m	1.14	1.41	1.62	1.79

TABLE 7.15
Design Results for Concrete Check Dam with Vertical U/S Face Height = 2.00 m and Stream Width = 15.0 m

Catchment Area, ha	50.00	100.00	150.00	200.00	250.00	300.00
A. WEIR						
Span of Weir W, m	12.00	12.00	12.00	12.00	12.00	12.00
Flood Discharge Q, m^3/s	6.54	11.00	14.91	18.50	21.87	25.07
Head Over Crest, m	0.44	0.63	0.77	0.89	0.99	1.09
Top Level, m	100.00	100.00	100.00	100.00	100.00	100.00
Foundation Level, m	98.00	98.00	98.00	98.00	98.00	98.00
Height of Weir H_{WEIR}, m	2.00	2.00	2.00	2.00	2.00	2.00
Top Width of Weir TW_{WEIR}, mm	500	500	500	500	500	500
Base Width of Weir B_{WEIR}, mm	1920	2040	2120	2200	2250	2300
Width B1', mm	1420	1540	1620	1700	1750	1800
D/S Slope – H_{DS}: 1.0(V)	0.710	0.770	0.810	0.850	0.875	0.900
Pressure at Toe, t/m^2	4.42	4.41	4.42	4.37	4.40	4.39
Pressure at Heel, t/m^2	0.03	0.02	0.00	0.04	0.00	0.01
Factor of Safety against Overturning	1.55	1.54	1.53	1.53	1.53	1.53
Shear Friction Factor F	2.83	2.70	2.61	2.56	2.50	2.45
Total Volume of Concrete, m^3	29.04	30.48	31.44	32.40	33.00	33.60
Length of D/S Apron Floor $L_{D/S\ APRON}$, m	6.10	6.00	5.90	5.80	5.80	5.70
B. NON-OVER FLOW PORTION						
Length, m	3.00	3.00	3.00	3.00	3.00	3.00
Top Level including Free Board, m	101.00	101.20	101.40	101.50	101.60	101.70
Foundation Level, m	98.00	98.00	98.00	98.00	98.00	98.00
Height of Non-Over Flow H_{NOF}, m	3.00	3.20	3.40	3.50	3.60	3.70
Height H1, mm	2000	2000	2000	2000	2000	2000
Height H2, mm	1000	1200	1400	1500	1600	1700
Top Width of Non-Over Flow TW_{NOF}, mm	500	500	500	500	500	500
Base Width of Non-Over Flow B_{NOF}, mm	1920	2040	2120	2200	2250	2300
Width B1, mm	1420	1540	1620	1700	1750	1800
D/S Slope – H_{DS}: 1.0(V)	0.710	0.770	0.810	0.850	0.875	0.900
Pressure at Toe, t/m^2	4.34	4.60	5.13	5.14	5.37	5.59
Pressure at Heel, t/m^2	1.26	1.09	0.70	0.70	0.52	0.34
Factor of Safety against Overturning	1.78	1.73	1.66	1.65	1.62	1.59
Shear Friction Factor F	2.75	2.57	2.37	2.31	2.23	2.16
Total Volume of Concrete, m^3	8.76	9.42	9.96	10.35	10.65	10.95
C. WING WALLS (ABUTMENTS) ON BOTH BANKS						
Length in U/S $L_{U/SABUT}$, m	1.50	1.50	1.50	1.50	1.50	1.50
Length in D/S $L_{D/SABUT}$, m	2.42	2.54	2.62	2.70	2.75	2.80
Top Level including Free Board, m	101.00	101.20	101.40	101.50	101.60	101.70

(Continued)

TABLE 7.15 (Continued)

Catchment Area, ha	50.00	100.00	150.00	200.00	250.00	300.00
Foundation Level, m	98.00	98.00	98.00	98.00	98.00	98.00
Total Height H_{ABUT}, m	3.00	3.20	3.40	3.50	3.60	3.70
D. D/S WING WALLS ON BOTH BANKS						
Length in D/S $L_{D/SWING}$, m	3.68	3.46	3.28	3.10	3.05	2.90
Top Level including Free Board, m	98.99	99.20	99.35	99.49	99.60	99.71
Foundation Level, m	98.00	98.00	98.00	98.00	98.00	98.00
Total Height $H_{D/SWING}$, m	0.99	1.20	1.35	1.49	1.60	1.71

TABLE 7.16

Design Results for Concrete Check Dam with Vertical U/S Face Height = 2.00 m and Stream Width = 20.0 m

Catchment Area, ha	50.00	100.00	150.00	200.00	250.00	300.00
A. WEIR						
Span of Weir W, m	16.00	16.00	16.00	16.00	16.00	16.00
Flood Discharge Q, m^3/s	6.54	11.00	14.91	18.50	21.87	25.07
Head Over Crest, m	0.37	0.52	0.64	0.73	0.82	0.90
Top Level, m	100.00	100.00	100.00	100.00	100.00	100.00
Foundation Level, m	98.00	98.00	98.00	98.00	98.00	98.00
Height of Weir H_{WEIR}, m	2.00	2.00	2.00	2.00	2.00	2.00
Top Width of Weir TW_{WEIR}, mm	500	500	500	500	500	500
Base Width of Weir B_{WEIR}, mm	1870	1970	2040	2100	2150	2200
Width B1', mm	1370	1470	1540	1600	1650	1700
D/S Slope – $H_{DS'}$: 1.0(V)	0.685	0.735	0.770	0.800	0.825	0.850
Pressure at Toe, t/m^2	4.46	4.42	4.43	4.40	4.40	4.39
Pressure at Heel, t/m^2	0.01	0.02	0.00	0.02	0.02	0.02
Factor of Safety against Overturning	1.55	1.54	1.54	1.54	1.53	1.53
Shear Friction Factor F	2.87	2.77	2.69	2.64	2.59	2.55
Total Volume of Concrete, m^3	37.92	39.52	40.64	41.60	42.40	43.20
Length of D/S Apron Floor $L_{D/S\ APRON}$, m	6.20	6.10	6.00	5.90	5.90	5.80
B. NON-OVER FLOW PORTION						
Length, m	4.00	4.00	4.00	4.00	4.00	4.00
Top Level including Free Board, m	100.90	101.10	101.20	101.30	101.40	101.50
Foundation Level, m	98.00	98.00	98.00	98.00	98.00	98.00
Height of Non-Over Flow H_{NOF}, m	2.90	3.10	3.20	3.30	3.40	3.50
Height H1, mm	2000	2000	2000	2000	2000	2000
Height H2, mm	900	1100	1200	1300	1400	1500
Top Width of Non-Over Flow TW_{NOF}, mm	500	500	500	500	500	500
Base Width of Non-Over Flow B_{NOF}, mm	1870	1970	2040	2100	2150	2200
Width B1, mm	1370	1470	1540	1600	1650	1700
D/S Slope – H_{DS}: 1.0(V)	0.685	0.735	0.770	0.800	0.825	0.850
Pressure at Toe, t/m^2	4.22	4.54	4.60	4.72	4.95	5.14

(Continued)

TABLE 7.16 (Continued)

Catchment Area, ha	50.00	100.00	150.00	200.00	250.00	300.00
Pressure at Heel, t/m²	1.32	1.13	1.09	1.02	0.84	0.70
Factor of Safety against Overturning	1.80	1.75	1.73	1.72	1.68	1.65
Shear Friction Factor F	2.87	2.65	2.57	2.49	2.40	2.31
Total Volume of Concrete, m³	11.28	12.08	12.56	13.00	13.40	13.80
C. WING WALLS (ABUTMENTS) ON BOTH BANKS						
Length in U/S $L_{U/SABUT}$, m	1.50	1.50	1.50	1.50	1.50	1.50
Length in D/S $L_{D/SABUT}$, m	2.37	2.47	2.54	2.60	2.65	2.70
Top Level including Free Board, m	100.90	101.10	101.20	101.30	101.40	101.50
Foundation Level, m	98.00	98.00	98.00	98.00	98.00	98.00
Total Height H_{ABUT}, m	2.90	3.10	3.20	3.30	3.40	3.50
D. D/S WING WALLS ON BOTH BANKS						
Length in D/S $L_{D/SWING}$, m	3.83	3.63	3.46	3.30	3.25	3.10
Top Level including Free Board, m	98.91	99.08	99.21	99.31	99.41	99.50
Foundation Level, m	98.00	98.00	98.00	98.00	98.00	98.00
Total Height $H_{D/SWING}$, m	0.91	1.08	1.21	1.31	1.41	1.50

7.3.2 Section with Vertical D/S Face

The tables developed after the rigorous design of the different components for 2.00 m high concrete check dams for different combinations are presented from Table 7.17 to Table 7.20.

TABLE 7.17
Design Results for Concrete Check Dam with Vertical D/S Face Height = 2.00 m and Stream Width = 5.0 m

Catchment Area, ha	50.00	100.00
A. WEIR		
Span of Weir W, m	4.00	4.00
Flood Discharge Q, m³/s	6.54	11.00
Head Over Crest, m	0.92	1.31
Top Level, m	100.00	100.00
Foundation Level, m	98.00	98.00
Height of Weir H_{WEIR}, m	2.00	2.00
Top Width of Weir TW_{WEIR}, mm	2330	2450
Base Width of Weir B_{WEIR}, mm	2330	2450
Pressure at Toe, t/m²	9.20	9.74
Pressure at Heel, t/m²	0.02	0.01
Factor of Safety against Overturning	2.17	2.17
Shear Friction Factor F	3.17	2.87

(Continued)

TABLE 7.17 (Continued)

Catchment Area, ha	50.00	100.00
Total Volume of Concrete, m^3	18.64	19.60
Length of D/S Apron Floor $L_{D/S\ APRON}$, m	5.70	5.60
B. NON-OVER FLOW PORTION		
Length, m	1.00	1.00
Top Level including Free Board, m	101.50	102.00
Foundation Level, m	98.00	98.00
Height of Non-Over Flow H_{NOF}, m	3500	4000
Height H1, mm	2800	3200
Height H2, mm	700	800
Top Width of Non-Over Flow TW_{NOF}, mm	1000	1000
Base Width of Non-Over Flow B_{NOF}, mm	2450	2570
Width B1, mm	1450	1570
Pressure at Toe, t/m^2	13.31	15.91
Pressure at Heel, t/m^2	0.36	0.01
Factor of Safety against Overturning	2.47	2.52
Shear Friction Factor F	3.29	2.77
Total Volume of Concrete, m^3	7.56	9.02
C. WING WALLS (ABUTMENTS) ON BOTH BANKS		
Length in U/S $L_{U/SABUT}$, m	3.45	3.57
Length in D/S $L_{D/SABUT}$, m	2.00	2.00
Top Level including Free Board, m	101.50	102.00
Foundation Level, m	98.00	98.00
Total Height H_{ABUT}, m	3.50	4.00
D. D/S WING WALLS ON BOTH BANKS		
Length in D/S $L_{D/SWING}$, m	3.70	3.60
Top Level including Free Board, m	99.52	99.95
Foundation Level, m	98.00	98.00
Total Height $H_{D/SWING}$, m	1.52	1.95

TABLE 7.18
Design Results for Concrete Check Dam with Vertical D/S Face Height = 2.00 m and Stream Width = 10.0 m

Catchment Area, ha	50.00	100.00	150.00	200.00
A. WEIR				
Span of Weir W, m	8.00	8.00	8.00	8.00
Flood Discharge Q, m^3/s	6.54	11.00	14.91	18.50

(Continued)

TABLE 7.18 (Continued)

Catchment Area, ha	**50.00**	**100.00**	**150.00**	**200.00**
Head Over Crest, m	0.58	0.82	1.01	1.16
Top Level, m	100.00	100.00	100.00	100.00
Foundation Level, m	98.00	98.00	98.00	98.00
Height of Weir H_{WEIR}, m	2.00	2.00	2.00	2.00
Top Width of Weir TW_{WEIR}, mm	2190	2290	2360	2410
Base Width of Weir B_{WEIR}, mm	2190	2290	2360	2410
Pressure at Toe, t/m^2	8.75	9.08	9.33	9.53
Pressure at Heel, t/m^2	0.02	0.01	0.02	0.02
Factor of Safety against Overturning	2.16	2.17	2.17	2.18
Shear Friction Factor F	3.48	3.25	3.09	2.98
Total Volume of Concrete, m^3	35.04	36.64	37.76	38.56
Length of D/S Apron Floor $L_{D/S\ APRON}$, m	5.90	5.80	5.70	5.60
B. NON-OVER FLOW PORTION				
Length, m	2.00	2.00	2.00	2.00
Top Level including Free Board, m	101.10	101.40	101.60	101.80
Foundation Level, m	98.00	98.00	98.00	98.00
Height of Non-Over Flow H_{NOF}, m	3.10	3.40	3.60	3.80
Height H1, mm	2330	2550	2700	2850
Height H2, mm	770	850	900	950
Top Width of Non-Over Flow TW_{NOF}, mm	1000	1000	1000	1000
Base Width of Non-Over Flow B_{NOF}, mm	2380	2560	2670	2770
Width B1, mm	1380	1560	1670	1770
Pressure at Toe, t/m^2	11.58	12.83	13.67	13.75
Pressure at Heel, t/m^2	0.00	0.00	0.01	0.79
Factor of Safety against Overturning	2.32	2.46	2.54	2.61
Shear Friction Factor F	3.86	3.54	3.36	3.18
Total Volume of Concrete, m^3	12.62	14.76	16.22	17.69
C. WING WALLS (ABUTMENTS) ON BOTH BANKS				
Length in U/S $L_{U/SABUT}$, m	3.38	3.56	3.67	3.77
Length in D/S $L_{D/SABUT}$, m	2.00	2.00	2.00	2.00
Top Level including Free Board, m	101.10	101.40	101.60	101.80
Foundation Level, m	98.00	98.00	98.00	98.00
Total Height H_{ABUT}, m	3.10	3.40	3.60	3.80
D. D/S WING WALLS ON BOTH BANKS				
Length in D/S $L_{D/SWING}$, m	3.90	3.80	3.70	3.60
Top Level including Free Board, m	99.14	99.41	99.62	99.79
Foundation Level, m	98.00	98.00	98.00	98.00
Total Height $H_{D/SWING}$, m	1.14	1.41	1.62	1.79

TABLE 7.19
Design Results for Concrete Check Dam with Vertical D/S Face Height = 2.00 m and Stream Width = 15.0 m

Catchment Area, ha	50.00	100.00	150.00	200.00	250.00	300.00
A. WEIR						
Span of Weir W, m	12.00	12.00	12.00	12.00	12.00	12.00
Flood Discharge Q, m^3/s	6.54	11.00	14.91	18.50	21.87	25.07
Head Over Crest, m	0.44	0.63	0.77	0.89	0.99	1.09
Top Level, m	100.00	100.00	100.00	100.00	100.00	100.00
Foundation Level, m	98.00	98.00	98.00	98.00	98.00	98.00
Height of Weir H_{WEIR}, m	2.00	2.00	2.00	2.00	2.00	2.00
Top Width of Weir TW_{WEIR}, mm	2120	2210	2270	2320	2350	2390
Base Width of Weir B_{WEIR}, mm	2120	2210	2270	2320	2350	2390
Pressure at Toe, t/m^2	8.57	8.83	9.01	9.16	9.31	9.42
Pressure at Heel, t/m^2	0.01	0.01	0.01	0.03	0.01	0.03
Factor of Safety against Overturning	2.16	2.16	2.17	2.17	2.17	2.18
Shear Friction Factor F	3.62	3.42	3.29	3.19	3.11	3.04
Total Volume of Concrete, m^3	50.88	53.04	54.48	55.68	56.40	57.36
Length of D/S Apron Floor $L_{D/S\ APRON}$, m	5.90	5.80	5.80	5.70	5.70	5.70
B. NON-OVER FLOW PORTION						
Length, m	3.00	3.00	3.00	3.00	3.00	3.00
Top Level including Free Board, m	101.00	101.20	101.40	101.50	101.60	101.70
Foundation Level, m	98.00	98.00	98.00	98.00	98.00	98.00
Height of Non-Over Flow H_{NOF}, m	3.00	3.20	3.40	3.50	3.60	3.70
Height H1, mm	2250	2400	2550	2630	2700	2780
Height H2, mm	750	800	850	870	900	920
Top Width of Non-Over Flow TW_{NOF}, mm	1000	1000	1000	1000	1000	1000
Base Width of Non-Over Flow B_{NOF}, mm	2320	2450	2560	2620	2670	2720
Width B1, mm	1320	1450	1560	1620	1670	1720
Pressure at Toe, t/m^2	11.14	11.97	12.83	13.24	13.67	14.10
Pressure at Heel, t/m^2	0.01	0.01	0.00	0.02	0.01	0.01
Factor of Safety against Overturning	2.27	2.38	2.46	2.50	2.54	2.58
Shear Friction Factor F	3.97	3.76	3.54	3.45	3.36	3.26
Total Volume of Concrete, m^3	17.91	20.04	22.13	23.26	24.33	25.42
C. WING WALLS (ABUTMENTS) ON BOTH BANKS						
Length in U/S $L_{U/SABUT}$, m	3.32	3.45	3.56	3.62	3.67	3.72
Length in D/S $L_{D/SABUT}$, m	2.00	2.00	2.00	2.00	2.00	2.00
Top Level including Free Board, m	101.00	101.20	101.40	101.50	101.60	101.70
Foundation Level, m	98.00	98.00	98.00	98.00	98.00	98.00
Total Height H_{ABUT}, m	3.00	3.20	3.40	3.50	3.60	3.70
D. D/S WING WALLS ON BOTH BANKS						
Length in D/S $L_{D/SWING}$, m	3.90	3.80	3.80	3.70	3.70	3.70
Top Level including Free Board, m	98.99	99.20	99.35	99.49	99.60	99.71
Foundation Level, m	98.00	98.00	98.00	98.00	98.00	98.00
Total Height $H_{D/SWING}$, m	0.99	1.20	1.35	1.49	1.60	1.71

TABLE 7.20
Design Results for Concrete Check Dam with Vertical D/S Face Height = 2.00 m and Stream Width = 20.0 m

Catchment Area, ha	50.00	100.00	150.00	200.00	250.00	300.00
A. WEIR						
Span of Weir W, m	16.00	16.00	16.00	16.00	16.00	16.00
Flood Discharge Q, m^3/s	6.54	11.00	14.91	18.50	21.87	25.07
Head Over Crest, m	0.37	0.52	0.64	0.73	0.82	0.90
Top Level, m	100.00	100.00	100.00	100.00	100.00	100.00
Foundation Level, m	98.00	98.00	98.00	98.00	98.00	98.00
Height of Weir H_{WEIR}, m	2.00	2.00	2.00	2.00	2.00	2.00
Top Width of Weir TW_{WEIR}, mm	2080	2160	2210	2250	2290	2320
Base Width of Weir B_{WEIR}, mm	2080	2160	2210	2250	2290	2320
Pressure at Toe, t/m^2	8.49	8.68	8.86	8.97	9.08	9.19
Pressure at Heel, t/m^2	0.00	0.01	0.00	0.00	0.01	0.01
Factor of Safety against Overturning	2.15	2.16	2.16	2.16	2.17	2.17
Shear Friction Factor F	3.69	3.53	3.41	3.32	3.25	3.18
Total Volume of Concrete, m^3	66.56	69.12	70.72	72.00	73.28	74.24
Length of D/S Apron Floor $L_{D/S\ APRON}$, m	6.00	5.90	5.80	5.80	5.80	5.70
B. NON-OVER FLOW PORTION						
Length, m	4.00	4.00	4.00	4.00	4.00	4.00
Top Level including Free Board, m	100.90	101.10	101.20	101.30	101.40	101.50
Foundation Level, m	98.00	98.00	98.00	98.00	98.00	98.00
Height of Non-Over Flow H_{NOF}, m	2.90	3.10	3.20	3.30	3.40	3.50
Height H1, mm	2180	2330	2400	2480	2550	2630
Height H2, mm	720	770	800	820	850	870
Top Width of Non-Over Flow TW_{NOF}, mm	1000	1000	1000	1000	1000	1000
Base Width of Non-Over Flow B_{NOF}, mm	2340	2380	2450	2500	2560	2620
Width B1, mm	1340	1380	1450	1500	1560	1620
Pressure at Toe, t/m^2	10.51	11.58	11.97	12.42	12.83	13.24
Pressure at Heel, t/m^2	0.18	0.00	0.01	0.00	0.00	0.02
Factor of Safety against Overturning	2.29	2.32	2.38	2.41	2.46	2.50
Shear Friction Factor F	4.24	3.86	3.76	3.64	3.54	3.45
Total Volume of Concrete, m^3	23.26	25.23	26.72	28.05	29.51	31.01
C. WING WALLS (ABUTMENTS) ON BOTH BANKS						
Length in U/S $L_{U/SABUT}$, m	3.34	3.38	3.45	3.50	3.56	3.62
Length in D/S $L_{D/SABUT}$, m	2.00	2.00	2.00	2.00	2.00	2.00
Top Level including Free Board, m	100.90	101.10	101.20	101.30	101.40	101.50
Foundation Level, m	98.00	98.00	98.00	98.00	98.00	98.00
Total Height H_{ABUT}, m	2.90	3.10	3.20	3.30	3.40	3.50
D. D/S WING WALLS ON BOTH BANKS						
Length in D/S $L_{D/SWING}$, m	4.00	3.90	3.80	3.80	3.80	3.70
Top Level including Free Board, m	98.91	99.08	99.21	99.31	99.41	99.50
Foundation Level, m	98.00	98.00	98.00	98.00	98.00	98.00
Total Height $H_{D/SWING}$, m	0.91	1.08	1.21	1.31	1.41	1.50

7.3.3 Section with Steps on Both Faces

The tables developed after the rigorous design of the different components for 2.00 m high concrete check dams for different combinations are presented from Table 7.21 to Table 7.24.

TABLE 7.21
Design Results for Concrete Check Dam with Steps on Both Faces Height = 2.00 m and Stream Width = 5.0 m

Catchment Area, ha	50.00	100.00
A. WEIR		
Span of Weir W, m	4.00	4.00
Flood Discharge Q, m^3/s	6.54	11.00
Head Over Crest, m	0.92	1.31
Top Level, m	100.00	100.00
Foundation Level, m	98.00	98.00
Height of Weir H_{WEIR}, m	2.00	2.00
Size of Concrete Steps (W, H), m		
Base Width B_{WEIR}, Thickness H1'	2.64, 1.00	2.96, 1.00
Width B1', Thickness H2'	–	–
Top Width TW_{WEIR}, Thickness H3'	1.64, 1.00	1.96, 1.00
Pressure at Toe, t/m^2	7.08	7.46
Pressure at Heel, t/m^2	0.01	0.02
Factor of Safety against Overturning	2.01	2.08
Shear Friction Factor F	2.36	2.24
Total Volume of Concrete, m^3	17.12	19.68
Length of D/S Apron Floor $L_{D/S\ APRON}$, m	4.25	5.51
B. NON-OVER FLOW PORTION		
Length, m	1.00	1.00
Top Level including Free Board, m	101.50	102.00
Foundation Level, m	98.00	98.00
Height of Non-Over Flow H_{NOF}, m	3.50	4.00
Size of Concrete Steps (W, H), m		
Base Width B_{NOF}, Thickness H1	2.20, 1.00	2.57, 1.00
Width B1, Thickness H2	1.20, 1.00	1.57, 1.00
Width B2, Thickness H3	–	–
Top Width TW_{NOF}, Thickness H4	0.50, 1.50	0.57, 2.00
Pressure at Toe, t/m^2	9.61	10.72
Pressure at Heel, t/m^2	0.04	0.04
Factor of Safety against Overturning	1.97	2.00
Shear Friction Factor F	3.61	3.36
Total Volume of Concrete, m^3	4.15	5.28
C. WING WALLS (ABUTMENTS) ON BOTH BANKS		
Length in U/S $L_{U/SABUT}$, m	2.40	2.50
Length in D/S $L_{D/SABUT}$, m	2.40	2.50
Top Level including Free Board, m	101.50	102.00
Foundation Level, m	98.00	98.00

(Continued)

TABLE 7.21 (Continued)

Catchment Area, ha	50.00	100.00
Total Height H_{ABUT}, m	3.50	4.00
D. D/S WING WALLS ON BOTH BANKS		
Length in D/S $L_{D/SWING}$, m	4.40	4.10
Top Level including Free Board, m	99.52	99.95
Foundation Level, m	98.00	98.00
Total Height $H_{D/SWING}$, m	1.52	1.95

TABLE 7.22
Design Results for Concrete Check Dam with Steps on Both Faces Height = 2.00 m and Stream Width = 10.0 m

Catchment Area, ha	50.00	100.00	150.00	200.00
A. WEIR				
Span of Weir W, m	8.00	8.00	8.00	8.00
Flood Discharge Q, m³/s	6.54	11.00	14.91	18.50
Head Over Crest, m	0.58	0.82	1.01	1.16
Top Level, m	100.00	100.00	100.00	100.00
Foundation Level, m	98.00	98.00	98.00	98.00
Height of Weir H_{WEIR}, m	2.00	2.00	2.00	2.00
Size of Concrete Steps (W, H), m				
Base Width B_{WEIR}, Thickness H1'	2.34, 1.00	2.55, 1.00	2.72, 1.00	2.84, 1.00
Width B1', Thickness H2'	–	–	–	–
Top Width TW_{WEIR}, Thickness H3'	1.34, 1.00	1.55, 1.00	1.72, 1.00	1.84, 1.00
Pressure at Toe, t/m²	6.73	6.98	7.14	7.31
Pressure at Heel, t/m²	0.02	0.00	0.04	0.02
Factor of Safety against Overturning	1.95	1.99	2.03	2.05
Shear Friction Factor F	2.49	2.40	2.34	2.29
Total Volume of Concrete, m³	29.44	32.80	35.52	37.44
Length of D/S Apron Floor $L_{D/S\ APRON}$, m	3.01	3.90	4.54	5.06
B. NON-OVER FLOW PORTION				
Length, m	2.00	2.00	2.00	2.00
Top Level including Free Board, m	101.10	101.40	101.60	101.80
Foundation Level, m	98.00	98.00	98.00	98.00
Height of Non-Over Flow H_{NOF}, m	3.10	3.40	3.60	3.80
Size of Concrete Steps (W, H), m				
Base Width B_{NOF}, Thickness H1	1.98, 1.00	2.15, 1.00	2.27, 1.00	2.43, 1.00
Width B1, Thickness H2	0.98, 1.00	1.15, 1.00	1.27, 1.00	1.43, 1.00
Width B2, Thickness H3	–	–	–	–
Top Width TW_{NOF}, Thickness H4	0.50, 1.10	0.50, 1.40	0.50, 1.60	0.50, 1.80
Pressure at Toe, t/m²	8.55	9.32	9.82	10.16
Pressure at Heel, t/m²	0.03	0.06	0.02	0.04
Factor of Safety against Overturning	1.95	1.97	1.97	1.98

(Continued)

TABLE 7.22 (Continued)

Catchment Area, ha	50.00	100.00	150.00	200.00
Shear Friction Factor F	3.98	3.70	3.54	3.45
Total Volume of Concrete, m^3	7.02	8.00	8.68	9.52
C. WING WALLS (ABUTMENTS) ON BOTH BANKS				
Length in U/S $L_{U/SABUT}$, m	2.20	2.30	2.40	2.50
Length in D/S $L_{D/SABUT}$, m	2.20	2.30	2.40	2.50
Top Level including Free Board, m	101.1	101.4	101.6	101.8
Foundation Level, m	98.00	98.00	98.00	98.00
Total Height H_{ABUT}, m	3.10	3.40	3.60	3.80
D. D/S WING WALLS ON BOTH BANKS				
Length in D/S $L_{D/SWING}$, m	4.70	4.50	4.30	4.20
Top Level including Free Board, m	99.14	99.41	99.62	99.79
Foundation Level, m	98.00	98.00	98.00	98.00
Total Height $H_{D/SWING}$, m	1.14	1.41	1.62	1.79

TABLE 7.23
Design Results for Concrete Check Dam with Steps on Both Faces Height = 2.00 m and Stream Width = 15.0 m

Catchment Area, ha	50.00	100.00	150.00	200.00	250.00	300.00
A. WEIR						
Span of Weir W, m	12.00	12.00	12.00	12.00	12.00	12.00
Flood Discharge Q, m^3/s	6.54	11.00	14.91	18.50	21.87	25.07
Head Over Crest, m	0.44	0.63	0.77	0.89	0.99	1.09
Top Level, m	100.00	100.00	100.00	100.00	100.00	100.00
Foundation Level, m	98.00	98.00	98.00	98.00	98.00	98.00
Height of Weir H_{WEIR}, m	2.00	2.00	2.00	2.00	2.00	2.00
Size of Concrete Steps (W, H), m						
Base Width B_{WEIR}, Thickness H1'	2.21, 1.00	2.38, 1.00	2.51, 1.00	2.61, 1.00	2.7, 1.00	2.78, 1.00
Width B1', Thickness H2'	–	–	–	–	–	–
Top Width TW_{WEIR}, Thickness H3'	1.21, 1.00	1.38, 1.00	1.51, 1.00	1.61, 1.00	1.70, 1.00	1.78, 1.00
Pressure at Toe, t/m^2	6.59	6.79	6.91	7.05	7.14	7.24
Pressure at Heel, t/m^2	0.02	0.00	0.02	0.00	0.01	0.02
Factor of Safety against Overturning	1.93	1.95	1.98	2.00	2.02	2.04
Shear Friction Factor F	2.55	2.47	2.42	2.37	2.34	2.31
Total Volume of Concrete, m^3	41.04	45.12	48.24	50.64	52.80	54.72
Length of D/S Apron Floor $L_{D/S APRON}$, m	2.46	3.18	3.71	4.13	4.49	4.81
B. NON-OVER FLOW PORTION						
Length, m	3.00	3.00	3.00	3.00	3.00	3.00

(Continued)

TABLE 7.23 (Continued)

Catchment Area, ha	50.00	100.00	150.00	200.00	250.00	300.00
Top Level including Free Board, m	101.00	101.20	101.40	101.50	101.60	101.70
Foundation Level, m	98.00	98.00	98.00	98.00	98.00	98.00
Height of Non-Over Flow H_{NOF}, m	3.00	3.20	3.40	3.50	3.60	3.70
Size of Concrete Steps (W, H), m						
Base Width B_{NOF}, Thickness H1	1.92, 1.00	2.03, 1.00	2.15, 1.00	2.20, 1.00	2.27, 1.00	2.35, 1.00
Width B1, Thickness H2	0.92, 1.00	1.03, 1.00	1.15, 1.00	1.20, 1.00	1.27, 1.00	1.35, 1.00
Width B2, Thickness H3	–	–	–	–	–	–
Top Width TW_{NOF}, Thickness H4	0.50, 1.00	0.50, 1.20	0.50, 1.40	0.50, 1.50	0.50, 1.60	0.50, 1.70
Pressure at Toe, t/m^2	8.28	8.84	9.32	9.61	9.82	10.02
Pressure at Heel, t/m^2	0.02	0.02	0.06	0.04	0.02	0.01
Factor of Safety against Overturning	1.94	1.95	1.97	1.97	1.97	1.97
Shear Friction Factor F	4.06	3.87	3.70	3.61	3.54	3.50
Total Volume of Concrete, m^3	10.02	10.98	12.00	12.45	13.02	13.65
C. WING WALLS (ABUTMENTS) ON BOTH BANKS						
Length in U/S $L_{U/SABUT}$, m	2.20	2.20	2.30	2.40	2.40	2.40
Length in D/S $L_{D/SABUT}$, m	2.20	2.20	2.30	2.40	2.40	2.40
Top Level including Free Board, m	101	101.2	101.4	101.5	101.6	101.7
Foundation Level, m	98.00	98.00	98.00	98.00	98.00	98.00
Total Height H_{ABUT}, m	3.00	3.20	3.40	3.50	3.60	3.70
D. D/S WING WALLS ON BOTH BANKS						
Length in D/S $L_{D/SWING}$, m	4.80	4.70	4.50	4.40	4.30	4.30
Top Level including Free Board, m	98.99	99.20	99.35	99.49	99.60	99.71
Foundation Level, m	98.00	98.00	98.00	98.00	98.00	98.00
Total Height $H_{D/SWING}$, m	0.99	1.20	1.35	1.49	1.60	1.71

TABLE 7.24
Design Results for Concrete Check Dam with Steps on Both Faces Height = 2.00 m and Stream Width = 20.0 m

Catchment Area, ha	50.00	100.00	150.00	200.00	250.00	300.00
A. WEIR						
Span of Weir W, m	16.00	16.00	16.00	16.00	16.00	16.00
Flood Discharge Q, m^3/s	6.54	11.00	14.91	18.50	21.87	25.07
Head Over Crest, m	0.37	0.52	0.64	0.73	0.82	0.90
Top Level, m	100.00	100.00	100.00	100.00	100.00	100.00
Foundation Level, m	98.00	98.00	98.00	98.00	98.00	98.00
Height of Weir H_{WEIR}, m	2.00	2.00	2.00	2.00	2.00	2.00

(Continued)

TABLE 7.24 (Continued)

Catchment Area, ha	50.00	100.00	150.00	200.00	250.00	300.00
Size of Concrete Steps (W, H), m						
Base Width B_{WEIR}, Thickness H1'	2.14, 1.00	2.28, 1.00	2.39, 1.00	2.48, 1.00	2.55, 1.00	2.62, 1.00
Width B1', Thickness H2'	–	–	–	–	–	–
Top Width TW_{WEIR}, Thickness H3'	1.14, 1.00	1.28, 1.00	1.39, 1.00	1.48, 1.00	1.55, 1.00	1.62, 1.00
Pressure at Toe, t/m^2	6.54	6.68	6.79	6.86	6.98	7.06
Pressure at Heel, t/m^2	0.00	0.00	0.01	0.04	0.00	0.00
Factor of Safety against Overturning	1.91	1.93	1.96	1.98	1.99	2.00
Shear Friction Factor F	2.58	2.51	2.46	2.44	2.40	2.37
Total Volume of Concrete, m^3	52.48	56.96	60.48	63.36	65.60	67.84
Length of D/S Apron Floor $L_{D/S\ APRON}$, m	2.13	2.76	3.21	3.58	3.89	4.16
B. NON-OVER FLOW PORTION						
Length, m	4.00	4.00	4.00	4.00	4.00	4.00
Top Level including Free Board, m	100.90	101.10	101.20	101.30	101.40	101.50
Foundation Level, m	98.00	98.00	98.00	98.00	98.00	98.00
Height of Non-Over Flow H_{NOF}, m	2.90	3.10	3.20	3.30	3.40	3.50
Size of Concrete Steps (W, H), m						
Base Width B_{NOF}, Thickness H1	1.88, 1.00	1.98, 1.00	2.03, 1.00	2.08, 1.00	2.15, 1.00	2.20, 1.00
Width B1, Thickness H2	0.88, 1.00	0.98, 1.00	1.03, 1.00	1.08, 1.00	1.15, 1.00	1.20, 1.00
Width B2, Thickness H3	–	–	–	–	–	–
Top Width TW_{NOF}, Thickness H4	0.50, 0.90	0.50, 1.10	0.50, 1.20	0.50, 1.30	0.50, 1.40	0.50, 1.50
Pressure at Toe, t/m^2	7.95	8.55	8.84	9.11	9.32	9.61
Pressure at Heel, t/m^2	0.08	0.03	0.02	0.01	0.06	0.04
Factor of Safety against Overturning	1.95	1.95	1.95	1.96	1.97	1.97
Shear Friction Factor F	4.21	3.98	3.87	3.77	3.70	3.61
Total Volume of Concrete, m^3	12.84	14.04	14.64	15.24	16.00	16.60
C. WING WALLS (ABUTMENTS) ON BOTH BANKS						
Length in U/S $L_{U/SABUT}$, m	2.10	2.20	2.20	2.30	2.30	2.40
Length in D/S $L_{D/SABUT}$, m	2.10	2.20	2.20	2.30	2.30	2.40
Top Level including Free Board, m	100.9	101.1	101.2	101.3	101.4	101.5
Foundation Level, m	98.00	98.00	98.00	98.00	98.00	98.00
Total Height H_{ABUT}, m	2.90	3.10	3.20	3.30	3.40	3.50
D. D/S WING WALLS ON BOTH BANKS						
Length in D/S $L_{D/SWING}$, m	4.90	4.80	4.70	4.60	4.50	4.40
Top Level including Free Board, m	98.91	99.08	99.21	99.31	99.41	99.50
Foundation Level, m	98.00	98.00	98.00	98.00	98.00	98.00
Total Height $H_{D/SWING}$, m	0.91	1.08	1.21	1.31	1.41	1.50

7.4 DESIGN OF 2.50 M HIGH CONCRETE CHECK DAMS

7.4.1 Section with Vertical U/S Face

The tables developed after the rigorous design of the different components for 2.50 m high concrete check dams for different combinations are presented from Table 7.25 to Table 7.28.

TABLE 7.25
Design Results for Concrete Check Dam with Vertical U/S Face Height = 2.50 m and Stream Width = 5.0 m

Catchment Area, ha	50.00	100.00
A. WEIR		
Span of Weir W, m	4.00	4.00
Flood Discharge Q, m^3/s	6.54	11.00
Head Over Crest, m	0.92	1.31
Top Level, m	100.00	100.00
Foundation Level, m	97.50	97.50
Height of Weir H_{WEIR}, m	2.50	2.50
Top Width of Weir TW_{WEIR}, mm	500	500
Base Width of Weir B_{WEIR}, mm	2700	2930
Width B1', mm	2200	2430
D/S Slope – H_{DS}: 1.0(V)	0.880	0.970
Pressure at Toe, t/m^2	5.11	5.10
Pressure at Heel, t/m^2	0.02	0.01
Factor of Safety against Overturning	1.51	1.51
Shear Friction Factor F	2.22	2.08
Total Volume of Concrete, m^3	16.00	17.15
Length of D/S Apron Floor $L_{D/S\ APRON}$, m	7.30	7.10
B. NON-OVER FLOW PORTION		
Length, m	1.00	1.00
Top Level including Free Board, m	101.50	102.00
Foundation Level, m	97.50	97.50
Height of Non-Over Flow H_{NOF}, m	4.00	4.50
Height H1, mm	2500	2448
Height H2, mm	1500	2052
Top Width of Non-Over Flow TW_{NOF}, mm	500	550
Base Width of Non-Over Flow B_{NOF}, mm	2700	2930
Width B1, mm	2200	2430
D/S Slope – H_{DS}: 1.0(V)	0.880	0.970
Pressure at Toe, t/m^2	5.68	6.69
Pressure at Heel, t/m^2	0.61	0.01
Factor of Safety against Overturning	1.60	1.52
Shear Friction Factor F	2.09	1.81
Total Volume of Concrete, m^3	4.75	5.39
C. WING WALLS (ABUTMENTS) ON BOTH BANKS		
Length in U/S $L_{U/SABUT}$, m	1.50	1.50

(Continued)

TABLE 7.25 (Continued)

Catchment Area, ha	50.00	100.00
Length in D/S $L_{D/SABUT}$, m	3.20	3.43
Top Level including Free Board, m	101.50	102.00
Foundation Level, m	97.50	97.50
Total Height H_{ABUT}, m	4.00	4.50
D. D/S WING WALLS ON BOTH BANKS		
Length in D/S $L_{D/SWING}$, m	4.10	3.67
Top Level including Free Board, m	99.02	99.45
Foundation Level, m	97.50	97.50
Total Height $H_{D/SWING}$, m	1.52	1.95

TABLE 7.26
Design Results for Concrete Check Dam with Vertical U/S Face Height = 2.50 m and Stream Width = 10.0 m

Catchment Area, ha	50.00	100.00	150.00	200.00
A. WEIR				
Span of Weir W, m	8.00	8.00	8.00	8.00
Flood Discharge Q, m^3/s	6.54	11.00	14.91	18.50
Head Over Crest, m	0.58	0.82	1.01	1.16
Top Level, m	100.00	100.00	100.00	100.00
Foundation Level, m	97.50	97.50	97.50	97.50
Height of Weir H_{WEIR}, m	2.50	2.50	2.50	2.50
Top Width of Weir TW_{WEIR}, mm	500	500	500	500
Base Width of Weir B_{WEIR}, mm	2470	2630	2750	2840
Width B1', mm	1970	2130	2250	2340
D/S Slope – H_{DS}: 1.0(V)	0.788	0.850	0.900	0.935
Pressure at Toe, t/m^2	5.16	5.13	5.13	5.12
Pressure at Heel, t/m^2	0.01	0.01	0.00	0.01
Factor of Safety against Overturning	1.52	1.51	1.51	1.51
Shear Friction Factor F	2.36	2.26	2.18	2.13
Total Volume of Concrete, m^3	29.70	31.30	32.50	33.40
Length of D/S Apron Floor $L_{D/S\ APRON}$, m	7.60	7.40	7.30	7.20
B. NON-OVER FLOW PORTION				
Length, m	2.00	2.00	2.00	2.00
Top Level including Free Board, m	101.10	101.40	101.60	101.80
Foundation Level, m	97.50	97.50	97.50	97.50
Height of Non-Over Flow H_{NOF}, m	3.60	3.90	4.10	4.30
Height H1, mm	2500	2500	2500	2500
Height H2, mm	1100	1400	1600	1800
Top Width of Non-Over Flow TW_{NOF}, mm	500	500	500	500
Base Width of Non-Over Flow B_{NOF}, mm	2470	2630	2750	2840
Width B1, mm	1970	2130	2250	2340

(Continued)

TABLE 7.26 (Continued)

Catchment Area, ha	50.00	100.00	150.00	200.00
D/S Slope – H_{DS}: 1.0(V)	0.788	0.850	0.900	0.935
Pressure at Toe, t/m^2	5.04	5.53	5.86	6.35
Pressure at Heel, t/m^2	1.09	0.74	0.48	0.08
Factor of Safety against Overturning	1.68	1.62	1.58	1.53
Shear Friction Factor F	2.34	2.14	2.03	1.91
Total Volume of Concrete, m^3	8.53	9.23	9.73	10.15
C. WING WALLS (ABUTMENTS) ON BOTH BANKS				
Length in U/S $L_{U/SABUT}$, m	1.50	1.50	1.50	1.50
Length in D/S $L_{D/SABUT}$, m	2.97	3.13	3.25	3.34
Top Level including Free Board, m	101.10	101.40	101.60	101.80
Foundation Level, m	97.50	97.50	97.50	97.50
Total Height H_{ABUT}, m	3.60	3.90	4.10	4.30
D. D/S WING WALLS ON BOTH BANKS				
Length in D/S $L_{D/SWING}$, m	4.63	4.27	4.05	3.86
Top Level including Free Board, m	98.64	98.91	99.12	99.29
Foundation Level, m	97.50	97.50	97.50	97.50
Total Height $H_{D/SWING}$, m	1.14	1.41	1.62	1.79

TABLE 7.27
Design Results for Concrete Check Dam with Vertical U/S Face Height = 2.50 m and Stream Width = 15.0 m

Catchment Area, ha	50.00	100.00	150.00	200.00	250.00	300.00
A. WEIR						
Span of Weir W, m	12.00	12.00	12.00	12.00	12.00	12.00
Flood Discharge Q, m^3/s	6.54	11.00	14.91	18.50	21.87	25.07
Head Over Crest, m	0.44	0.63	0.77	0.89	0.99	1.09
Top Level, m	100.00	100.00	100.00	100.00	100.00	100.00
Foundation Level, m	97.50	97.50	97.50	97.50	97.50	97.50
Height of Weir H_{WEIR}, m	2.50	2.50	2.50	2.50	2.50	2.50
Top Width of Weir TW_{WEIR}, mm	500	500	500	500	500	500
Base Width of Weir B_{WEIR}, mm	2370	2510	2600	2690	2750	2810
Width B1', mm	1870	2010	2100	2190	2250	2310
D/S Slope – $H_{DS'}$: 1.0(V)	0.746	0.805	0.840	0.875	0.900	0.922
Pressure at Toe, t/m^2	5.16	5.12	5.13	5.07	5.08	5.07
Pressure at Heel, t/m^2	0.03	0.04	0.02	0.07	0.04	0.05
Factor of Safety against Overturning	1.52	1.52	1.51	1.52	1.51	1.51
Shear Friction Factor F	2.42	2.34	2.28	2.24	2.20	2.16
Total Volume of Concrete, m^3	43.05	45.15	46.50	47.85	48.75	49.65
Length of D/S Apron Floor $L_{D/S\ APRON}$, m	7.70	7.50	7.40	7.40	7.30	7.20

(Continued)

TABLE 7.27 (Continued)

Catchment Area, ha	50.00	100.00	150.00	200.00	250.00	300.00
B. NON-OVER FLOW PORTION						
Length, m	3.00	3.00	3.00	3.00	3.00	3.00
Top Level including Free Board, m	101.00	101.20	101.40	101.50	101.60	101.70
Foundation Level, m	97.50	97.50	97.50	97.50	97.50	97.50
Height of Non-Over Flow H_{NOF}, m	3.50	3.70	3.90	4.00	4.10	4.20
Height H1, mm	2500	2500	2500	2500	2500	2500
Height H2, mm	1000	1200	1400	1500	1600	1700
Top Width of Non-Over Flow TW_{NOF}, mm	500	500	500	500	500	500
Base Width of Non-Over Flow B_{NOF}, mm	2370	2510	2600	2690	2750	2810
Width B1, mm	1870	2010	2100	2190	2250	2310
D/S Slope – H_{DS}: 1.0(V)	0.746	0.805	0.840	0.875	0.900	0.922
Pressure at Toe, t/m^2	5.09	5.27	5.67	5.71	5.86	6.00
Pressure at Heel, t/m^2	1.03	0.92	0.62	0.60	0.48	0.38
Factor of Safety against Overturning	1.68	1.65	1.61	1.60	1.58	1.57
Shear Friction Factor F	2.38	2.26	2.12	2.08	2.03	1.97
Total Volume of Concrete, m^3	12.26	13.09	13.73	14.21	14.59	14.96
C. WING WALLS (ABUTMENTS) ON BOTH BANKS						
Length in U/S $L_{U/SABUT}$, m	1.50	1.50	1.50	1.50	1.50	1.50
Length in D/S $L_{D/SABUT}$, m	2.87	3.01	3.10	3.19	3.25	3.31
Top Level including Free Board, m	101.00	101.20	101.40	101.50	101.60	101.70
Foundation Level, m	97.50	97.50	97.50	97.50	97.50	97.50
Total Height H_{ABUT}, m	3.50	3.70	3.90	4.00	4.10	4.20
D. D/S WING WALLS ON BOTH BANKS						
Length in D/S $L_{D/SWING}$, m	4.83	4.49	4.30	4.21	4.05	3.89
Top Level including Free Board, m	98.49	98.70	98.85	98.99	99.10	99.21
Foundation Level, m	97.50	97.50	97.50	97.50	97.50	97.50
Total Height $H_{D/SWING}$, m	0.99	1.20	1.35	1.49	1.60	1.71

TABLE 7.28

Design Results for Concrete Check Dam with Vertical U/S Face Height = 2.50 m and Stream Width = 20.0 m

Catchment Area, ha	50.00	100.00	150.00	200.00	250.00	300.00
A. WEIR						
Span of Weir W, m	16.00	16.00	16.00	16.00	16.00	16.00
Flood Discharge Q, m^3/s	6.54	11.00	14.91	18.50	21.87	25.07
Head Over Crest, m	0.37	0.52	0.64	0.73	0.82	0.90
Top Level, m	100.00	100.00	100.00	100.00	100.00	100.00
Foundation Level, m	97.50	97.50	97.50	97.50	97.50	97.50
Height of Weir H_{WEIR}, m	2.50	2.50	2.50	2.50	2.50	2.50
Top Width of Weir TW_{WEIR}, mm	500	500	500	500	500	500
Base Width of Weir B_{WEIR}, mm	2320	2430	2510	2580	2630	2690

(Continued)

TABLE 7.28 (Continued)

Catchment Area, ha	50.00	100.00	150.00	200.00	250.00	300.00
Width B1', mm	1820	1930	2010	2080	2130	2190
D/S Slope – H_{DS},: 1.0(V)	0.726	0.770	0.805	0.830	0.850	0.875
Pressure at Toe, t/m^2	5.19	5.15	5.14	5.12	5.13	5.09
Pressure at Heel, t/m^2	0.01	0.03	0.02	0.02	0.01	0.05
Factor of Safety against Overturning	1.52	1.52	1.52	1.51	1.51	1.52
Shear Friction Factor F	2.45	2.39	2.33	2.30	2.26	2.23
Total Volume of Concrete, m^3	56.40	58.60	60.20	61.60	62.60	63.80
Length of D/S Apron Floor $L_{D/S\ APRON}$, m	7.70	7.60	7.50	7.50	7.40	7.40
B. NON-OVER FLOW PORTION						
Length, m	4.00	4.00	4.00	4.00	4.00	4.00
Top Level including Free Board, m	100.90	101.10	101.20	101.30	101.40	101.50
Foundation Level, m	97.50	97.50	97.50	97.50	97.50	97.50
Height of Non-Over Flow H_{NOF}, m	3.40	3.60	3.70	3.80	3.90	4.00
Height H1, mm	2500	2500	2500	2500	2500	2500
Height H2, mm	900	1100	1200	1300	1400	1500
Top Width of Non-Over Flow TW_{NOF}, mm	500	500	500	500	500	500
Base Width of Non-Over Flow B_{NOF}, mm	2320	2430	2510	2580	2630	2690
Width B1, mm	1820	1930	2010	2080	2130	2190
D/S Slope – H_{DS}: 1.0(V)	0.726	0.770	0.805	0.830	0.850	0.875
Pressure at Toe, t/m^2	4.97	5.21	5.27	5.37	5.53	5.71
Pressure at Heel, t/m^2	1.10	0.95	0.92	0.85	0.74	0.60
Factor of Safety against Overturning	1.69	1.66	1.65	1.64	1.62	1.60
Shear Friction Factor F	2.46	2.31	2.26	2.20	2.14	2.08
Total Volume of Concrete, m^3	15.90	16.85	17.45	18.00	18.45	18.95
C. WING WALLS (ABUTMENTS) ON BOTH BANKS						
Length in U/S $L_{U/SABUT}$, m	1.50	1.50	1.50	1.50	1.50	1.50
Length in D/S $L_{D/SABUT}$, m	2.82	2.93	3.01	3.08	3.13	3.19
Top Level including Free Board, m	100.90	101.10	101.20	101.30	101.40	101.50
Foundation Level, m	97.50	97.50	97.50	97.50	97.50	97.50
Total Height H_{ABUT}, m	3.40	3.60	3.70	3.80	3.90	4.00
D. D/S WING WALLS ON BOTH BANKS						
Length in D/S $L_{D/SWING}$, m	4.88	4.67	4.49	4.42	4.27	4.21
Top Level including Free Board, m	98.41	98.58	98.71	98.81	98.91	99.00
Foundation Level, m	97.50	97.50	97.50	97.50	97.50	97.50
Total Height $H_{/SWING}$, m	0.91	1.08	1.21	1.31	1.41	1.50

7.4.2 Section with Vertical D/S Face

The tables developed after the rigorous design of the different components for 2.50 m high concrete check dams for different combinations are presented from Table 7.29 to Table 7.32.

TABLE 7.29
Design Results for Concrete Check Dam with Vertical D/S Face Height = 2.50 m and Stream Width = 5.0 m

Catchment Area, ha	50.00	100.00
A. WEIR		
Span of Weir W, m	4.00	4.00
Flood Discharge Q, m^3/s	6.54	11.00
Head Over Crest, m	0.92	1.31
Top Level, m	100.00	100.00
Foundation Level, m	97.50	97.50
Height of Weir H_{WEIR}, m	2.50	2.50
Top Width of Weir TW_{WEIR}, mm	2820	2960
Base Width of Weir B_{WEIR}, mm	2820	2960
Pressure at Toe, t/m^2	11.21	11.75
Pressure at Heel, t/m^2	0.01	0.00
Factor of Safety against Overturning	2.16	2.17
Shear Friction Factor F	2.85	2.63
Total Volume of Concrete, m^3	28.20	29.60
Length of D/S Apron Floor $L_{D/S\ APRON}$, m	7.20	7.10
B. NON-OVER FLOW PORTION		
Length, m	1.00	1.00
Top Level including Free Board, m	101.50	102.00
Foundation Level, m	97.50	97.50
Height of Non-Over Flow H_{NOF}, m	4.00	4.50
Height H1, mm	3000	3375
Height H2, mm	1000	1125
Top Width of Non-Over Flow TW_{NOF}, mm	1000	1000
Base Width of Non-Over Flow B_{NOF}, mm	2870	3000
Width B1, mm	1870	2000
Pressure at Toe, t/m^2	14.89	16.71
Pressure at Heel, t/m^2	0.01	0.37
Factor of Safety against Overturning	2.35	2.43
Shear Friction Factor F	2.85	2.48
Total Volume of Concrete, m^3	9.61	11.25
C. WING WALLS (ABUTMENTS) ON BOTH BANKS		
Length in U/S $L_{U/SABUT}$, m	3.87	4.00
Length in D/S $L_{D/SABUT}$, m	2.00	2.00
Top Level including Free Board, m	101.50	102.00
Foundation Level, m	97.50	97.50
Total Height H_{ABUT}, m	4.00	4.50
D. D/S WING WALLS ON BOTH BANKS		
Length in D/S $L_{D/SWING}$, m	5.20	5.10
Top Level including Free Board, m	99.02	99.45
Foundation Level, m	97.50	97.50
Total Height $H_{D/SWING}$, m	1.52	1.95

TABLE 7.30
Design Results for Concrete Check Dam with Vertical D/S Face Height = 2.50 m and Stream Width = 10.0 m

Catchment Area, ha	50.00	100.00	150.00	200.00
A. WEIR				
Span of Weir W, m	8.00	8.00	8.00	8.00
Flood Discharge Q, m^3/s	6.54	11.00	14.91	18.50
Head Over Crest, m	0.58	0.82	1.01	1.16
Top Level, m	100.00	100.00	100.00	100.00
Foundation Level, m	97.50	97.50	97.50	97.50
Height of Weir H_{WEIR}, m	2.50	2.50	2.50	2.50
Top Width of Weir TW_{WEIR}, mm	2660	2780	2860	2910
Base Width of Weir B_{WEIR}, mm	2660	2780	2860	2910
Pressure at Toe, t/m^2	10.78	11.07	11.32	11.54
Pressure at Heel, t/m^2	0.00	0.02	0.03	0.01
Factor of Safety against Overturning	2.15	2.16	2.17	2.17
Shear Friction Factor F	3.07	2.92	2.80	2.71
Total Volume of Concrete, m^3	53.20	55.60	57.20	58.20
Length of D/S Apron Floor $L_{D/S\ APRON}$, m	7.40	7.30	7.20	7.10
B. NON-OVER FLOW PORTION				
Length, m	2.00	2.00	2.00	2.00
Top Level including Free Board, m	101.10	101.40	101.60	101.80
Foundation Level, m	97.50	97.50	97.50	97.50
Height of Non-Over Flow H_{NOF}, m	3.60	3.90	4.10	4.30
Height H1, mm	2880	3120	3280	3440
Height H2, mm	720	780	820	860
Top Width of Non-Over Flow TW_{NOF}, mm	1000	1000	1000	1000
Base Width of Non-Over Flow B_{NOF}, mm	2660	2770	2840	2910
Width B1, mm	1660	1770	1840	1910
Pressure at Toe, t/m^2	13.51	14.85	15.75	16.64
Pressure at Heel, t/m^2	0.02	0.02	0.01	0.01
Factor of Safety against Overturning	2.31	2.40	2.45	2.50
Shear Friction Factor F	3.17	2.92	2.77	2.63
Total Volume of Concrete, m^3	16.76	18.84	20.27	21.74
C. WING WALLS (ABUTMENTS) ON BOTH BANKS				
Length in U/S $L_{U/SABUT}$, m	3.66	3.77	3.84	3.91
Length in D/S $L_{D/SABUT}$, m	2.00	2.00	2.00	2.00
Top Level including Free Board, m	101.10	101.40	101.60	101.80
Foundation Level, m	97.50	97.50	97.50	97.50
Total Height H_{ABUT}, m	3.60	3.90	4.10	4.30
D. D/S WING WALLS ON BOTH BANKS				
Length in D/S $L_{D/SWING}$, m	5.40	5.30	5.20	5.10
Top Level including Free Board, m	98.64	98.91	99.12	99.29
Foundation Level, m	97.50	97.50	97.50	97.50
Total Height $H_{D/SWING}$, m	1.14	1.41	1.62	1.79

TABLE 7.31
Design Results for Concrete Check Dam with Vertical D/S Face Height = 2.50 m and Stream Width = 15.0 m

Catchment Area, ha	50.00	100.00	150.00	200.00	250.00	300.00
A. WEIR						
Span of Weir W, m	12.00	12.00	12.00	12.00	12.00	12.00
Flood Discharge Q, m^3/s	6.54	11.00	14.91	18.50	21.87	25.07
Head Over Crest, m	0.44	0.63	0.77	0.89	0.99	1.09
Top Level, m	100.00	100.00	100.00	100.00	100.00	100.00
Foundation Level, m	97.50	97.50	97.50	97.50	97.50	97.50
Height of Weir H_{WEIR}, m	2.50	2.50	2.50	2.50	2.50	2.50
Top Width of Weir TW_{WEIR}, mm	2590	2690	2750	2810	2850	2890
Base Width of Weir B_{WEIR}, mm	2590	2690	2750	2810	2850	2890
Pressure at Toe, t/m^2	10.57	10.83	11.03	11.16	11.30	11.43
Pressure at Heel, t/m^2	0.01	0.01	0.00	0.02	0.02	0.03
Factor of Safety against Overturning	2.15	2.16	2.16	2.17	2.17	2.17
Shear Friction Factor F	3.18	3.04	2.94	2.87	2.81	2.76
Total Volume of Concrete, m^3	77.70	80.70	82.50	84.30	85.50	86.70
Length of D/S Apron Floor $L_{D/S\ APRON}$, m	7.50	7.40	7.30	7.20	7.20	7.20
B. NON-OVER FLOW PORTION						
Length, m	3.00	3.00	3.00	3.00	3.00	3.00
Top Level including Free Board, m	101.00	101.20	101.40	101.50	101.60	101.70
Foundation Level, m	97.50	97.50	97.50	97.50	97.50	97.50
Height of Non-Over Flow H_{NOF}, m	3.50	3.70	3.90	4.00	4.10	4.20
Height H1, mm	2800	2960	3120	3200	3280	3360
Height H2, mm	700	740	780	800	820	840
Top Width of Non-Over Flow TW_{NOF}, mm	1000	1000	1000	1000	1000	1000
Base Width of Non-Over Flow B_{NOF}, mm	2620	2690	2770	2800	2840	2870
Width B1, mm	1620	1690	1770	1800	1840	1870
Pressure at Toe, t/m^2	13.07	13.98	14.85	15.31	15.75	16.21
Pressure at Heel, t/m^2	0.02	0.00	0.02	0.00	0.01	0.00
Factor of Safety against Overturning	2.28	2.34	2.40	2.42	2.45	2.48
Shear Friction Factor F	3.26	3.07	2.92	2.83	2.77	2.69
Total Volume of Concrete, m^3	24.11	26.11	28.27	29.28	30.41	31.45
C. WING WALLS (ABUTMENTS) ON BOTH BANKS						
Length in U/S $L_{U/SABUT}$, m	3.62	3.69	3.77	3.80	3.84	3.87
Length in D/S $L_{D/SABUT}$, m	2.00	2.00	2.00	2.00	2.00	2.00
Top Level including Free Board, m	101.00	101.20	101.40	101.50	101.60	101.70
Foundation Level, m	97.50	97.50	97.50	97.50	97.50	97.50
Total Height H_{ABUT}, m	3.50	3.70	3.90	4.00	4.10	4.20
D. D/S WING WALLS ON BOTH BANKS						
Length in D/S $L_{D/SWING}$, m	5.50	5.40	5.30	5.20	5.20	5.20
Top Level including Free Board, m	98.49	98.70	98.85	98.99	99.10	99.21
Foundation Level, m	97.50	97.50	97.50	97.50	97.50	97.50
Total Height $H_{D/SWING}$, m	0.99	1.20	1.35	1.49	1.60	1.71

TABLE 7.32
Design Results for Concrete Check Dam with Vertical D/S Face Height = 2.50 m and Stream Width = 20.0 m

Catchment Area, ha	50.00	100.00	150.00	200.00	250.00	300.00
A. WEIR						
Span of Weir W, m	16.00	16.00	16.00	16.00	16.00	16.00
Flood Discharge Q, m^3/s	6.54	11.00	14.91	18.50	21.87	25.07
Head Over Crest, m	0.37	0.52	0.64	0.73	0.82	0.90
Top Level, m	100.00	100.00	100.00	100.00	100.00	100.00
Foundation Level, m	97.50	97.50	97.50	97.50	97.50	97.50
Height of Weir H_{WEIR}, m	2.50	2.50	2.50	2.50	2.50	2.50
Top Width of Weir TW_{WEIR}, mm	2550	2630	2690	2740	2780	2810
Base Width of Weir B_{WEIR}, mm	2550	2630	2690	2740	2780	2810
Pressure at Toe, t/m^2	10.48	10.69	10.85	10.95	11.07	11.19
Pressure at Heel, t/m^2	0.01	0.00	0.00	0.02	0.02	0.01
Factor of Safety against Overturning	2.15	2.15	2.16	2.16	2.16	2.16
Shear Friction Factor F	3.23	3.12	3.03	2.98	2.92	2.86
Total Volume of Concrete, m^3	102.00	105.20	107.60	109.60	111.20	112.40
Length of D/S Apron Floor $L_{D/S\ APRON}$, m	7.50	7.40	7.40	7.30	7.30	7.20
B. NON-OVER FLOW PORTION						
Length, m	4.00	4.00	4.00	4.00	4.00	4.00
Top Level including Free Board, m	100.90	101.10	101.20	101.30	101.40	101.50
Foundation Level, m	97.50	97.50	97.50	97.50	97.50	97.50
Height of Non-Over Flow H_{NOF}, m	3.40	3.60	3.70	3.80	3.90	4.00
Height H1, mm	2720	2880	2960	3040	3120	3200
Height H2, mm	680	720	740	760	780	800
Top Width of Non-Over Flow TW_{NOF}, mm	1000	1000	1000	1000	1000	1000
Base Width of Non-Over Flow B_{NOF}, mm	2580	2660	2690	2730	2770	2800
Width B1, mm	1580	1660	1690	1730	1770	1800
Pressure at Toe, t/m^2	12.62	13.51	13.98	14.41	14.85	15.31
Pressure at Heel, t/m^2	0.02	0.02	0.00	0.01	0.02	0.00
Factor of Safety against Overturning	2.25	2.31	2.34	2.37	2.40	2.42
Shear Friction Factor F	3.36	3.17	3.07	2.99	2.92	2.83
Total Volume of Concrete, m^3	30.79	33.52	34.81	36.24	37.69	39.04
C. WING WALLS (ABUTMENTS) ON BOTH BANKS						
Length in U/S $L_{U/SABUT}$, m	3.58	3.66	3.69	3.73	3.77	3.80
Length in D/S $L_{D/SABUT}$, m	2.00	2.00	2.00	2.00	2.00	2.00
Top Level including Free Board, m	100.90	101.10	101.20	101.30	101.40	101.50
Foundation Level, m	97.50	97.50	97.50	97.50	97.50	97.50
Total Height H_{ABUT}, m	3.40	3.60	3.70	3.80	3.90	4.00
D. D/S WING WALLS ON BOTH BANKS						
Length in D/S $L_{D/SWING}$, m	5.50	5.40	5.40	5.30	5.30	5.20
Top Level including Free Board, m	98.41	98.58	98.71	98.81	98.91	99.00
Foundation Level, m	97.50	97.50	97.50	97.50	97.50	97.50
Total Height $H_{D/SWING}$, m	0.91	1.08	1.21	1.31	1.41	1.50

7.4.3 Section with Steps on Both Faces

The tables developed after the rigorous design of the different components for 2.50 m high concrete check dams for different combinations are presented from Table 7.33 to Table 7.36.

TABLE 7.33
Design Results for Concrete Check Dam with Steps on Both Faces Height = 2.50 m and Stream Width = 5.0 m

Catchment Area, ha	50.00	100.00
A. WEIR		
Span of Weir W, m	4.00	4.00
Flood Discharge Q, m^3/s	6.54	11.00
Head Over Crest, m	0.92	1.31
Top Level, m	100.00	100.00
Foundation Level, m	97.50	97.50
Height of Weir H_{WEIR}, m	2.50	2.50
Size of Concrete Steps (W, H), m		
Base Width B_{WEIR}, Thickness H1'	3.09, 1.00	3.40, 1.00
Width B1', Thickness H2'	2.09, 1.00	2.40, 1.00
Top Width TW_{WEIR}, Thickness H3'	1.09, 0.50	1.40, 0.50
Pressure at Toe, t/m^2	8.35	8.83
Pressure at Heel, t/m^2	0.01	0.00
Factor of Safety against Overturning	1.94	2.00
Shear Friction Factor F	1.93	1.83
Total Volume of Concrete, m^3	22.90	26.00
Length of D/S Apron Floor $L_{D/S\ APRON}$, m	4.75	6.17
B. NON-OVER FLOW PORTION		
Length, m	1.000	1.00
Top Level including Free Board, m	101.50	102.00
Foundation Level, m	97.50	97.50
Height of Non-Over Flow H_{NOF}, m	4.00	4.50
Size of Concrete Steps (W, H), m		
Base Width B_{NOF}, Thickness H1	2.74, 1.00	3.09, 1.00
Width B1, Thickness H2	1.74, 1.00	2.09, 1.00
Width B2, Thickness H3	0.74, 1.00	1.09, 1.00
Top Width TW_{NOF}, Thickness H4	0.50, 1.00	0.50, 1.50
Pressure at Toe, t/m^2	10.29	11.44
Pressure at Heel, t/m^2	0.02	0.00
Factor of Safety against Overturning	1.91	1.95
Shear Friction Factor F	3.39	3.17
Total Volume of Concrete, m^3	5.72	7.02
C. WING WALLS (ABUTMENTS) ON BOTH BANKS		
Length in U/S $L_{U/SABUT}$, m	2.60	2.70
Length in D/S $L_{D/SABUT}$, m	2.60	2.70
Top Level including Free Board, m	101.50	102.00

(Continued)

TABLE 7.33 (Continued)

Catchment Area, ha	**50.00**	**100.00**
Foundation Level, m	97.50	97.50
Total Height H_{ABUT}, m	4.00	4.50
D. D/S WING WALLS ON BOTH BANKS		
Length in D/S $L_{D/SWING}$, m	6.00	5.60
Top Level including Free Board, m	99.02	99.45
Foundation Level, m	97.50	97.50
Total Height $H_{D/SWING}$, m	1.52	1.95

TABLE 7.34
Design Results for Concrete Check Dam with Steps on Both Faces Height = 2.50 m and Stream Width = 10.0 m

Catchment Area, ha	**50.00**	**100.00**	**150.00**	**200.00**
A. WEIR				
Span of Weir W, m	8.00	8.00	8.00	8.00
Flood Discharge Q, m^3/s	6.54	11.00	14.91	18.50
Head Over Crest, m	0.58	0.82	1.01	1.16
Top Level, m	100.00	100.00	100.00	100.00
Foundation Level, m	97.50	97.50	97.50	97.50
Height of Weir H_{WEIR}, m	2.50	2.50	2.50	2.50
Size of Concrete Steps (W, H), m				
Base Width B_{WEIR}, Thickness H1'	2.81, 1.00	3.01, 1.00	3.17, 1.00	3.29, 1.00
Width B1', Thickness H2'	1.81, 1.00	2.01, 1.00	2.17, 1.00	2.29, 1.00
Top Width TW_{WEIR}, Thickness H3'	0.81, 0.50	1.01, 0.50	1.17, 0.50	1.29, 0.50
Pressure at Toe, t/m^2	7.89	8.24	8.46	8.63
Pressure at Heel, t/m^2	0.05	0.01	0.02	0.03
Factor of Safety against Overturning	1.90	1.93	1.96	1.99
Shear Friction Factor F	2.04	1.96	1.91	1.87
Total Volume of Concrete, m^3	40.20	44.20	47.40	49.80
Length of D/S Apron Floor $L_{D/S\ APRON}$, m	3.36	4.36	5.08	5.65
B. NON-OVER FLOW PORTION				
Length, m	2.00	2.00	2.00	2.00
Top Level including Free Board, m	101.10	101.40	101.60	101.80
Foundation Level, m	97.50	97.50	97.50	97.50
Height of Non-Over Flow H_{NOF}, m	3.60	3.90	4.10	4.30
Size of Concrete Steps (W, H), m				
Base Width B_{NOF}, Thickness H1	2.45, 1.00	2.67, 1.00	2.80, 1.00	2.95, 1.00
Width B1, Thickness H2	1.45, 1.00	1.67, 1.00	1.80, 1.00	1.95, 1.00
Width B2, Thickness H3	0.50, 1.60	0.50, 1.90	0.80, 1.00	0.95, 1.00
Top Width TW_{NOF}, Thickness H4			0.50, 1.10	0.50, 1.30
Pressure at Toe, t/m^2	9.31	9.86	10.54	11.01
Pressure at Heel, t/m^2	0.03	0.01	0.00	0.00

(Continued)

TABLE 7.34 (Continued)

Catchment Area, ha	50.00	100.00	150.00	200.00
Factor of Safety against Overturning	1.88	1.89	1.92	1.93
Shear Friction Factor F	3.58	3.41	3.34	3.26
Total Volume of Concrete, m^3	9.40	10.58	11.90	13.00
C. WING WALLS (ABUTMENTS) ON BOTH BANKS				
Length in U/S $L_{U/SABUT}$, m	2.50	2.60	2.60	2.70
Length in D/S $L_{D/SABUT}$, m	2.50	2.60	2.60	2.70
Top Level including Free Board, m	101.1	101.4	101.6	101.8
Foundation Level, m	97.50	97.50	97.50	97.50
Total Height H_{ABUT}, m	3.60	3.90	4.10	4.30
D. D/S WING WALLS ON BOTH BANKS				
Length in D/S $L_{D/SWING}$, m	6.20	6.00	5.90	5.80
Top Level including Free Board, m	98.64	98.91	99.12	99.29
Foundation Level, m	97.50	97.50	97.50	97.50
Total Height $H_{D/SWING}$, m	1.14	1.41	1.62	1.79

TABLE 7.35
Design Results for Concrete Check Dam with Steps on Both Faces Height = 2.50 m and Stream Width = 15.0 m

Catchment Area, ha	50.00	100.00	150.00	200.00	250.00	300.00
A. WEIR						
Span of Weir W, m	12.00	12.00	12.00	12.00	12.00	12.00
Flood Discharge Q, m^3/s	6.54	11.00	14.91	18.50	21.87	25.07
Head Over Crest, m	0.44	0.63	0.77	0.89	0.99	1.09
Top Level, m	100.00	100.00	100.00	100.00	100.00	100.00
Foundation Level, m	97.50	97.50	97.50	97.50	97.50	97.50
Height of Weir H_{WEIR}, m	2.50	2.50	2.50	2.50	2.50	2.50
Size of Concrete Steps (W, H), m						
Base Width B_{WEIR}, Thickness H1'	2.68, 1.00	2.85, 1.00	2.97, 1.00	3.08, 1.00	3.15, 1.00	3.23, 1.00
Width B1', Thickness H2'	1.68, 1.00	1.85, 1.00	1.97, 1.00	2.08, 1.00	2.15, 1.00	2.23, 1.00
Top Width TW_{WEIR}, Thickness H3'	0.68, 0.50	0.85, 0.50	0.97, 0.50	1.08, 0.50	1.15, 0.50	1.23, 0.50
Pressure at Toe, t/m^2	7.74	7.99	8.17	8.29	8.44	8.56
Pressure at Heel, t/m^2	0.02	0.02	0.01	0.05	0.01	0.01
Factor of Safety against Overturning	1.87	1.90	1.92	1.95	1.95	1.97
Shear Friction Factor F	2.09	2.02	1.98	1.95	1.91	1.88
Total Volume of Concrete, m^3	56.40	61.50	65.10	68.40	70.50	72.90
Length of D/S Apron Floor $L_{D/S\ APRON}$, m	2.74	3.56	4.14	4.62	5.02	5.37
B. NON-OVER FLOW PORTION						
Length, m	3.00	3.00	3.00	3.00	3.00	3.00

(Continued)

TABLE 7.35 (Continued)

Catchment Area, ha	50.00	100.00	150.00	200.00	250.00	300.00
Top Level including Free Board, m	101.00	101.20	101.40	101.50	101.60	101.70
Foundation Level, m	97.50	97.50	97.50	97.50	97.50	97.50
Height of Non-Over Flow H_{NOF}, m	3.50	3.70	3.90	4.00	4.10	4.20
Size of Concrete Steps (W, H), m						
Base Width B_{NOF}, Thickness H1	2.38, 1.00	2.52, 1.00	2.66, 1.00	2.73, 1.00	2.80, 1.00	2.87, 1.00
Width B1, Thickness H2	1.38, 1.00	1.52, 1.00	1.66, 1.00	1.73, 1.00	1.80, 1.00	1.87, 1.00
Width B2, Thickness H3	0.50, 1.50	0.52, 1.70	0.66, 1.00	0.73, 1.00	0.80, 1.00	0.87, 1.00
Top Width TW_{NOF}, Thickness H4			0.50, 0.90	0.50, 1.00	0.50, 1.10	0.50, 1.20
Pressure at Toe, t/m²	9.15	9.56	10.01	10.31	10.54	10.76
Pressure at Heel, t/m²	0.00	0.01	0.04	0.00	0.00	0.02
Factor of Safety against Overturning	1.87	1.89	1.91	1.91	1.92	1.93
Shear Friction Factor F	3.65	3.53	3.43	3.38	3.34	3.29
Total Volume of Concrete, m³	13.53	14.77	16.29	17.07	17.85	18.63
C. WING WALLS (ABUTMENTS) ON BOTH BANKS						
Length in U/S $L_{U/SABUT}$, m	2.40	2.50	2.50	2.60	2.60	2.70
Length in D/S $L_{D/SABUT}$, m	2.40	2.50	2.50	2.60	2.60	2.70
Top Level including Free Board, m	101	101.2	101.4	101.5	101.6	101.7
Foundation Level, m	97.50	97.50	97.50	97.50	97.50	97.50
Total Height H_{ABUT}, m	3.50	3.70	3.90	4.00	4.10	4.20
D. D/S WING WALLS ON BOTH BANKS						
Length in D/S $L_{D/SWING}$, m	6.40	6.20	6.10	6.00	5.90	5.80
Top Level including Free Board, m	98.49	98.70	98.85	98.99	99.10	99.21
Foundation Level, m	97.50	97.50	97.50	97.50	97.50	97.50
Total Height $H_{D/SWING}$, m	0.99	1.20	1.35	1.49	1.60	1.71

TABLE 7.36
Design Results for Concrete Check Dam with Steps on Both Faces Height = 2.50 m and Stream Width = 20.0 m

Catchment Area, ha	50.00	100.00	150.00	200.00	250.00	300.00
A. WEIR						
Span of Weir W, m	16.00	16.00	16.00	16.00	16.00	16.00
Flood Discharge Q, m³/s	6.54	11.00	14.91	18.50	21.87	25.07
Head Over Crest, m	0.37	0.52	0.64	0.73	0.82	0.90
Top Level, m	100.00	100.00	100.00	100.00	100.00	100.00
Foundation Level, m	97.50	97.50	97.50	97.50	97.50	97.50
Height of Weir H_{WEIR}, m	2.50	2.50	2.50	2.50	2.50	2.50

(Continued)

TABLE 7.36 (Continued)

Catchment Area, ha	50.00	100.00	150.00	200.00	250.00	300.00
Size of Concrete Steps (W, H), m						
Base Width B_{WEIR}, Thickness H1'	2.61, 1.00	2.75, 1.00	2.86, 1.00	2.93, 1.00	3.01, 1.00	3.08, 1.00
Width B1', Thickness H2'	1.61, 1.00	1.75, 1.00	1.86, 1.00	1.93, 1.00	2.01, 1.00	2.08, 1.00
Top Width TW_{WEIR}, Thickness H3'	0.61, 0.50	0.75, 0.50	0.86, 0.50	0.93, 0.50	1.01, 0.50	1.08, 0.50
Pressure at Toe, t/m^2	7.64	7.83	8.00	8.12	8.24	8.31
Pressure at Heel, t/m^2	0.03	0.03	0.01	0.01	0.01	0.03
Factor of Safety against Overturning	1.86	1.89	1.90	1.92	1.93	1.95
Shear Friction Factor F	2.11	2.06	2.02	1.98	1.96	1.94
Total Volume of Concrete, m^3	72.40	78.00	82.40	85.20	88.40	91.20
Length of D/S Apron Floor $L_{D/S\ APRON}$, m	2.38	3.08	3.59	4.00	4.35	4.65
B. NON-OVER FLOW PORTION						
Length, m	4.00	4.00	4.00	4.00	4.00	4.00
Top Level including Free Board, m	100.90	101.10	101.20	101.30	101.40	101.50
Foundation Level, m	97.50	97.50	97.50	97.50	97.50	97.50
Height of Non-Over Flow H_{NOF}, m	3.40	3.60	3.70	3.80	3.90	4.00
Size of Concrete Steps (W, H), m						
Base Width B_{NOF}, Thickness H1	2.32, 1.00	2.45, 1.00	2.52, 1.00	2.58, 1.00	2.63, 1.00	2.73, 1.00
Width B1, Thickness H2	1.32, 1.00	1.45, 1.00	1.52, 1.00	1.58, 1.00	1.63, 1.00	1.73, 1.00
Width B2, Thickness H3	0.50, 1.40	0.50, 1.60	0.52, 1.70	0.58, 1.80	0.63, 1.90	0.73, 1.00
Top Width TW_{NOF}, Thickness H4						0.50, 1.00
Pressure at Toe, t/m^2	8.93	9.31	9.56	9.89	10.23	10.31
Pressure at Heel, t/m^2	0.03	0.03	0.01	0.02	0.00	0.00
Factor of Safety against Overturning	1.87	1.88	1.89	1.90	1.91	1.91
Shear Friction Factor F	3.73	3.58	3.53	3.48	3.42	3.38
Total Volume of Concrete, m^3	17.36	18.80	19.70	20.82	21.83	22.76
C. WING WALLS (ABUTMENTS) ON BOTH BANKS						
Length in U/S $L_{U/SABUT}$, m	2.40	2.40	2.50	2.50	2.60	2.60
Length in D/S $L_{D/SABUT}$, m	2.40	2.40	2.50	2.50	2.60	2.60
Top Level including Free Board, m	100.9	101.1	101.2	101.3	101.4	101.5
Foundation Level, m	97.50	97.50	97.50	97.50	97.50	97.50
Total Height H_{ABUT}, m	3.40	3.60	3.70	3.80	3.90	4.00
D. D/S WING WALLS ON BOTH BANKS						
Length in D/S $L_{D/SWING}$, m	6.40	6.30	6.20	6.10	6.00	6.00
Top Level including Free Board, m	98.41	98.58	98.71	98.81	98.91	99.00
Foundation Level, m	97.50	97.50	97.50	97.50	97.50	97.50
Total Height $H_{D/SWING}$, m	0.91	1.08	1.21	1.31	1.41	1.50

7.5 DESIGN OF 3.00 M HIGH CONCRETE CHECK DAMS

7.5.1 Section with Vertical U/S Face

The tables developed after the rigorous design of the different components for 3.00 m high concrete check dams for different combinations are presented from Table 7.37 to Table 7.40.

TABLE 7.37
Design Results for Concrete Check Dam with Vertical U/S Face Height = 3.00 m and Stream Width = 5.0 m

Catchment Area, ha	50.00	100.00
A. WEIR		
Span of Weir W, m	4.00	4.00
Flood Discharge Q, m^3/s	6.54	11.00
Head Over Crest, m	0.92	1.31
Top Level, m	100.00	100.00
Foundation Level, m	97.00	97.00
Height of Weir H_{WEIR}, m	3.00	3.00
Top Width of Weir TW_{WEIR}, mm	500	500
Base Width of Weir B_{WEIR}, mm	3180	3440
Width B1', mm	2680	2940
D/S Slope – $H_{DS'}$: 1.0(V)	0.892	0.980
Pressure at Toe, t/m^2	5.86	5.82
Pressure at Heel, t/m^2	0.01	0.02
Factor of Safety against Overturning	1.50	1.49
Shear Friction Factor F	1.99	1.89
Total Volume of Concrete, m^3	22.08	23.64
Length of D/S Apron Floor $L_{D/S\ APRON}$, m	8.90	8.60
B. NON-OVER FLOW PORTION		
Length, m	1.00	1.00
Top Level including Free Board, m	101.50	102.00
Foundation Level, m	97.00	97.00
Height of Non-Over Flow H_{NOF}, m	4.50	5.00
Height H1, mm	3000	2939
Height H2, mm	1500	2061
Top Width of Non-Over Flow TW_{NOF}, mm	500	560
Base Width of Non-Over Flow B_{NOF}, mm	3180	3440
Width B1, mm	2680	2940
D/S Slope – H_{DS}: 1.0(V)	0.892	0.980
Pressure at Toe, t/m^2	6.31	7.19
Pressure at Heel, t/m^2	0.55	0.03
Factor of Safety against Overturning	1.57	1.51
Shear Friction Factor F	1.90	1.69
Total Volume of Concrete, m^3	6.27	7.03
C. WING WALLS (ABUTMENTS) ON BOTH BANKS		
Length in U/S $L_{U/SABUT}$, m	1.50	1.50
Length in D/S $L_{D/SABUT}$, m	3.68	3.94
Top Level including Free Board, m	101.50	102.00
Foundation Level, m	97.00	97.00
Total Height H_{ABUT}, m	4.50	5.00
D. D/S WING WALLS ON BOTH BANKS		
Length in D/S $L_{D/SWING}$, m	5.22	4.66
Top Level including Free Board, m	98.52	98.95
Foundation Level, m	97.00	97.00
Total Height $H_{D/SWING}$, m	1.52	1.95

TABLE 7.38
Design Results for Concrete Check Dam with Vertical U/S Face Height = 3.00 m and Stream Width = 10.0 m

Catchment Area, ha	50.00	100.00	150.00	200.00
A. WEIR				
Span of Weir W, m	8.00	8.00	8.00	8.00
Flood Discharge Q, m^3/s	6.54	11.00	14.91	18.50
Head Over Crest, m	0.58	0.82	1.01	1.16
Top Level, m	100.00	100.00	100.00	100.00
Foundation Level, m	97.00	97.00	97.00	97.00
Height of Weir H_{WEIR}, m	3.00	3.00	3.00	3.00
Top Width of Weir TW_{WEIR}, mm	500	500	500	500
Base Width of Weir B_{WEIR}, mm	2940	3110	3250	3350
Width B1', mm	2440	2610	2750	2850
D/S Slope – $H_{DS'}$: 1.0(V)	0.812	0.870	0.915	0.950
Pressure at Toe, t/m^2	5.88	5.87	5.84	5.81
Pressure at Heel, t/m^2	0.03	0.01	0.02	0.04
Factor of Safety against Overturning	1.50	1.50	1.50	1.50
Shear Friction Factor F	2.10	2.02	1.97	1.93
Total Volume of Concrete, m^3	41.28	43.32	45.00	46.20
Length of D/S Apron Floor $L_{D/S\ APRON}$, m	9.10	8.90	8.80	8.70
B. NON-OVER FLOW PORTION				
Length, m	2.00	2.00	2.00	2.00
Top Level including Free Board, m	101.10	101.40	101.60	101.80
Foundation Level, m	97.00	97.00	97.00	97.00
Height of Non-Over Flow H_{NOF}, m	4.10	4.40	4.60	4.80
Height H1, mm	3000	3000	3000	3000
Height H2, mm	1100	1400	1600	1800
Top Width of Non-Over Flow TW_{NOF}, mm	500	500	500	500
Base Width of Non-Over Flow B_{NOF}, mm	2940	3110	3250	3350
Width B1, mm	2440	2610	2750	2850
D/S Slope – H_{DS}: 1.0(V)	0.812	0.870	0.915	0.950
Pressure at Toe, t/m^2	5.75	6.21	6.46	6.82
Pressure at Heel, t/m^2	0.96	0.62	0.42	0.14
Factor of Safety against Overturning	1.63	1.58	1.55	1.52
Shear Friction Factor F	2.09	1.93	1.86	1.77
Total Volume of Concrete, m^3	11.42	12.23	12.85	13.35
C. WING WALLS (ABUTMENTS) ON BOTH BANKS				
Length in U/S $L_{U/SABUT}$, m	1.50	1.50	1.50	1.50
Length in D/S $L_{D/SABUT}$, m	3.44	3.61	3.75	3.85
Top Level including Free Board, m	101.10	101.40	101.60	101.80
Foundation Level, m	97.00	97.00	97.00	97.00
Total Height H_{ABUT}, m	4.10	4.40	4.60	4.80
D. D/S WING WALLS ON BOTH BANKS				
Length in D/S $L_{D/SWING}$, m	5.66	5.29	5.05	4.85
Top Level including Free Board, m	98.14	98.41	98.62	98.79
Foundation Level, m	97.00	97.00	97.00	97.00
Total Height $H_{D/SWING}$, m	1.14	1.41	1.62	1.79

TABLE 7.39
Design Results for Concrete Check Dam with Vertical U/S Face Height = 3.00 m and Stream Width = 15.0 m

Catchment Area, ha	50.00	100.00	150.00	200.00	250.00	300.00
A. WEIR						
Span of Weir W, m	12.00	12.00	12.00	12.00	12.00	12.00
Flood Discharge Q, m^3/s	6.54	11.00	14.91	18.50	21.87	25.07
Head Over Crest, m	0.44	0.63	0.77	0.89	0.99	1.09
Top Level, m	100.00	100.00	100.00	100.00	100.00	100.00
Foundation Level, m	97.00	97.00	97.00	97.00	97.00	97.00
Height of Weir H_{WEIR}, m	3.00	3.00	3.00	3.00	3.00	3.00
Top Width of Weir TW_{WEIR}, mm	500	500	500	500	500	500
Base Width of Weir B_{WEIR}, mm	2830	2980	3080	3170	3230	3300
Width B1', mm	2330	2480	2580	2670	2730	2800
D/S Slope – H_{DS}.: 1.0(V)	0.775	0.825	0.860	0.890	0.910	0.932
Pressure at Toe, t/m^2	5.91	5.88	5.85	5.82	5.85	5.84
Pressure at Heel, t/m^2	0.01	0.02	0.03	0.05	0.02	0.02
Factor of Safety against Overturning	1.50	1.50	1.50	1.50	1.50	1.50
Shear Friction Factor F	2.14	2.08	2.04	2.01	1.97	1.95
Total Volume of Concrete, m^3	59.94	62.64	64.44	66.06	67.14	68.40
Length of D/S Apron Floor $L_{D/S\ APRON}$, m	9.20	9.10	9.00	8.90	8.80	8.70
B. NON-OVER FLOW PORTION						
Length, m	3.00	3.00	3.00	3.00	3.00	3.00
Top Level including Free Board, m	101.00	101.20	101.40	101.50	101.60	101.70
Foundation Level, m	97.00	97.00	97.00	97.00	97.00	97.00
Height of Non-Over Flow H_{NOF}, m	4.00	4.20	4.40	4.50	4.60	4.70
Height H1, mm	3000	3000	3000	3000	3000	3000
Height H2, mm	1000	1200	1400	1500	1600	1700
Top Width of Non-Over Flow TW_{NOF}, mm	500	500	500	500	500	500
Base Width of Non-Over Flow B_{NOF}, mm	2830	2980	3080	3170	3230	3300
Width B1, mm	2330	2480	2580	2670	2730	2800
D/S Slope – H_{DS}: 1.0(V)	0.775	0.825	0.860	0.890	0.910	0.932
Pressure at Toe, t/m^2	5.86	5.97	6.35	6.34	6.54	6.64
Pressure at Heel, t/m^2	0.85	0.80	0.51	0.52	0.36	0.28
Factor of Safety against Overturning	1.61	1.60	1.57	1.57	1.55	1.54
Shear Friction Factor F	2.11	2.02	1.92	1.89	1.85	1.81
Total Volume of Concrete, m^3	16.49	17.46	18.21	18.77	19.19	19.65
C. WING WALLS (ABUTMENTS) ON BOTH BANKS						
Length in U/S $L_{U/SABUT}$, m	1.50	1.50	1.50	1.50	1.50	1.50
Length in D/S $L_{D/SABUT}$, m	3.33	3.48	3.58	3.67	3.73	3.80
Top Level including Free Board, m	101.00	101.20	101.40	101.50	101.60	101.70
Foundation Level, m	97.00	97.00	97.00	97.00	97.00	97.00
Total Height H_{ABUT}, m	4.00	4.20	4.40	4.50	4.60	4.70
D. D/S WING WALLS ON BOTH BANKS						
Length in D/S $L_{D/SWING}$, m	5.87	5.62	5.42	5.23	5.07	4.90
Top Level including Free Board, m	97.99	98.20	98.35	98.49	98.60	98.71
Foundation Level, m	97.00	97.00	97.00	97.00	97.00	97.00
Total Height $H_{D/SWING}$, m	0.99	1.20	1.35	1.49	1.60	1.71

TABLE 7.40
Design Results for Concrete Check Dam with Vertical U/S Face Height = 3.00 m and Stream Width = 20.0 m

Catchment Area, ha	50.00	100.00	150.00	200.00	250.00	300.00
A. WEIR						
Span of Weir W, m	16.00	16.00	16.00	16.00	16.00	16.00
Flood Discharge Q, m^3/s	6.54	11.00	14.91	18.50	21.87	25.07
Head Over Crest, m	0.37	0.52	0.64	0.73	0.82	0.90
Top Level, m	100.00	100.00	100.00	100.00	100.00	100.00
Foundation Level, m	97.00	97.00	97.00	97.00	97.00	97.00
Height of Weir H_{WEIR}, m	3.00	3.00	3.00	3.00	3.00	3.00
Top Width of Weir TW_{WEIR}, mm	500	500	500	500	500	500
Base Width of Weir B_{WEIR}, mm	2780	2900	2990	3050	3110	3170
Width B1', mm	2280	2400	2490	2550	2610	2670
D/S Slope – $H_{DS'}$: 1.0(V)	0.760	0.800	0.830	0.850	0.870	0.890
Pressure at Toe, t/m^2	5.87	5.86	5.85	5.86	5.87	5.85
Pressure at Heel, t/m^2	0.07	0.06	0.05	0.03	0.01	0.02
Factor of Safety against Overturning	1.51	1.51	1.51	1.50	1.50	1.50
Shear Friction Factor F	2.17	2.12	2.08	2.05	2.02	2.00
Total Volume of Concrete, m^3	78.72	81.60	83.76	85.20	86.64	88.08
Length of D/S Apron Floor $L_{D/S\ APRON}$, m	9.30	9.10	9.10	9.00	8.90	8.90
B. NON-OVER FLOW PORTION						
Length, m	4.00	4.00	4.00	4.00	4.00	4.00
Top Level including Free Board, m	100.90	101.10	101.20	101.30	101.40	101.50
Foundation Level, m	97.00	97.00	97.00	97.00	97.00	97.00
Height of Non-Over Flow H_{NOF}, m	3.90	4.10	4.20	4.30	4.40	4.50
Height H1, mm	3000	3000	3000	3000	3000	3000
Height H2, mm	900	1100	1200	1300	1400	1500
Top Width of Non-Over Flow TW_{NOF}, mm	500	500	500	500	500	500
Base Width of Non-Over Flow B_{NOF}, mm	2780	2900	2990	3050	3110	3170
Width B1, mm	2280	2400	2490	2550	2610	2670
D/S Slope – H_{DS}: 1.0(V)	0.760	0.800	0.830	0.850	0.870	0.890
Pressure at Toe, t/m^2	5.63	5.92	5.91	6.08	6.21	6.34
Pressure at Heel, t/m^2	1.03	0.82	0.85	0.72	0.62	0.52
Factor of Safety against Overturning	1.64	1.61	1.61	1.59	1.58	1.57
Shear Friction Factor F	2.17	2.06	2.03	1.98	1.93	1.89
Total Volume of Concrete, m^3	21.48	22.60	23.34	23.90	24.46	25.02
C. WING WALLS (ABUTMENTS) ON BOTH BANKS						
Length in U/S $L_{U/SABUT}$, m	1.50	1.50	1.50	1.50	1.50	1.50
Length in D/S $L_{D/SABUT}$, m	3.28	3.40	3.49	3.55	3.61	3.67
Top Level including Free Board, m	100.90	101.10	101.20	101.30	101.40	101.50
Foundation Level, m	97.00	97.00	97.00	97.00	97.00	97.00
Total Height H_{ABUT}, m	3.90	4.10	4.20	4.30	4.40	4.50
D. D/S WING WALLS ON BOTH BANKS						
Length in D/S $L_{D/SWING}$, m	6.02	5.70	5.61	5.45	5.29	5.23
Top Level including Free Board, m	97.91	98.08	98.21	98.31	98.41	98.50
Foundation Level, m	97.00	97.00	97.00	97.00	97.00	97.00
Total Height $H_{D/SWING}$, m	0.91	1.08	1.21	1.31	1.41	1.50

7.5.2 Section with Vertical D/S Face

The tables developed after the rigorous design of the different components for 3.00 m high concrete check dams for different combinations are presented from Table 7.41 to Table 7.44.

TABLE 7.41
Design Results for Concrete Check Dam with Vertical D/S Face Height = 3.00 m and Stream Width = 5.0 m

Catchment Area, ha	**50.00**	**100.00**
A. WEIR		
Span of Weir W, m	4.00	4.00
Flood Discharge Q, m^3/s	6.54	11.00
Head Over Crest, m	0.92	1.31
Top Level, m	100.00	100.00
Foundation Level, m	97.00	97.00
Height of Weir H_{WEIR}, m	3.00	3.00
Top Width of Weir TW_{WEIR}, mm	3300	3460
Base Width of Weir B_{WEIR}, mm	3300	3460
Pressure at Toe, t/m^2	13.23	13.75
Pressure at Heel, t/m^2	0.00	0.00
Factor of Safety against Overturning	2.16	2.16
Shear Friction Factor F	2.62	2.45
Total Volume of Concrete, m^3	39.60	41.52
Length of D/S Apron Floor $L_{D/S\ APRON}$, m	8.70	8.60
B. NON-OVER FLOW PORTION		
Length, m	1.00	1.00
Top Level including Free Board, m	101.50	102.00
Foundation Level, m	97.00	97.00
Height of Non-Over Flow H_{NOF}, m	4.50	5.00
Height H1, mm	3600	4000
Height H2, mm	900	1000
Top Width of Non-Over Flow TW_{NOF}, mm	1000	1000
Base Width of Non-Over Flow B_{NOF}, mm	3030	4360
Width B1, mm	2030	3360
Pressure at Toe, t/m^2	17.01	18.63
Pressure at Heel, t/m^2	0.01	0.28
Factor of Safety against Overturning	2.31	2.38
Shear Friction Factor F	2.42	2.95
Total Volume of Concrete, m^3	11.81	18.44
C. WING WALLS (ABUTMENTS) ON BOTH BANKS		
Length in U/S $L_{U/SABUT}$, m	4.03	5.36
Length in D/S $L_{D/SABUT}$, m	2.00	2.00
Top Level including Free Board, m	101.50	102.00
Foundation Level, m	97.00	97.00
Total Height H_{ABUT}, m	4.50	5.00

(Continued)

TABLE 7.41 (Continued)

Catchment Area, ha	50.00	100.00
D. D/S WING WALLS ON BOTH BANKS		
Length in D/S $L_{D/SWING}$, m	6.70	6.60
Top Level including Free Board, m	98.52	98.95
Foundation Level, m	97.00	97.00
Total Height $H_{D/SWING}$, m	1.52	1.95

TABLE 7.42
Design Results for Concrete Check Dam with Vertical D/S Face Height = 3.00 m and Stream Width = 10.0 m

Catchment Area, ha	50.00	100.00	150.00	200.00
A. WEIR				
Span of Weir W, m	8.00	8.00	8.00	8.00
Flood Discharge Q, m^3/s	6.54	11.00	14.91	18.50
Head Over Crest, m	0.58	0.82	1.01	1.16
Top Level, m	100.00	100.00	100.00	100.00
Foundation Level, m	97.00	97.00	97.00	97.00
Height of Weir H_{WEIR}, m	3.00	3.00	3.00	3.00
Top Width of Weir TW_{WEIR}, mm	3140	3260	3340	3410
Base Width of Weir B_{WEIR}, mm	3140	3260	3340	3410
Pressure at Toe, t/m^2	12.75	13.07	13.35	13.52
Pressure at Heel, t/m^2	0.02	0.02	0.00	0.03
Factor of Safety against Overturning	2.16	2.16	2.16	2.17
Shear Friction Factor F	2.80	2.67	2.58	2.52
Total Volume of Concrete, m^3	75.36	78.24	80.16	81.84
Length of D/S Apron Floor $L_{D/S\ APRON}$, m	8.90	8.80	8.70	8.60
B. NON-OVER FLOW PORTION				
Length, m	2.00	2.00	2.00	2.00
Top Level including Free Board, m	101.10	101.40	101.60	101.80
Foundation Level, m	97.00	97.00	97.00	97.00
Height of Non-Over Flow H_{NOF}, m	4.10	4.40	4.60	4.80
Height H1, mm	3075	3080	2944	3120
Height H2, mm	1025	1320	1656	1680
Top Width of Non-Over Flow TW_{NOF}, mm	1000	1000	1000	1000
Base Width of Non-Over Flow B_{NOF}, mm	3250	3430	3570	3640
Width B1, mm	2250	2430	2570	2640
Pressure at Toe, t/m^2	14.23	15.48	15.58	16.45
Pressure at Heel, t/m^2	0.45	0.00	0.12	0.17
Factor of Safety against Overturning	2.30	2.28	2.19	2.27
Shear Friction Factor F	2.90	2.70	2.55	2.45
Total Volume of Concrete, m^3	22.04	23.77	24.33	26.07

(Continued)

TABLE 7.42 (Continued)

Catchment Area, ha	50.00	100.00	150.00	200.00
C. WING WALLS (ABUTMENTS) ON BOTH BANKS				
Length in U/S $L_{U/SABUT}$, m	4.25	4.43	4.57	4.64
Length in D/S $L_{D/SABUT}$, m	2.00	2.00	2.00	2.00
Top Level including Free Board, m	101.10	101.40	101.60	101.80
Foundation Level, m	97.00	97.00	97.00	97.00
Total Height H_{ABUT}, m	4.10	4.40	4.60	4.80
D. D/S WING WALLS ON BOTH BANKS				
Length in D/S $L_{D/SWING}$, m	6.90	6.80	6.70	6.60
Top Level including Free Board, m	98.14	98.41	98.62	98.79
Foundation Level, m	97.00	97.00	97.00	97.00
Total Height $H_{D/SWING}$, m	1.14	1.41	1.62	1.79

TABLE 7.43
Design Results for Concrete Check Dam with Vertical D/S Face Height = 3.00 m and Stream Width = 15.0 m

Catchment Area, ha	50.00	100.00	150.00	200.00	250.00	300.00
A. WEIR						
Span of Weir W, m	12.00	12.00	12.00	12.00	12.00	12.00
Flood Discharge Q, m^3/s	6.54	11.00	14.91	18.50	21.87	25.07
Head Over Crest, m	0.44	0.63	0.77	0.89	0.99	1.09
Top Level, m	100.00	100.00	100.00	100.00	100.00	100.00
Foundation Level, m	97.00	97.00	97.00	97.00	97.00	97.00
Height of Weir H_{WEIR}, m	3.00	3.00	3.00	3.00	3.00	3.00
Top Width of Weir TW_{WEIR}, mm	3060	3160	3230	3290	3340	3380
Base Width of Weir B_{WEIR}, mm	3060	3160	3230	3290	3340	3380
Pressure at Toe, t/m^2	12.57	12.84	13.03	13.18	13.29	13.43
Pressure at Heel, t/m^2	0.02	0.00	0.00	0.01	0.03	0.02
Factor of Safety against Overturning	2.15	2.15	2.16	2.16	2.16	2.16
Shear Friction Factor F	2.87	2.76	2.69	2.64	2.59	2.55
Total Volume of Concrete, m^3	110.16	113.76	116.28	118.44	120.24	121.68
Length of D/S Apron Floor $L_{D/S\ APRON}$, m	9.00	8.90	8.80	8.80	8.70	8.70
B. NON-OVER FLOW PORTION						
Length, m	3.00	3.00	3.00	3.00	3.00	3.00
Top Level including Free Board, m	101.00	101.20	101.40	101.50	101.60	101.70
Foundation Level, m	97.00	97.00	97.00	97.00	97.00	97.00
Height of Non-Over Flow H_{NOF}, m	4.00	4.20	4.40	4.50	4.60	4.70
Height H1, mm	2760	2898	2860	2925	2714	2632
Height H2, mm	1240	1302	1540	1575	1886	2068
Top Width of Non-Over Flow TW_{NOF}, mm	1000	1000	1000	1000	1000	1000
Base Width of Non-Over Flow B_{NOF}, mm	3320	3450	3500	3600	3840	3910
Width B1, mm	2320	2450	2500	2600	2840	2910

(Continued)

TABLE 7.43 (Continued)

Catchment Area, ha	50.00	100.00	150.00	200.00	250.00	300.00
Pressure at Toe, t/m^2	13.67	14.52	14.95	15.36	14.68	15.14
Pressure at Heel, t/m^2	0.03	0.00	0.03	0.00	0.40	0.00
Factor of Safety against Overturning	2.16	2.24	2.17	2.23	2.15	2.09
Shear Friction Factor F	2.99	2.89	2.69	2.67	2.67	2.59
Total Volume of Concrete, m^3	31.21	33.90	34.65	36.32	36.92	37.08
C. WING WALLS (ABUTMENTS) ON BOTH BANKS						
Length in U/S $L_{U/SABUT}$, m	4.32	4.45	4.50	4.60	4.84	4.91
Length in D/S $L_{D/SABUT}$, m	2.00	2.00	2.00	2.00	2.00	2.00
Top Level including Free Board, m	101.00	101.20	101.40	101.50	101.60	101.70
Foundation Level, m	97.00	97.00	97.00	97.00	97.00	97.00
Total Height H_{ABUT}, m	4.00	4.20	4.40	4.50	4.60	4.70
D. D/S WING WALLS ON BOTH BANKS						
Length in D/S $L_{D/SWING}$, m	7.00	6.90	6.80	6.80	6.70	6.70
Top Level including Free Board, m	97.99	98.20	98.35	98.49	98.60	98.71
Foundation Level, m	97.00	97.00	97.00	97.00	97.00	97.00
Total Height $H_{D/SWING}$, m	0.99	1.20	1.35	1.49	1.60	1.71

TABLE 7.44

Design Results for Concrete Check Dam with Vertical D/S Face Height = 3.00 m and Stream Width = 20.0 m

Catchment Area, ha	50.00	100.00	150.00	200.00	250.00	300.00
A. WEIR						
Span of Weir W, m	16.00	16.00	16.00	16.00	16.00	16.00
Flood Discharge Q, m^3/s	6.54	11.00	14.91	18.50	21.87	25.07
Head Over Crest, m	0.37	0.52	0.64	0.73	0.82	0.90
Top Level, m	100.00	100.00	100.00	100.00	100.00	100.00
Foundation Level, m	97.00	97.00	97.00	97.00	97.00	97.00
Height of Weir H_{WEIR}, m	3.00	3.00	3.00	3.00	3.00	3.00
Top Width of Weir TW_{WEIR}, mm	3020	3100	3170	3220	3260	3300
Base Width of Weir B_{WEIR}, mm	3020	3100	3170	3220	3260	3300
Pressure at Toe, t/m^2	12.47	12.69	12.84	12.94	13.07	13.17
Pressure at Heel, t/m^2	0.02	0.00	0.02	0.03	0.02	0.03
Factor of Safety against Overturning	2.15	2.15	2.16	2.16	2.16	2.16
Shear Friction Factor F	2.91	2.82	2.76	2.72	2.67	2.64

(Continued)

TABLE 7.44 (Continued)

Catchment Area, ha	50.00	100.00	150.00	200.00	250.00	300.00
Total Volume of Concrete, m^3	144.96	148.80	152.16	154.56	156.48	158.40
Length of D/S Apron Floor $L_{D/S\ APRON}$, m	9.00	8.90	8.90	8.80	8.80	8.70
B. NON-OVER FLOW PORTION						
Length, m	4.00	4.00	4.00	4.00	4.00	4.00
Top Level including Free Board, m	100.90	101.10	101.20	101.30	101.40	101.50
Foundation Level, m	97.00	97.00	97.00	97.00	97.00	97.00
Height of Non-Over Flow H_{NOF}, m	3.90	4.10	4.20	4.30	4.40	4.50
Height H1, mm	2691	2829	2898	2752	2684	2745
Height H2, mm	1209	1271	1302	1548	1716	1755
Top Width of Non-Over Flow TW_{NOF}, mm	1000	1000	1000	1000	1000	1000
Base Width of Non-Over Flow B_{NOF}, mm	3300	3380	3450	3620	3660	3700
Width B1, mm	2300	2380	2450	2620	2660	2700
Pressure at Toe, t/m^2	13.18	14.10	14.52	14.41	14.33	14.83
Pressure at Heel, t/m^2	0.10	0.01	0.00	0.00	0.20	0.09
Factor of Safety against Overturning	2.13	2.20	2.24	2.18	2.12	2.15
Shear Friction Factor F	3.09	2.94	2.89	2.86	2.75	2.69
Total Volume of Concrete, m^3	40.36	43.33	45.20	46.04	46.16	47.65
C. WING WALLS (ABUTMENTS) ON BOTH BANKS						
Length in U/S $L_{U/SABUT}$, m	4.30	4.38	4.45	4.62	4.66	4.70
Length in D/S $L_{D/SABUT}$, m	2.00	2.00	2.00	2.00	2.00	2.00
Top Level including Free Board, m	100.90	101.10	101.20	101.30	101.40	101.50
Foundation Level, m	97.00	97.00	97.00	97.00	97.00	97.00
Total Height H_{ABUT}, m	3.90	4.10	4.20	4.30	4.40	4.50
D. D/S WING WALLS ON BOTH BANKS						
Length in D/S $L_{D/SWING}$, m	7.00	6.90	6.90	6.80	6.80	6.70
Top Level including Free Board, m	97.91	98.08	98.21	98.31	98.41	98.50
Foundation Level, m	97.00	97.00	97.00	97.00	97.00	97.00
Total Height $H_{D/SWING}$, m	0.91	1.08	1.21	1.31	1.41	1.50

7.5.3 Section with Steps on Both Faces

The tables developed after the rigorous design of the different components for 3.00 m high concrete check dams for different combinations are presented from Table 7.45 to Table 7.48.

TABLE 7.45
Design Results for Concrete Check Dam with Steps on Both Faces Height = 3.00 m and Stream Width = 5.0 m

Catchment Area, ha	50.00	100.00
A. WEIR		
Span of Weir W, m	4.00	4.00
Flood Discharge Q, m^3/s	6.54	11.00
Head Over Crest, m	0.92	1.31
Top Level, m	100.00	100.00
Foundation Level, m	97.00	97.00
Height of Weir H_{WEIR}, m	3.00	3.00
Size of Concrete Steps (W, H), m		
Base Width B_{WEIR}, Thickness H1’	3.52, 1.00	3.83, 1.00
Width B1’, Thickness H2’	2.52, 1.00	2.83, 1.00
Top Width TW_{WEIR}, Thickness H3’	1.52, 1.00	1.83, 1.00
Pressure at Toe, t/m^2	9.81	10.30
Pressure at Heel, t/m^2	0.00	0.03
Factor of Safety against Overturning	1.91	1.96
Shear Friction Factor F	1.62	1.54
Total Volume of Concrete, m^3	30.24	33.96
Length of D/S Apron Floor $L_{D/S\ APRON}$, m	5.21	6.75
B. NON-OVER FLOW PORTION		
Length, m	1.00	1.00
Top Level including Free Board, m	101.50	102.00
Foundation Level, m	97.00	97.00
Height of Non-Over Flow H_{NOF}, m	4.50	5.00
Size of Concrete Steps (W, H), m		
Base Width B_{NOF}, Thickness H1	3.28, 1.00	3.67, 1.00
Width B1, Thickness H2	2.28, 1.00	2.67, 1.00
Width B2, Thickness H3	1.28, 1.00	1.67, 1.00
Top Width TW_{NOF}, Thickness H4	0.50, 1.50	0.50, 2.00
Pressure at Toe, t/m^2	11.06	12.02
Pressure at Heel, t/m^2	0.00	0.02
Factor of Safety against Overturning	1.88	1.91
Shear Friction Factor F	3.26	3.10
Total Volume of Concrete, m^3	7.59	9.01
C. WING WALLS (ABUTMENTS) ON BOTH BANKS		
Length in U/S $L_{U/SABUT}$, m	2.80	3.00
Length in D/S $L_{D/SABUT}$, m	2.80	3.00
Top Level including Free Board, m	101.50	102.00
Foundation Level, m	97.00	97.00
Total Height H_{ABUT}, m	4.50	5.00
D. D/S WING WALLS ON BOTH BANKS		
Length in D/S $L_{D/SWING}$, m	7.50	7.20
Top Level including Free Board, m	98.52	98.95
Foundation Level, m	97.00	97.00
Total Height $H_{D/SWING}$, m	1.52	1.95

TABLE 7.46
Design Results for Concrete Check Dam with Steps on Both Faces Height = 3.00 m and Stream Width = 10.0 m

Catchment Area, ha	50.00	100.00	150.00	200.00
A. WEIR				
Span of Weir W, m	8.00	8.00	8.00	8.00
Flood Discharge Q, m^3/s	6.54	11.00	14.91	18.50
Head Over Crest, m	0.58	0.82	1.01	1.16
Top Level, m	100.00	100.00	100.00	100.00
Foundation Level, m	97.00	97.00	97.00	97.00
Height of Weir H_{WEIR}, m	3.00	3.00	3.00	3.00
Size of Concrete Steps (W, H), m				
Base Width B_{WEIR}, Thickness H1'	3.23, 1.00	3.44, 1.00	3.59, 1.00	3.72, 1.00
Width B1', Thickness H2'	2.23, 1.00	2.44, 1.00	2.59, 1.00	2.72, 1.00
Top Width TW_{WEIR}, Thickness H3'	1.23, 1.00	1.44, 1.00	1.59, 1.00	1.72, 1.00
Pressure at Toe, t/m^2	9.32	9.65	9.92	10.10
Pressure at Heel, t/m^2	0.00	0.01	0.00	0.03
Factor of Safety against Overturning	1.86	1.90	1.92	1.95
Shear Friction Factor F	1.70	1.64	1.59	1.57
Total Volume of Concrete, m^3	53.52	58.56	62.16	65.28
Length of D/S Apron Floor $L_{D/S\ APRON}$, m	3.68	4.78	5.56	6.19
B. NON-OVER FLOW PORTION				
Length, m	2.00	2.00	2.00	2.00
Top Level including Free Board, m	101.10	101.40	101.60	101.80
Foundation Level, m	97.00	97.00	97.00	97.00
Height of Non-Over Flow H_{NOF}, m	4.10	4.40	4.60	4.80
Size of Concrete Steps (W, H), m				
Base Width B_{NOF}, Thickness H1	3.00, 1.00	3.21, 1.00	3.37, 1.00	3.51, 1.00
Width B1, Thickness H2	2.00, 1.00	2.21, 1.00	2.37, 1.00	2.51, 1.00
Width B2, Thickness H3	1.00, 1.00	1.21, 1.00	1.37, 1.00	1.51, 1.00
Top Width TW_{NOF}, Thickness H4	0.50, 1.10	0.50, 1.40	0.50, 1.60	0.51, 1.80
Pressure at Toe, t/m^2	10.21	10.85	11.25	11.65
Pressure at Heel, t/m^2	0.01	0.00	0.02	0.03
Factor of Safety against Overturning	1.85	1.87	1.89	1.90
Shear Friction Factor F	3.42	3.29	3.24	3.16
Total Volume of Concrete, m^3	13.10	14.66	15.82	16.90
C. WING WALLS (ABUTMENTS) ON BOTH BANKS				
Length in U/S $L_{U/SABUT}$, m	2.70	2.80	2.80	2.90
Length in D/S $L_{D/SABUT}$, m	2.70	2.80	2.80	2.90
Top Level including Free Board, m	101.1	101.4	101.6	101.8
Foundation Level, m	97.00	97.00	97.00	97.00
Total Height H_{ABUT}, m	4.10	4.40	4.60	4.80
D. D/S WING WALLS ON BOTH BANKS				
Length in D/S $L_{D/SWING}$, m	7.80	7.60	7.50	7.30
Top Level including Free Board, m	98.14	98.41	98.62	98.79
Foundation Level, m	97.00	97.00	97.00	97.00
Total Height $H_{D/SWING}$, m	1.14	1.41	1.62	1.79

TABLE 7.47
Design Results for Concrete Check Dam with Steps on Both Faces Height = 3.00 m and Stream Width = 15.0 m

Catchment Area, ha	50.00	100.00	150.00	200.00	250.00	300.00
A. WEIR						
Span of Weir W, m	12.00	12.00	12.00	12.00	12.00	12.00
Flood Discharge Q, m^3/s	6.54	11.00	14.91	18.50	21.87	25.07
Head Over Crest, m	0.44	0.63	0.77	0.89	0.99	1.09
Top Level, m	100.00	100.00	100.00	100.00	100.00	100.00
Foundation Level, m	97.00	97.00	97.00	97.00	97.00	97.00
Height of Weir H_{WEIR}, m	3.00	3.00	3.00	3.00	3.00	3.00
Size of Concrete Steps (W, H), m						
Base Width B_{WEIR}, Thickness H1'	3.11, 1.00	3.28, 1.00	3.40, 1.00	3.50, 1.00	3.58, 1.00	3.66, 1.00
Width B1', Thickness H2'	2.11, 1.00	2.28, 1.00	2.40, 1.00	2.50, 1.00	2.58, 1.00	2.66, 1.00
Top Width TW_{WEIR}, Thickness H3'	1.11, 1.00	1.28, 1.00	1.40, 1.00	1.50, 1.00	1.58, 1.00	1.66, 1.00
Pressure at Toe, t/m^2	9.09	9.39	9.60	9.74	9.90	10.03
Pressure at Heel, t/m^2	0.02	0.01	0.00	0.02	0.00	0.01
Factor of Safety against Overturning	1.85	1.87	1.89	1.91	1.92	1.93
Shear Friction Factor F	1.74	1.69	1.65	1.63	1.60	1.58
Total Volume of Concrete, m^3	75.96	82.08	86.40	90.00	92.88	95.76
Length of D/S Apron Floor $L_{D/S\ APRON}$, m	3.01	3.90	4.54	5.06	5.50	5.89
B. NON-OVER FLOW PORTION						
Length, m	3.00	3.00	3.00	3.00	3.00	3.00
Top Level including Free Board, m	101.00	101.20	101.40	101.50	101.60	101.70
Foundation Level, m	97.00	97.00	97.00	97.00	97.00	97.00
Height of Non-Over Flow H_{NOF}, m	4.00	4.20	4.40	4.50	4.60	4.70
Size of Concrete Steps (W, H), m						
Base Width B_{NOF}, Thickness H1	2.93, 1.00	3.07, 1.00	3.21, 1.00	3.28, 1.00	3.37, 1.00	3.44, 1.00
Width B1, Thickness H2	1.93, 1.00	2.07, 1.00	2.21, 1.00	2.28, 1.00	2.37, 1.00	2.44, 1.00
Width B2, Thickness H3	0.93, 1.00	1.07, 1.00	1.21, 1.00	1.28, 1.00	1.37, 1.00	1.44, 1.00
Top Width TW_{NOF}, Thickness H4	0.50, 1.00	0.50, 1.20	0.50, 1.40	0.50, 1.50	0.50, 1.60	0.50, 1.70
Pressure at Toe, t/m^2	9.99	10.42	10.85	11.06	11.25	11.43
Pressure at Heel, t/m^2	0.00	0.01	0.00	0.00	0.02	0.04
Factor of Safety against Overturning	1.84	1.86	1.87	1.88	1.89	1.89
Shear Friction Factor F	3.46	3.38	3.29	3.26	3.24	3.20
Total Volume of Concrete, m^3	18.87	20.43	21.99	22.77	23.73	24.51
C. WING WALLS (ABUTMENTS) ON BOTH BANKS						
Length in U/S $L_{U/SABUT}$, m	2.60	2.70	2.70	2.80	2.80	2.90
Length in D/S $L_{D/SABUT}$, m	2.60	2.70	2.70	2.80	2.80	2.90
Top Level including Free Board, m	101	101.2	101.4	101.5	101.6	101.7
Foundation Level, m	97.00	97.00	97.00	97.00	97.00	97.00
Total Height H_{ABUT}, m	4.00	4.20	4.40	4.50	4.60	4.70
D. D/S WING WALLS ON BOTH BANKS						
Length in D/S $L_{D/SWING}$, m	7.90	7.80	7.60	7.50	7.50	7.40
Top Level including Free Board, m	97.99	98.20	98.35	98.49	98.60	98.71
Foundation Level, m	97.00	97.00	97.00	97.00	97.00	97.00
Total Height $H_{D/SWING}$, m	0.99	1.20	1.35	1.49	1.60	1.71

TABLE 7.48
Design Results for Concrete Check Dam with Steps on Both Faces Height = 3.00 m and Stream Width = 20.0 m

Catchment Area, ha	50.00	100.00	150.00	200.00	250.00	300.00
A. WEIR						
Span of Weir W, m	16.00	16.00	16.00	16.00	16.00	16.00
Flood Discharge Q, m^3/s	6.54	11.00	14.91	18.50	21.87	25.07
Head Over Crest, m	0.37	0.52	0.64	0.73	0.82	0.90
Top Level, m	100.00	100.00	100.00	100.00	100.00	100.00
Foundation Level, m	97.00	97.00	97.00	97.00	97.00	97.00
Height of Weir H_{WEIR}, m	3.00	3.00	3.00	3.00	3.00	3.00
Size of Concrete Steps (W, H), m						
Base Width B_{WEIR}, Thickness H1'	3.05, 1.00	3.18, 1.00	3.29, 1.00	3.37, 1.00	3.44, 1.00	3.50, 1.00
Width B1', Thickness H2'	2.05, 1.00	2.18, 1.00	2.29, 1.00	2.37, 1.00	2.44, 1.00	2.50, 1.00
Top Width TW_{WEIR}, Thickness H3'	1.05, 1.00	1.18, 1.00	1.29, 1.00	1.37, 1.00	1.44, 1.00	1.50, 1.00
Pressure at Toe, t/m^2	8.97	9.23	9.38	9.53	9.65	9.77
Pressure at Heel, t/m^2	0.05	0.01	0.03	0.01	0.01	0.00
Factor of Safety against Overturning	1.84	1.86	1.88	1.89	1.90	1.91
Shear Friction Factor F	1.76	1.72	1.69	1.66	1.64	1.62
Total Volume of Concrete, m^3	98.40	104.64	109.92	113.76	117.12	120.00
Length of D/S Apron Floor $L_{D/S\ APRON}$, m	2.60	3.38	3.93	4.38	4.76	5.10
B. NON-OVER FLOW PORTION						
Length, m	4.00	4.00	4.00	4.00	4.00	4.00
Top Level including Free Board, m	100.90	101.10	101.20	101.30	101.40	101.50
Foundation Level, m	97.00	97.00	97.00	97.00	97.00	97.00
Height of Non-Over Flow H_{NOF}, m	3.90	4.10	4.20	4.30	4.40	4.50
Size of Concrete Steps (W, H), m						
Base Width B_{NOF}, Thickness H1	2.87, 1.00	3.00, 1.00	3.07, 1.00	3.14, 1.00	3.21, 1.00	3.28, 1.00
Width B1, Thickness H2	1.87, 1.00	2.00, 1.00	2.07, 1.00	2.14, 1.00	2.21, 1.00	2.28, 1.00
Width B2, Thickness H3	0.87, 1.00	1.00, 1.00	1.07, 1.00	1.14, 1.00	1.21, 1.00	1.28, 1.00
Top Width TW_{NOF}, Thickness H4	0.50, 0.90	0.50, 1.10	0.50, 1.20	0.50, 1.30	0.50, 1.40	0.50, 1.50
Pressure at Toe, t/m^2	9.71	10.21	10.42	10.65	10.85	11.06
Pressure at Heel, t/m^2	0.05	0.01	0.01	0.00	0.00	0.00
Factor of Safety against Overturning	1.84	1.85	1.86	1.87	1.87	1.88
Shear Friction Factor F	3.52	3.42	3.38	3.34	3.29	3.26
Total Volume of Concrete, m^3	24.24	26.20	27.24	28.28	29.32	30.36
C. WING WALLS (ABUTMENTS) ON BOTH BANKS						
Length in U/S $L_{U/SABUT}$, m	2.60	2.60	2.70	2.70	2.80	2.80
Length in D/S $L_{D/SABUT}$, m	2.60	2.60	2.70	2.70	2.80	2.80
Top Level including Free Board, m	100.9	101.1	101.2	101.3	101.4	101.5
Foundation Level, m	97.00	97.00	97.00	97.00	97.00	97.00
Total Height H_{ABUT}, m	3.90	4.10	4.20	4.30	4.40	4.50
D. D/S WING WALLS ON BOTH BANKS						
Length in D/S $L_{D/SWING}$, m	8.00	7.90	7.80	7.70	7.60	7.50
Top Level including Free Board, m	97.91	98.08	98.21	98.31	98.41	98.50
Foundation Level, m	97.00	97.00	97.00	97.00	97.00	97.00
Total Height $H_{/SWING}$, m	0.91	1.08	1.21	1.31	1.41	1.50

7.6 DESIGN OF CONCRETE WING WALLS

The following forces are considered, in general, in the design of concrete check dams:

- Self weight of wall;
- Water pressure;
- Uplift pressures; and
- Active earth pressures in the design of abutment and wing walls.

The following parameters were varied for developing the design tables for concrete walls:

Height of Concrete Wing Walls	:	1.00 m to 6.00 m
Unit Weight of Water, γ_{water}	:	1.00 t/m³ or 9.81 kN/m³
Unit Weight of Concrete, $\gamma_{Concrete}$	:	2.50 t/m³ or 24.52 kN/m³
Effect of Water for Uplift Pressure	:	30% of Total Height
Unit Weight of Saturated Earthfill, γ_{sat}	:	1.80 t/m³ or 17.66 kN/m³
Angle of Repose of Backfill, ϕ	:	30°
Angle of Friction between the Concrete Wall and Backfill, δ	:	⅓ ϕ
Angle of Surcharge of Backfill, ι	:	0°
Coefficient of Friction between the Concrete and Soil/rock mass μ	:	0.55

The base width of the concrete wing wall/abutment was worked out on the principles discussed in Chapter 3 on design philosophy.

The generalised cross section of concrete wing wall (abutment) and downstream wing wall with vertical water face and vertical earth face (Figure 7.4) for heights varying from 1.00 m to 6.00 m have been adopted for carrying out the analysis.

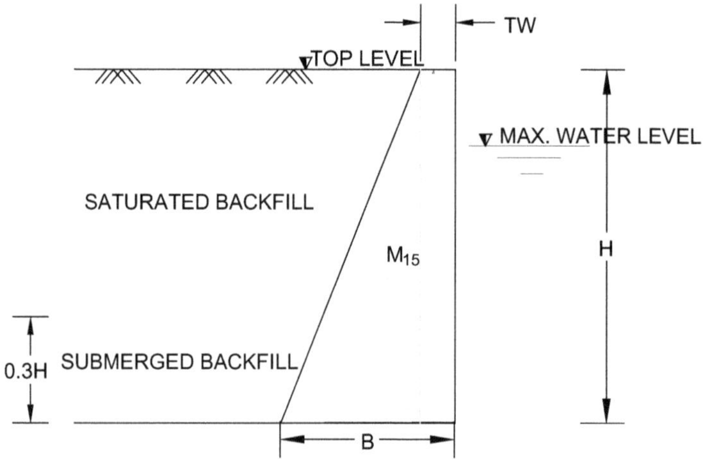

FIGURE 7.4A, 7.4B Typical Cross Sections of Concrete Wing Wall (Abutment) and Downstream Wing Wall.

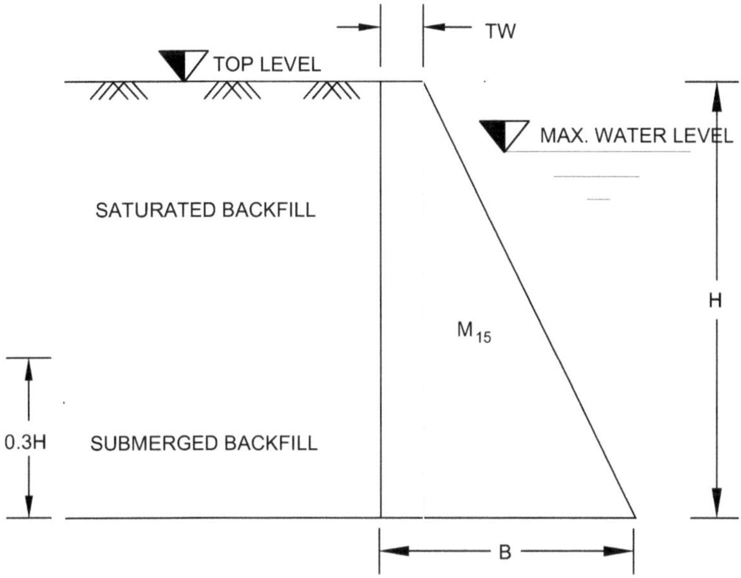

B. Vertical Earth Face

FIGURE 7.4A, 7.4B (Continued)

7.6.1 Section with Vertical Water Face

The design details of the concrete wing walls with vertical water face, computed for heights from 1.00 m to 6.00 m, are presented in Table 7.49, and the variation of the base width and the volume of the concrete with the height of wall have been depicted in Figure 7.5.

TABLE 7.49
Design Details for Concrete Wing Wall with Vertical Water Face

Height m	Top Width TW mm	Base Width B mm	Pressure at Toe t/m²	Pressure at Heel t/m²	F_{OVT}	F_{SLD}	Volume of Concrete per m run m³	B/H Ratio
1.00	470	470	4.81	0.08	2.60	1.73	0.47	0.47
1.25	500	970	5.07	0.01	3.20	1.80	0.92	0.78
1.50	500	1280	5.85	0.02	3.29	1.78	1.34	0.85
1.75	500	1540	6.70	0.01	3.29	1.75	1.79	0.88
2.00	500	1790	7.57	0.00	3.28	1.72	2.29	0.90
2.25	500	2040	8.43	0.01	3.27	1.71	2.86	0.91
2.50	500	2280	9.31	0.01	3.26	1.69	3.48	0.91
2.75	500	2520	10.18	0.01	3.25	1.68	4.15	0.92
3.00	500	2760	11.05	0.01	3.25	1.67	4.89	0.92
3.25	500	3000	11.92	0.02	3.24	1.66	5.69	0.92
3.50	500	3230	12.81	0.01	3.23	1.65	6.53	0.92

(Continued)

TABLE 7.49 (Continued)

Height m	Top Width TW mm	Base Width B mm	Pressure at Toe t/m²	Pressure at Heel t/m²	F_{OVT}	F_{SLD}	Volume of Concrete per m run m³	B/H Ratio
3.75	500	3470	13.68	0.02	3.23	1.65	7.44	0.93
4.00	500	3700	14.56	0.01	3.22	1.64	8.40	0.93
4.25	500	3940	15.43	0.02	3.22	1.63	9.44	0.93
4.50	500	4170	16.31	0.01	3.21	1.63	10.51	0.93
4.75	500	4400	17.20	0.01	3.20	1.62	11.64	0.93
5.00	500	4630	18.08	0.00	3.19	1.62	12.83	0.93
5.25	500	4870	18.95	0.02	3.19	1.62	14.10	0.93
5.50	500	5100	19.83	0.01	3.19	1.61	15.40	0.93
5.75	500	5330	20.71	0.00	3.18	1.61	16.76	0.93
6.00	500	5570	21.57	0.02	3.18	1.61	18.21	0.93

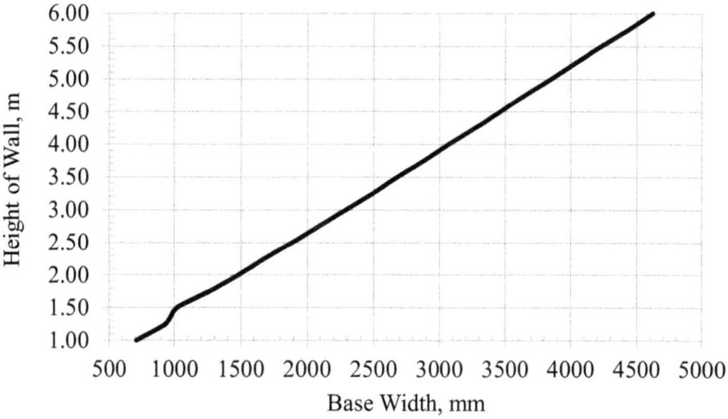

A. Variation of Base Width with Height

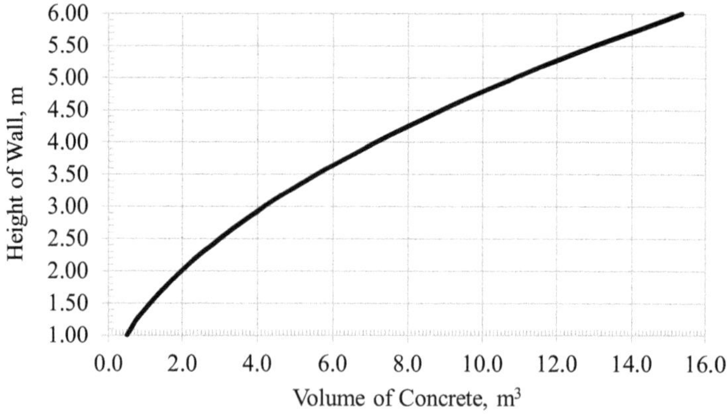

B. Variation of Volume of Concrete with Height

FIGURE 7.5A, 7.5B Variation of Base Width and Volume for Vertical Water Face Concrete Wing Wall.

7.6.2 Section with Vertical Earth Face

The design details of the concrete wing walls with vertical earth face, computed for heights from 1.00 m to 6.00 m, are presented in Table 7.50, and the variation of the base width and the volume of concrete with the height of wall have been depicted in Figure 7.6.

TABLE 7.50
Design Details for Concrete Wing Wall with Vertical Earth Face

Height m	Top Width TW mm	Base Width B mm	Pressure at Toe t/m²	Pressure at Heel t/m²	FOVT	FSLD	Depth Shear Key for μ = 0.55 mm	Volume of Concrete m³	B/H Ratio
1.00	300	430	4.00	0.15	2.20	1.48	70	0.37	0.43
1.25	300	520	4.77	0.05	2.03	1.33	140	0.51	0.42
1.50	500	640	6.53	0.01	2.20	1.54	70	0.86	0.43
1.75	500	740	7.04	0.13	2.13	1.44	150	1.09	0.42
2.00	500	830	7.81	0.03	2.03	1.35	210	1.33	0.42
2.25	500	930	8.43	0.04	1.98	1.29	270	1.61	0.41
2.50	500	1040	8.90	0.14	1.97	1.25	320	1.93	0.42
2.75	500	1140	9.58	0.08	1.93	1.21	380	2.26	0.41
3.00	500	1250	10.12	0.13	1.91	1.18	420	2.63	0.42
3.25	500	1350	10.84	0.02	1.88	1.15	480	3.01	0.42
3.50	500	1460	11.40	0.04	1.87	1.13	530	3.43	0.42
3.75	500	1570	11.99	0.05	1.86	1.11	580	3.88	0.42
4.00	500	1680	12.58	0.05	1.85	1.10	620	4.36	0.42
4.25	500	1790	13.18	0.04	1.84	1.09	670	4.87	0.42
4.50	500	1900	13.79	0.02	1.83	1.07	720	5.40	0.42
4.75	500	2010	14.41	0.00	1.82	1.06	770	5.96	0.42
5.00	500	2130	14.88	0.10	1.83	1.06	810	6.58	0.43
5.25	500	2240	15.51	0.07	1.82	1.05	850	7.19	0.43
5.50	500	2350	16.14	0.04	1.81	1.04	900	7.84	0.43
5.75	500	2460	16.77	0.00	1.81	1.03	960	8.51	0.43
6.00	500	2580	17.27	0.09	1.81	1.03	990	9.24	0.43

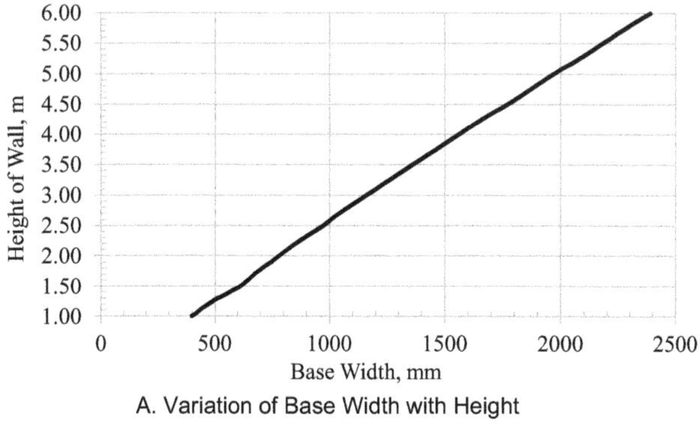

FIGURE 7.6A, 7.6B Variation of Base Width and Volume for Vertical Earth Face Concrete Wing Wall.

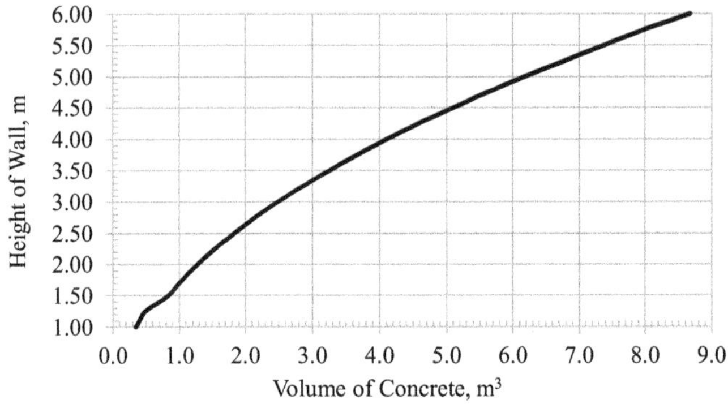

B. Variation of Volume of Concrete with Height

FIGURE 7.6A, 7.6B (Continued)

7.7 APPLICATION OF DESIGN TABLES

Steps to carry out the design of concrete check dams include the following:

Step 1: For the chosen shape of the cross section, given the value of the height of a check dam H and the width of the stream, select the appropriate table.

Step 2: From the appropriate table, for the known catchment area, select the appropriate column of the table and read from section A, top width, base width for the weir section, top level, foundation level and factor of safety against overturning, base pressures at toe and heel, length of downstream apron for check dams greater than 1.50 m high, total volume of concrete, etc.

Step 3: From section B, read the top width, base width for non-over flow section, top level, foundation level and factor of safety against overturning, base pressures at toe and pressure at heel, total volume of concrete, etc.

Step 4: Read the design details for concrete wing walls (abutments) from Table 7.49 or Table 7.50 for the height of the wing wall. For intermediate values, the data can be interpolated and partly read from Figure 7.5 or Figure 7.6.

Step 5: Read the design details for downstream concrete wing walls from Table 7.49 or Table 7.50 for the height of the D/S wing wall. For intermediate values, the data can be interpolated and partly read from Figure 7.5 or Figure 7.6.

Step 6: Engineering drawing for the concrete check dam can be prepared by selecting the appropriate figure (from 9.14 to 9.19) on the basis of the design data obtained from Step 2 to Step 5.

Step 7: Specifications and general guidelines to be followed during the construction of concrete check dams can be read from Figure 9.20.

7.8 DESIGN EXAMPLES

To illustrate the use of tables, the following examples have been chosen:

7.8.1 Example 1

Design a concrete check dam for following parameters:

i. Shape of the concrete section: vertical U/S face
ii. Height of a check dam H = 2.50m
iii. Width of stream/gadhera = 10.0 m
iv. Catchment area = 150 ha

Solution:

Step 1: For a chosen shape of the cross section as vertical U/S face, the given value of the height of a check dam as 2.50 m and width of stream as 10.0 m, select Table 7.26.

Step 2: From Table 7.26, for the known catchment area, select the appropriate column of the table. The values given in section A for the design of concrete weir are as follows:

Span of Weir W, m	8.00
Flood Discharge Q, m^3/s	14.91
Head Over Crest, m	1.01
Top Level, m	100.00
Foundation Level, m	97.50
Height of Weir H_{WEIR}, m	2.50
Top Width of Weir TW_{WEIR}, mm	500
Base Width of Weir B_{WEIR}, mm	2750
Width B1', mm	2250
D/S Slope – $H_{DS'}$: 1.0(V)	0.900
Pressure at Toe, t/m^2	5.13
Pressure at Heel, t/m^2	0.00
Factor of Safety against Overturning	1.51
Shear Friction Factor F	2.18

Total Volume of Concrete, m^3	32.50
Length of D/S Apron Floor $L_{D/S\ APRON}$, m	7.30

Step 3: From Table 7.26, read the values given in section B for the design of the non-over flow section:

Length, m	2.00
Top Level including Free Board, m	101.60
Foundation Level, m	97.50
Height of Non-Over Flow H_{NOF}, m	4.10
Height H1, mm	2500
Height H2, mm	1600
Top Width of Non-Over Flow TW_{NOF}, mm	500
Base Width of Non-Over Flow B_{NOF}, mm	2750
Width B1, mm	2250
D/S Slope – H_{DS}: 1.0(V)	0.900
Pressure at Toe, t/m^2	5.86
Pressure at Heel, t/m^2	0.48
Factor of Safety against Overturning	1.58
Shear Friction Factor F	2.03
Total Volume of Concrete, m^3	9.73

Step 4: From Table 7.26, read the values given in section **C** for the design of wing walls (abutments):

Length in U/S $L_{U/SABUT}$, m	1.50
Length in D/S $L_{D/SABUT}$, m	3.25
Top Level including Free Board, m	101.60
Foundation Level, m	97.50
Total Height H_{ABUT}, m	4.10

From Figure 7.5, for the concrete wall with vertical water face, read the values for a height of 4.10 m or interpolate the values from Table 7.49:

Top Width TW_{ABUT}, mm	500
Base Width B_{ABUT}, mm	3796
Pressure at Toe, t/m^2	14.91
Pressure at Heel, t/m^2	0.01
F_{OVT}	3.22
F_{SLD}	1.64
Volume of Concrete per m run, m^3	8.81
Total Volume of Concrete, m^3	41.85

Step 5: From Table 7.26, read the values given in section D for the design of D/S wing walls:

Length in D/S $L_{D/SWING}$, m	4.05
Top Level including Free Board, m	99.12
Foundation Level, m	97.50
Total Height $H_{D/SWING}$, m	1.62

From Figure 7.5, for the concrete wall with vertical water face, read the values for a height of 1.62 m or interpolate the values from Table 7.49:

Top Width $TW_{D/SWING}$, mm	500
Base Width $W_{D/SWING}$, mm	1405
Pressure at Toe, t/m^2	6.26
Pressure at Heel, t/m^2	0.01
F_{OVT}	3.29
F_{SLD}	1.76
Volume of Concrete per m run, m^3	1.55
Total Volume of Concrete, m^3	6.28

Step 6: Based on the design data obtained from Step 2 to Step 5, an engineering drawing for the 2.50 m high concrete check dam with vertical U/S face can be prepared with the help of Figure 9.17.

Step 7: Specifications and general guidelines for the construction of concrete check dams can be read from Figure 9.20.

7.8.2 EXAMPLE 2

Design a concrete check dam for the following parameters:

i. Shape of concrete section: vertical D/S face
ii. All the other data are the same, as given in example 1.

Solution:

Step 1: For a chosen shape of the cross section as vertical D/S face, the given value of the height of a check dam as 2.50 m and width of stream as 10.0 m, select Table 7.30.

Step 2: From Table 7.30, for the known catchment area, select the appropriate column of the table. The values given in section A for the design of concrete weir are as follows:

Span of Weir W, m	8.00
Flood Discharge Q, m^3/s	14.91

Head Over Crest, m	1.01
Top Level, m	100.00
Foundation Level, m	97.50
Height of Weir H_{WEIR}, m	2.50
Top Width of Weir TW_{WEIR}, mm	2860
Base Width of Weir B_{WEIR}, mm	2860
Pressure at Toe, t/m^2	11.32
Pressure at Heel, t/m^2	0.03
Factor of Safety against Overturning	2.17
Shear Friction Factor F	2.80
Total Volume of Concrete, m^3	57.20
Length of D/S Apron Floor $L_{D/S\ APRON}$, m	7.20

Step 3: From Table 7.30, read the values given in section B for the design of the non-over flow section:

Length, m	2.00
Top Level including Free Board, m	101.60
Foundation Level, m	97.50
Height of Non-Over Flow H_{NOF}, m	4.10
Height H1, mm	3280
Height H2, mm	820
Top Width of Non-Over Flow TW_{NOF}, mm	1000
Base Width of Non-Over Flow B_{NOF}, mm	2840
Width B1, mm	1740
Pressure at Toe, t/m^2	15.75
Pressure at Heel, t/m^2	0.01
Factor of Safety against Overturning	2.45
Shear Friction Factor F	2.77
Total Volume of Concrete, m^3	20.27

Step 4: From Table 7.30, read the values given in section C for the design of wing walls (abutments):

Length in U/S $L_{U/SABUT}$, m	3.84
Length in D/S $L_{D/SABUT}$, m	2.00
Top Level including Free Board, m	101.60
Foundation Level, m	97.50
Total Height H_{ABUT}, m	4.10

From Figure 7.5, for the concrete wall with vertical water fence, read the values for a height of 4.10 m or interpolate the values from Table 7.49:

Top Width TW_{ABUT}, mm	500
Base Width B_{ABUT}, mm	3796
Pressure at Toe, t/m^2	14.91
Pressure at Heel, t/m^2	0.01
F_{OVT}	3.22
F_{SLD}	1.64
Volume of Concrete per m run, m^3	8.81
Total Volume of Concrete, m^3	51.45

Step 5: From Table 7.30, read the values given in section D for the design of D/S wing walls:

Length in D/S $L_{D/SWING}$, m	5.20
Top Level including Free Board, m	99.12
Foundation Level, m	97.50
Total Height $H_{D/SWING}$, m	1.62

From Figure 7.5, for the concrete wall with vertical water face, read the values for a height of 1.62 m or interpolate the values from Table 7.49:

Top Width $TW_{D/SWING}$, mm	500
Base Width $W_{D/SWING}$, mm	1405
Pressure at Toe, t/m^2	6.26
Pressure at Heel, t/m^2	0.01
F_{OVT}	3.29
F_{SLD}	1.76
Volume of Concrete per m run, m^3	1.55
Total Volume of Concrete, m^3	8.06

Step 6: Based on the design data obtained from Step 2 to Step 5, an engineering drawing for the 2.50 m high concrete check dam with vertical D/S face can be prepared with the help of Figure 9.18.

Step 7: Specifications and general guidelines for the construction of concrete check dams can be read from Figure 9.20.

7.8.3 Example 3

Design a concrete check dam for the following parameters:

i. Shape of concrete section: steps on both faces
ii. All the other data are the same, as given in example 1.

Solution:

Step 1: For a chosen shape of the cross section as steps on both faces, the given value of the height of a check dam as 2.50 m and width of stream as 10.0 m, select Table 7.34.

Step 2: From Table 7.34, for the known catchment area, select the appropriate column of the table. The values given in section A for the design of concrete weir are as follows:

Span of Weir W, m	8.00
Flood Discharge Q, m^3/s	14.91
Head Over Crest, m	1.01
Top Level, m	100.00
Foundation Level, m	97.50
Height of Weir H_{WEIR}, m	2.50
Size of Concrete Steps (W, H), m	
Base Width B_{WEIR}, Thickness H1’	3.17, 1.00
Width B1’, Thickness H2’	2.17, 1.00
Top Width TW_{WEIR}, Thickness H3’	1.17, 0.50
Pressure at Toe, t/m^2	8.46
Pressure at Heel, t/m^2	0.02
Factor of Safety against Overturning	1.96
Shear Friction Factor F	1.91
Total Volume of Concrete, m^3	47.40
Length of D/S Apron Floor $L_{D/S\ APRON}$, m	5.08

Step 3: From Table 7.34, read the values given in section B for the design of the non-over flow section:

Length, m	2.00
Top Level including Free Board, m	101.60
Foundation Level, m	97.50
Height of Non-Over Flow H_{NOF}, m	4.10
Size of Concrete Steps (W, H), m	
Base Width B_{NOF}, Thickness H1	2.80, 1.00
Width B1, Thickness H2	1.80, 1.00

Width B2, Thickness H3	0.80, 1.00
Top Width TW_{NOF}, Thickness H4	0.50, 1.10
Pressure at Toe, t/m^2	10.54
Pressure at Heel, t/m^2	0.00
Factor of Safety against Overturning	1.92
Shear Friction Factor F	3.34
Total Volume of Concrete, m^3	11.90

Step 4: From Table 7.34, read the values given in section C for the design of wing walls (abutments):

Length in U/S $L_{U/SABUT}$, m	2.60
Length in D/S $L_{D/SABUT}$, m	2.60
Top Level including Free Board, m	101.60
Foundation Level, m	97.50
Total Height H_{ABUT}, m	4.10

From Figure 7.5, for the concrete wall with vertical water face, read the values for a height of 4.10 m or interpolate the values from Table 7.49:

Top Width TW_{ABUT}, mm	500
Base Width B_{ABUT}, mm	3796
Pressure at Toe, t/m^2	14.91
Pressure at Heel, t/m^2	0.01
F_{OVT}	3.22
F_{SLD}	1.64
Volume of Concrete per m run, m^3	8.81
Total Volume of Concrete, m^3	45.81

Step 5: From Table 7.34, read the values given in section D for the design of D/S wing walls:

Length in D/S $L_{D/SWING}$, m	5.90
Top Level including Free Board, m	99.12
Foundation Level, m	97.50
Total Height $H_{D/SWING}$, m	1.62

From Figure 7.5, for the concrete wall with vertical water face, read the values for a height of 1.62 m or interpolate the values from Table 7.49:

Top Width $TW_{D/SWING}$, mm	500
Base Width $B_{D/SWING}$, mm	1405
Pressure at Toe, t/m^2	6.26

Pressure at Heel, t/m^2	0.01
F_{OVT}	3.29
F_{SLD}	1.76
Volume of Concrete per m run, m^3	1.55
Total Volume of Concrete, m^3	9.15

Step 6: Based on the design data obtained from Step 2 to Step 5, an engineering drawing for the 2.50 m high concrete check dam with steps on both faces can be prepared with the help of Figure 9.19.

Step 7: Specifications and general guidelines for the construction of concrete check dams can be read from Figure 9.20.

REFERENCE

IS:6512 (2019) *Indian Standard Criteria for Design of Solid Gravity Dams*, IS:6512, Bureau of Indian Standards, New Delhi.

8 Construction and Maintenance Aspects

8.1 SITE PREPARATION FOR CHECK DAMS

- The alignment of the structure shall be marked before excavation.
- The excavated area shall be compacted using a roller.
- In areas where the vibro roller cannot reach, a plate compactor/vibro tamper should be used.
- Weak pockets shall be checked and must be replaced with compacted granular fill.
- The density of compacted ground shall be more than 95% modified Proctor density.

8.2 CONSTRUCTION ASPECTS OF GABION CHECK DAMS

Gabions delivered at the work site in bundles along with the lacing wire should be stored close to the site. While unloading, care shall be taken that gabion bundles are not damaged, and a maximum of four bundles shall be stacked over each other on a levelled ground with wooden rafters/sleepers in between them for easy handling, preventing any damage to the mesh and other components. The following steps may be followed at the site for the installation of gabion structures (IRC:SP:116, 2018):

i The surface of the formation should be free from any deleterious material and unwanted foreign objects. Loose pockets, if any, should be excavated and filled with suitable granular or backfill material.
ii The depth of the foundation shall be decided on the type of soil/rock available at the site. However, it is recommended that the foundation depth be kept at least 500 mm.
iii The place should be cleared of standing water to ensure the proper placement of gabions. In case dewatering is not possible, then gabions shall be installed underwater.
iv The density of the compacted foundation surface should be greater than or equal to 95% of the modified Proctor value.
v The outer alignment of the gabion facia should be marked as per the requirements in the drawings.
vi The gabions should be unfolded and laid over an even, level, and hard surface (compact ground). The gabions should be properly assembled and wired together so that all four corners match and form the designed rectangular shape and size.

DOI: 10.1201/9781003605621-8

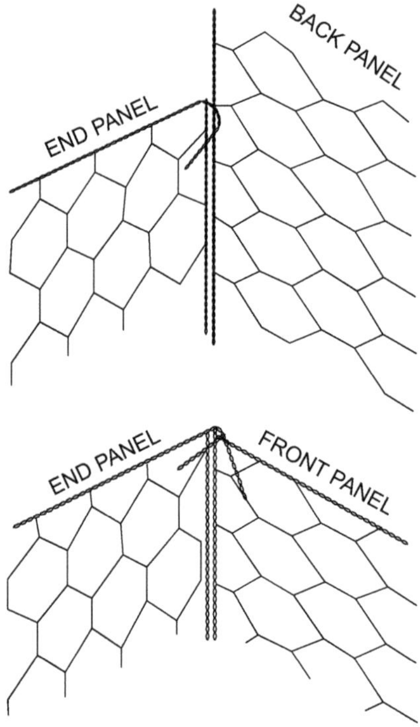

FIGURE 8.1 Assembling Details.

Source: IRC:SP:116 (2018)

vii The side panels should be opened up to form a box by tying the upper corner of each pair of side panels using the thick selvedge wires (Figure 8.1).

viii The edges should be laced together, starting from the top corner, in a continuous operation using alternate single and double loops at the spacing of one mesh length, i.e. 100 mm to 125 mm. The ends of the lacing wire should be secured at each corner and turned in to the gabions. Individual ties of lacing wire should not be used. Alternatively, steel rings can be used (Figure 8.2).

ix The gabions should be filled with stones of dimensions larger than that of the mesh opening, making sure that there are minimum voids.

x The average porosity of the gabion boxes can be found out by weighing the filled boxes. The weight of the box can be compared with the calculated weight of box.

xi In case of 1.0 m high boxes, gabions should be filled to one-third height and the tie wires be fixed (Figure 8.3). The same sequence is repeated at two-third height. For 0.5 m high, gabion requires only one row of tie wires at half the height of the gabion.

xii The filled layer should never be more than 300 mm higher than any adjoining cell and the filling pattern shown in Figure 8.4 should be followed. The

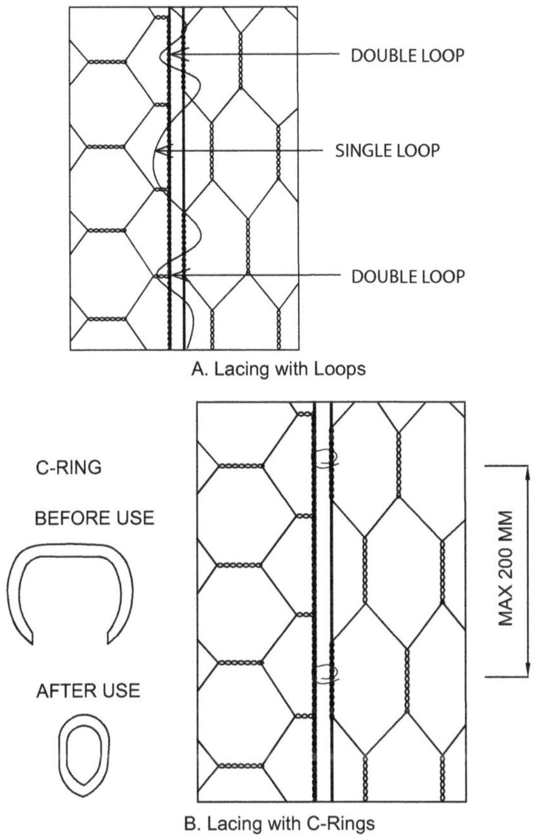

FIGURE 8.2A, 8.2B Assembling Details with C-Rings.

***Source*: IRC:SP:116 (2018)**

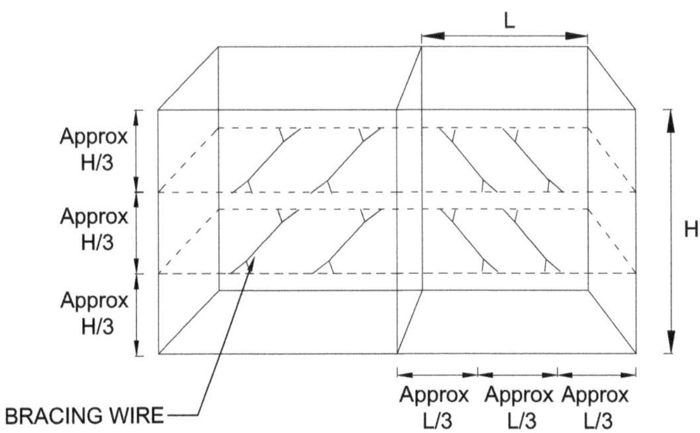

FIGURE 8.3 Bracing Wire Details.

***Source*: IS:16014 (2018)**

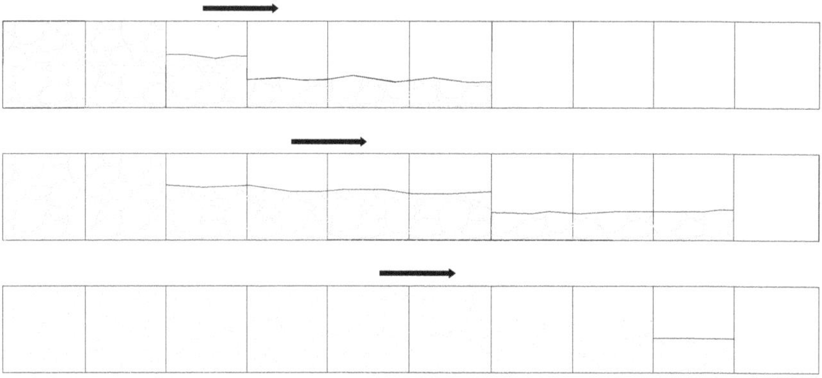

FIGURE 8.4 Filling Pattern of Gabion Boxes.

Source: IRC:SP:116 (2018)

Gabion should be over-filled by approximately 25 mm to 50 mm, to allow for settlement due to self weight/infill.

xiii The lids should be folded back and laced first to the front panels, then to the side ones and to any existing diaphragm to close the gabions. The gabions in the upper layer shall be connected to the top of gabions in the lower layer along the front and back edges of the contact surface using the same connecting procedure. Facia of adjoining units must be securely joined together along the vertical facing and top edges of their contact surfaces.

xiv Other boxes should also be placed and packed in the position following the same procedure to complete the gabion structure.

xv The gabions should be covered on the inner face with geo-synthetic clay liner or geo-membrane to prevent seepage from the U/S face and bottom of the gabion check dam.

xvi In case of gabion abutments/wing walls, the soil for structural fill shall be placed and compacted to 95% of modified Proctor density.

8.3 SEEPAGE PREVENTION THROUGH UPSTREAM FACE AND FOUNDATION

Seepage through dams and foundation is the most common cause of its deterioration and structural damage. Geo-synthetic products viz. the geotextile filters/liners, HDPE geo-membrane, or geo-synthetic clay liner are provided on the upstream face and bottom layer of gabion boxes to control seepage through the U/S face and bed. They are installed in vertical strips of 2 m to 4 m wide so as to cover the entire upstream face of the dam and foundation.

The geo-synthetic clay liner, though marginally costlier, is recommended based on the effectiveness, operation, and maintenance considerations. Further, the requirement of the subgrade preparation for the geo-synthetic clay liner is much less stringent compared to the HDPE geo-membrane and is likely to offset the additional cost to a certain extent.

8.3.1 Comparison between Geo-Synthetic Clay Liner and Geo-Membrane

Geo-synthetic clay liners and geo-membranes are frequently used in various lining applications, and many times, they are used together in a composite liner system. A comparison between geo-synthetic clay liner and geo-membrane is given in Table 8.1.

8.3.2 Geo-Synthetic Clay Liner

Geo-synthetic clay liner (Figure 8.5) is a composite geotextile material, usually made of the following three components:

i) geotextile top layer;
ii) granulated bentonite; and
iii) geotextile base layer.

TABLE 8.1
Comparison between Geo-Synthetic Clay Liner and Geo-Membrane

Item	Geo-Synthetic Clay Liner (GCL)	Geo-Membrane (GM)
Installation	• can be installed by unskilled labour • minimum training required • installation can be done in phases as per site conditions	• requires a specially trained and skilled team • installation needs to be performed all at once to minimise mobilisation of labour
	• seams are simply overlapped, providing same hydraulic performance as that of the mat itself • details such as penetrations and attachments are handled without mechanical fasteners • does not expand or contract due to temperature fluctuations	• seams must be welded together, which is time-consuming, labour-intensive, and expensive • details are difficult to seal properly, mechanical fastening to attachments is required • expands and contracts with temperature fluctuations, causing tensile stresses and wrinkles in the geo-membrane
Quality assurance	• minimal on-site quality assurance required because seams are overlapped	• extensive quality assurance required because seams are welded
	• minor puncture and other flaws are self-sealing • any damage can be easily repaired with simple overlapping patch	• minor punctures can cause major performance problems • damaged areas must be patched, welded, inspected, and tested
Installation Cost	• usually lower than GM • provides superior cost-performance ratio	• cost is higher than GCL • low cost-performance ratio compared to GCL
	• rapid installation	• slower installation compared to GCL

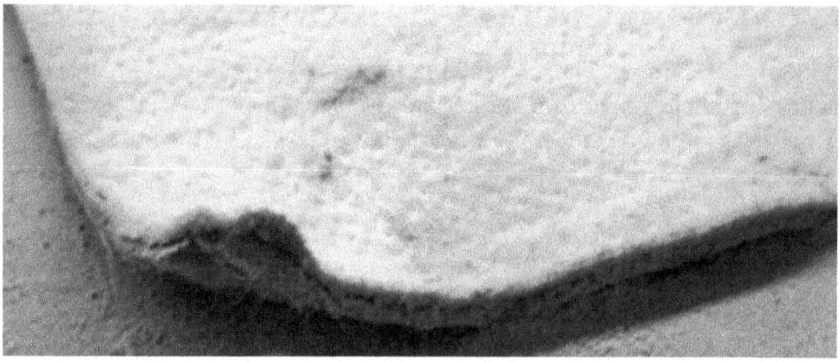

A. Geo-Synthetic Clay Liner

B. Cross Section of GCL

FIGURE 8.5A, 8.5B Geo-Synthetic Clay Liner.

Source: www.ashapura.com

In general, the geotextiles consist of a mechanically-bonded, non-woven fabric on one side and a woven geotextile on the other side. All the components of the layers are mechanically bonded by needle-punching over their entire area and are filled uniformly with bentonite granules. The use of bentonite granules instead of the conventional bentonite powder avoids the problem of dust production.

Other names used for GCLs are *Bentomat Geosynthetic Clay Liner*, *GCL Liner*, *Clay Blankets*, *Bentonite Blankets*, *Bentonite Mats*, *Clay Geosynthetic Barriers*, etc. The main component of the geo-synthetic clay liner is the mineral sodium bentonite, which is 90% montmorillonite. Bentonite in granular form is used for the manufacture of geo-synthetic clay liners. The excellent swelling capacity of bentonite when it comes into contact with moisture is responsible for the quick and reliable sealing effect of these liners. This special property of bentonite allows the sealing layer to repair itself securely if it gets mechanically damaged during installation or later. A 10 mm thick geo-synthetic clay liner can achieve a sealing effect comparable to 100 cm of mineral clay.

The use of bentonite mats is thus more environment friendly and cost-effective than the conventional method of construction. They are quick and easy to install, are supplied in rolls and are laid with an overlap.

8.3.3 General Installation Procedure for the Geo-Synthetic Clay Liner

- i Lifting equipment with a cross beam and suitable for lifting loads of at least 1,300 kg is required for unloading and laying the clay liners. The rolls, even if still packaged, must be stored in dry and weatherproof conditions.
- ii The material may be laid only in dry weather and by qualified personnel.
- iii The liners must always be laid without creasing and in the direction of the incline, and the subgrade must be flat, even and free of foreign bodies, surface water, rubble, roots, and pointed or sharp-edged stones.
- iv The overlap areas must be installed neatly with due care.
- v The bentonite granules must be distributed evenly in the overlap areas.
- vi The bentonite mat at the top edge of the incline be secured properly.
- vii Extra care must be taken in the area of pipe penetrations and connections.
- viii To prevent the material from swelling before the overburden is applied, it must be covered with suitable material immediately after laying.
- ix Vehicles must not drive directly onto areas that have already been laid.
- x The required overburden must cover the entire area of the Bentomat.

8.4 CONSTRUCTION ASPECTS OF MASONRY CHECK DAMS

Masonry check dams/walls should be made in random rubble masonry consisting of hammer-dressed, hard stones brought to course approx. every 600 mm. Masonry courses must be normal to the face batter, and the back of the wall can be left rough. Masonry work should proceed in a uniform level.

The minimum dimension of the stone should be 200 mm × 150 mm × 100 mm. Approximately half of the stones should tail into the wall by twice their height. Stones must break joint by half the height of the course.

8.4.1 General Guidelines for Masonry Construction

General guidelines for the construction of random rubble masonry construction works include the following:

8.4.1.1 Design

Before starting the construction, it should be ensured that the design and structural requirements for the random rubble masonry are prepared. The design should include the specific dimensions, mortar specifications and other additional construction considerations.

8.4.1.2 Material Selection

High-quality materials for both the rubble stones and the mortar should be used. Stones that are strong, durable, and suitable for rubble masonry should be chosen.

8.4.1.3 Stone Selection

Select random rubble stones of various sizes and shapes. The stones should be clean, free from any loose or flaky material, and reasonably square or rectangular in shape.

Avoid using irregular or excessively large stones that may compromise the stability of the masonry.

8.4.1.4 Foundation

Strong and stable foundation should be used to bear the weight of the masonry check dam. The foundation should be built on a compacted soil base, designed as per structural requirements and reinforced with appropriate reinforcement, if necessary.

8.4.1.5 Verticality and Alignment

Verticality and alignment of the masonry structure should be maintained using plumb lines, string lines, or laser-levelling devices. Regular checks for any deviations should be made to carry out necessary adjustments during construction.

8.4.1.6 Laying and Levelling

Begin building the random rubble masonry by laying the mortar bed using a trowel on the foundation or previous course. Place the rubble stones carefully, ensuring proper alignment and stability. Use smaller stones to fill gaps and provide support.

8.4.1.7 Bonding

Good bonding between masonry units should be achieved by applying mortar on the horizontal and vertical surfaces of the units. Provide proper bonding between the rubble stones by staggering the joints in adjacent courses. Interlock the stones and ensure that they are tightly packed to enhance the overall stability and strength of the masonry.

8.4.1.8 Curing

After completing each course of random rubble masonry, allow it to cure properly. Keep the masonry moist by sprinkling water regularly for a specific duration, typically around 7 to 14 days. Curing helps in achieving the desired strength and durability of the structure.

8.4.1.9 Quality Control

Regular inspections should be undertaken to ensure the quality control of the random rubble masonry work. It involves checking the alignment, levelness, mortar consistency and overall structural integrity. Ensure the stones are securely placed and adequately embedded in the mortar. Non-destructive tests and load tests may also be conducted if required.

8.4.1.10 Safety Precautions

Adhere to safety precautions during construction, such as wearing the appropriate Personal Protective Equipment (PPE) and following safe work practices. Ensure that scaffolding or support structures are properly installed and secure.

8.4.2 Through-Stones/Bonding Element

Through-stones shall be provided at every 1.0 m interval both horizontally and vertically throughout the length and height of the wall. In the absence of through-stones, reinforced concrete blocks of 100 mm × 100 mm shall be used. The through-stones or bonding element shall be staggered both horizontally and vertically.

8.4.3 Cement Mortar

Suitable cement mortar mix with the right proportions of cement, sand, and water should be used. The mortar should have good workability and sufficient strength for the specific application. It is suggested that Cement mortar ratio of 1:4 (cement: sand) or richer should be used in construction of Masonry check dams.

8.4.3.1 Mortar Coverage

An adequate amount of mortar should be applied on the horizontal and vertical surfaces of the rubble stones. Maintain consistent mortar joints, usually around 10–15 mm thick. Ensure proper mortar coverage to enhance bonding and structural stability.

8.4.4 Drainage in Downstream Walls

Provision should be made to prevent water accumulation behind the wing walls. Accumulated water causes increased pressure, seepage and, in areas subject to frost action, a force of considerable magnitude near the top of the wall. Drainage is an important aspect of wing wall construction, and a build-up of water behind the wing wall is the principal cause of failure. Good piped drainage that disposes of water away from behind the wall to limit this build-up of pressure is essential, except for walls that retain free-draining sand/gravel/boulder.

The drainage system should consist of free-draining granular material placed immediately behind the wall for a minimum width of 300 mm and for the full height of the wall. A perforated or slotted drainage pipe, protected with a filter cloth to prevent it from being clogged by silt, must be installed behind the base of the wall to collect and remove water, which is then drained to a storm water system. The typical drainage arrangement behind the wing wall has been depicted in Figure 8.6.

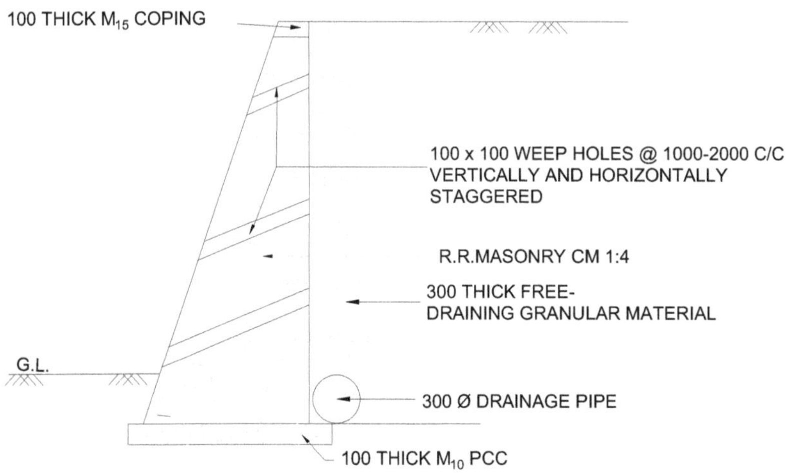

FIGURE 8.6 Typical Vertical Drainage Arrangement Behind Masonry/Concrete Wing Wall.

8.4.5 Weep Holes in Downstream Wing Walls

Weep holes provide an opening that allows the drainage of any moisture that may come from the back of the wall through penetration, capillary action, or leakage. From an engineering point of view, weep holes relieve hydrostatic pressure on walls. This reduces the structural design demand of the water and earth pressure by reducing thickness. With this, weep holes also reduce the buoyancy and uplift on the structure, making it possible to construct a lighter structure without any uplift-related stability issues.

Weep holes may be provided in cement stone masonry/concrete walls at a spacing of about 1.0–1.5 m centre-to-centre vertically and horizontally. 75–100 mm square-shape (Figure 8.7) or 100–150 mm PVC (flexible) pipes (Figure 8.8) embedded at a 10° angle down from the horizontal towards the valley side should be provided to drain the water. The lowest weep hole is generally kept 300 mm above the ground level.

8.5 CONSTRUCTION ASPECTS OF CONCRETE CHECK DAMS

Concrete check dams are recommended when local construction materials are not available for constructing masonry check dams.

8.5.1 General Guidelines for Concrete Construction

General guidelines for the construction of concrete check dams include the following:

8.5.1.1 Design

Before starting the construction, it should be ensured that the design and structural requirements for the concrete check dam are prepared. The design should include the specific dimensions, concrete mix design, and other additional construction considerations.

8.5.1.2 Material Selection

High-quality aggregates, cement, and admixtures should be used that comply with relevant standards and specifications.

8.5.1.3 Concrete Mix Design

Suitable concrete mix based on the project requirements and local conditions should be designed. It involves determining the optimal proportions of aggregates, cement, water, and admixtures to achieve the desired strength, workability, and durability.

8.5.1.4 Formwork

Construct the necessary formwork or moulds to give shape and support to the concrete check dam. Suitable materials, such as timber or steel, should be used for formwork, and it should be ensured that the formwork is properly braced and aligned.

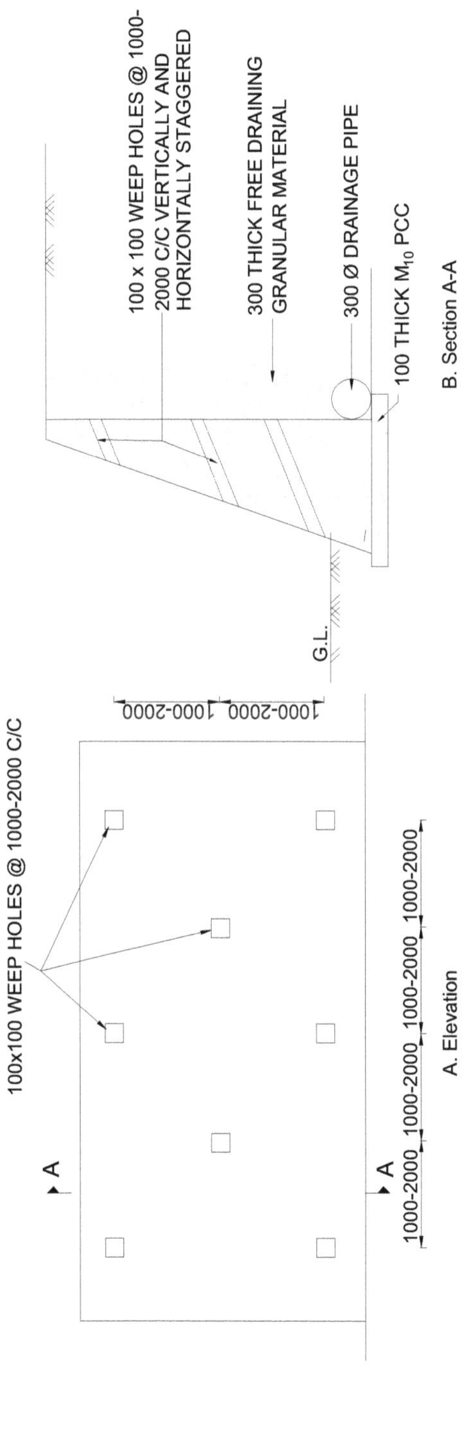

FIGURE 8.7A, 8.7B Elevation and Section of the Wall with Square Weep Holes.

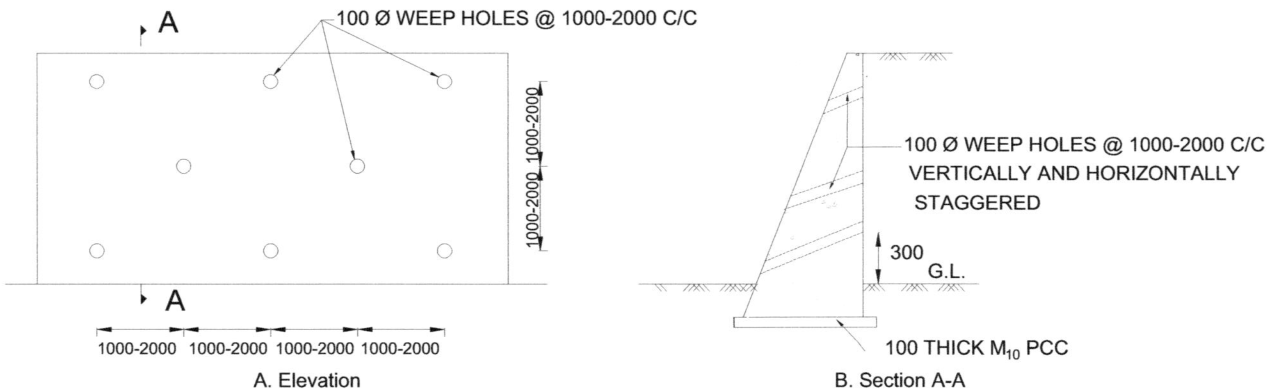

FIGURE 8.8A, 8.8B Elevation and Section of the Wall with Circular Pipe Weep Holes.

8.5.1.5 Concrete Placement

Pour the prepared concrete into the formwork in a controlled manner. Avoid segregation and ensure proper compaction of the concrete using vibration techniques. Pay attention to corners, edges, and hard-to-reach areas to ensure complete filling of the formwork.

8.5.1.6 Curing

After concrete placement, it is crucial to provide adequate curing to promote hydration and achieve the desired strength and durability. Curing of the concrete can be done by water, curing compounds, or insulating blankets, depending on the project requirements. The curing period specified in the design should be followed.

8.5.1.7 Quality Control

Regular inspection of concreting works should be done to ensure the quality control by monitoring concrete mix proportions, curing procedures, etc. Necessary tests should be conducted, such as compressive strength, slump, etc., to ensure compliance with design standards.

8.5.1.8 Safety Precautions

Adhere to safety precautions during construction, such as wearing the appropriate Personal Protective Equipment (PPE), ensuring proper scaffolding and formwork support and following safe work practices. Standard codes and regulations related to safety precautions in concrete construction should also be followed.

8.5.2 Wall Joints

8.5.2.1 Construction Joints in Concrete Walls

Vertical or horizontal joints should be placed between two successive pours of concrete. To increase shear resistance at the joints, keys may be provided as shown in Figure 8.9.

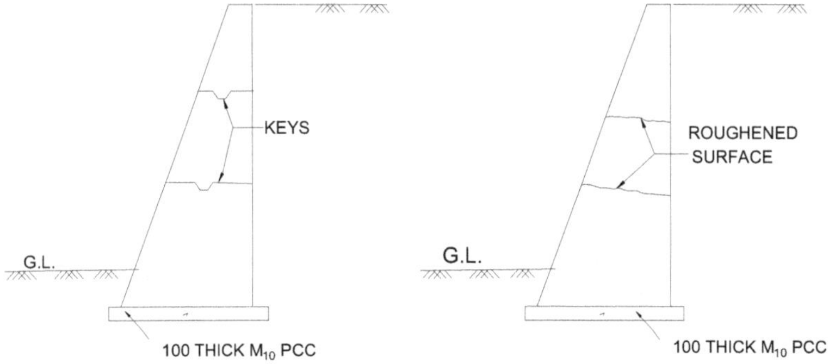

FIGURE 8.9 Construction Joints in Concrete Wall.

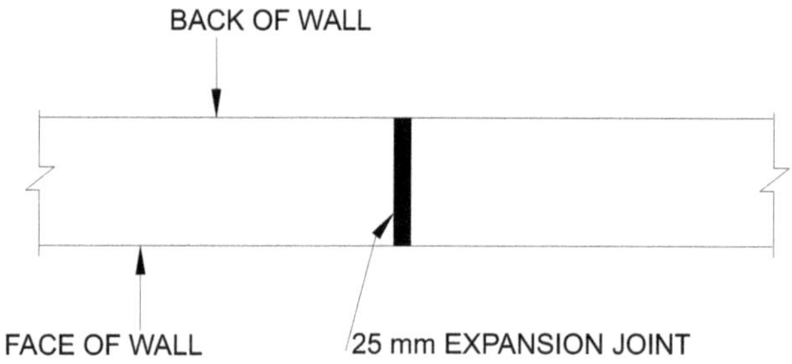

FIGURE 8.10 Expansion Joint in Concrete Wall.

8.5.2.2 Contraction Joint

A separation gap of 50–100 mm should be provided as vertical joints placed in the wall (from base to the top of wall) that allows the concrete to shrink without noticeable harm. Also, gaps shall be provided at bends and junctions.

8.5.2.3 Expansion Joint

These vertical joints are provided in large retaining walls to allow for the expansion of concrete due to temperature changes, and they are usually extended from the top to the bottom of the wall. These joints may be filled with flexible joint fillers.

8.6 GENERAL MAINTENANCE GUIDELINES

Check dam structures need to be properly operated and maintained so that they can continue to operate and function effectively and efficiently during their service life. Some general guidelines on the types of maintenance works required are as follows:

i Inspect check dam structures periodically and after each rainfall event for any damages. Some of the possible damages include the following:

a. Damages or deterioration of the main materials
b. Excessive seepage resulting in weak water level control
c. Bank erosion or scouring
d. Undermining of base of structure due to seepage/scouring
e. Bed degradation and erosion in downstream of structure
f. Accumulation of sediments and debris

ii Repair, replace, or reshape the damaged component of check dams immediately/periodically.
iii Remove any excessive sediments accumulation if necessary. Sediment removed should be disposed of properly.
iv Erosion caused by high flows along the bed and around the edges of the check dam should be repaired immediately.
v Remove the check dam structure when no longer needed.

All these normal maintenance works may not be applicable to all types of check dams, as it depends on many factors, such as the types, materials used and site conditions and location of the check dam structures, etc.

REFERENCES

IS:16014 (2018) *Indian Standard Mechanically Woven, Double-Twisted, Hexagonal Wire Mesh Gabions, Revet Mattresses, Rock Fall Netting and Other Products for Civil Engineering Purposes* (Galvanized Steel Wire or Galvanized Steel Wire With Polymer Coating) – Specification (First Revision), IS:16014, Bureau of Indian Standards, New Delhi, December.

IRC:SP:116 (2018) *Guidelines for Design and Installation of Gabion Structures* (IRC:SP:116), Indian Road Congress, New Delhi.

9 Engineering Drawings for Check Dams

9.1 ENGINEERING DRAWINGS

The completion of any construction project involves the participation of many stakeholders. Engineering drawings, if properly prepared, can help make sure all the different professionals working on a construction site know before the project begins what they are working on and what the outcome would be. Engineering drawings are the backbone of the construction engineering industry and precise, in-depth plans or diagrams convey the way a structure is to be built. These blueprints are used as a reference by engineers and contractors during construction. Since a picture is worth a thousand words, an engineering drawing is a much more effective tool for engineers than a written plan. The visual framework provided by engineering drawings give a deeper understanding to create a complete structure. Developing prototypes relies heavily on engineering drawings, especially when things are mass-produced. It helps streamline the construction process. The precise details of the technical drawings help executers create objects/elements of the project without errors and delays. There's no room for misinterpretation in engineering drawings and that's why they're so important. The engineering drawing should include the following aspects:

- i Engineering drawings should serve as a communication network between designers and executers.
- ii Engineering drawings should be so prepared that the field engineers and executers can understand them easily.

Considering the earlier principles, an attempt has been made on the basis of design tables, presented in Chapter 5, Chapter 6, and Chapter 7, to prepare easy-to-use templates for engineering drawings for the construction of gabion, masonry, and concrete check dams.

9.2 ENGINEERING DRAWINGS FOR GABION CHECK DAMS

Based on the design tables presented in Chapter 5, an attempt has been made to prepare the engineering drawings for 1.50 m high gabion check dams.

9.2.1 Gabion Check Dams with a Height of 1.50 M

The height being small, no abutment, wing walls, and D/S apron has been provided for 1.50 m high check dams. Further, to ensure outflanking, it is suggested that the check dam section be embedded to a minimum distance of 500 mm on both banks.

 DOI: 10.1201/9781003605621-9

a. *Geotextile filters on upstream face and base*
PVC geo-membranes or geo-synthetic clay liners are recommended to be provided on the U/S face of gabion boxes and below the bottom most layer of gabion boxes to prevent water infiltration and degradation of material. Improper liner selection may result in the failure of the structure or piping and erosion around the structure.

b. *Lining/jacketing of the crest of weir*
A 100 mm thick $M_{20}A_{20}$ concrete jacketing of the upper face of gabion boxes is recommended to be done on the crest of weir and D/S steps after the construction of the gabion weir, i.e. after most of the settlement has taken place, to protect the weir from the damages due to heavy bed load. The number of joints shall also be provided in the concrete jacketing to prevent the occurrence of cracks due to the uneven settlement of gabions at a later stage.

9.2.1.1 Vertical D/S Face

Figure 9.1 shows the generalised drawing for gabion check dams having vertical D/S face. The dimensions for various elements, shown in the template, can be directly read from Table 5.1 to Table 5.4 for the appropriate width of the river and catchment area.

9.2.1.2 Vertical U/S Face

Figure 9.2 shows the generalised drawing for gabion check dams having vertical U/S face. The dimensions for various elements, shown in the template, can be directly read from Table 5.5 to Table 5.8 for the appropriate width of the river and catchment area.

9.2.2 Gabion Check Dams with a Height of More than 1.50 M

For check dams having a height of more than 1.50 m, abutment and D/S wing walls have also been designed with the D/S apron, and results are presented in the design tables from Table 5.9 to Table 5.44. Further, to ensure outflanking, it is suggested that the check dam section be embedded to a minimum distance of 500 mm on both banks. It is also recommended to provide a revet/gabion mattress of 500 mm thickness as apron in the downstream, followed by a 300 mm wide and 1,500 mm deep cutoff of R. R. Masonry.

a. *Geotextile filters on upstream face and base*
PVC geo-membranes or geo-synthetic clay liners are recommended to be provided on U/S face of gabion boxes and below the bottom most layer of gabion boxes to prevent water infiltration and degradation of material. Improper liner selection may result in the failure of the structure or piping and erosion around the structure.

b. Lining/jacketing of the crest of weir
A 100 mm thick $M_{20}A_{20}$ concrete jacketing of the upper face of gabion boxes is recommended to be done on the crest of weir and D/S steps after the

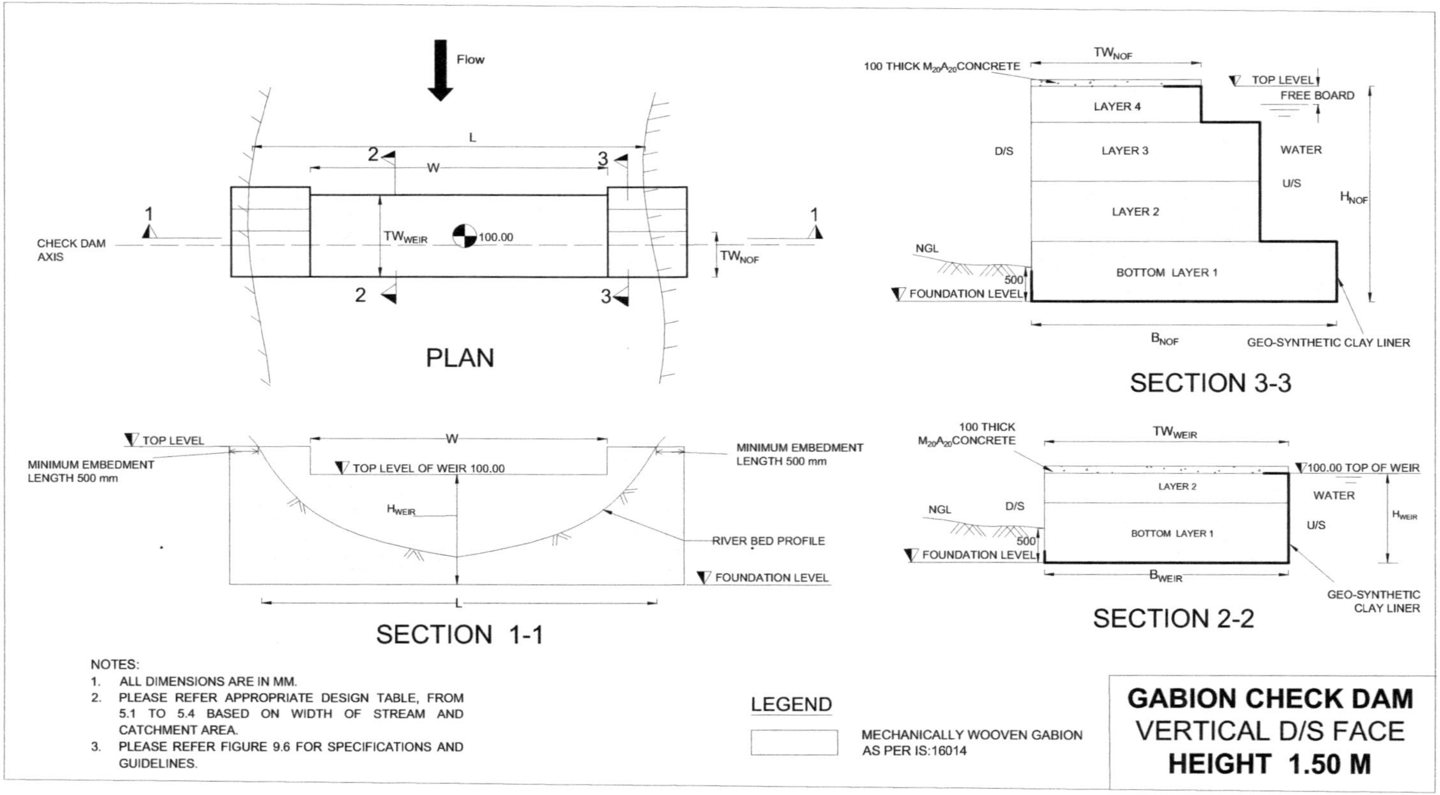

FIGURE 9.1 Gabion Check Dam 1.50 M High with Vertical D/S Face.

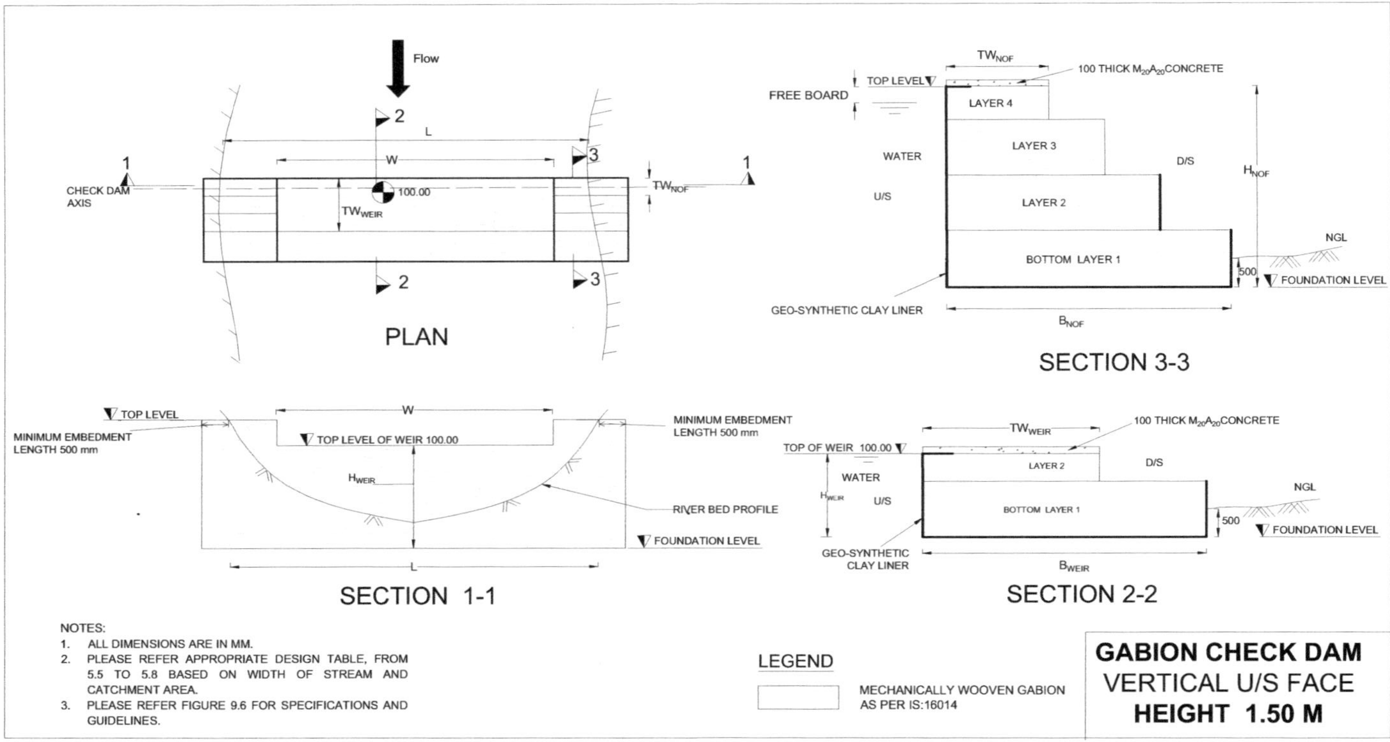

FIGURE 9.2 Gabion Check Dam 1.50 M High with Vertical U/S Face.

construction of the gabion weir, i.e. after most of the settlement has taken place, to protect the weir from the damages due to heavy bed load. The number of joints shall also be provided in the concrete jacketing to prevent the occurrence of cracks due to the uneven settlement of gabions at a later stage.

9.2.2.1 Vertical D/S Face

Figure 9.3 shows the generalised drawing for gabion check dams having vertical D/S face. The dimensions for various elements, shown in the template, can be directly read from Table 5.9 to Table 5.12 for 2.0 m high check dams, Table 5.21 to Table 5.24 for 2.50 m high check dams and Table 5.33 to Table 5.36 for 3.0 m high check dams for the appropriate width of the river and catchment area.

9.2.2.2 Vertical U/S Face

Figure 9.4 shows the generalised drawing for gabion check dams having vertical U/S face. The dimensions for various elements, shown in the template, can be directly read from Table 5.13 to Table 5.16 for 2.0 m high check dams, Table 5.25 to Table 5.28 for 2.50 m high check dams and Table 5.37 to Table 5.40 for 3.0 m high check dams for the appropriate width of the river and catchment area.

9.2.2.3 Steps on Both Faces

Figure 9.5 shows the generalised drawing for gabion check dams having steps on both faces. The dimensions for various elements, shown in the template, can be directly read from Table 5.17 to Table 5.20 for 2.0 m high check dams, Table 5.29 to Table 5.32 for 2.50 m high check dams and Table 5.41 to Table 5.44 for 3.0 m high check dams for the appropriate width of the river and catchment area.

9.2.3 Construction Specifications for Gabion Check Dams

Construction specification is a well-structured, detailed description of the quality, standards, workmanship, materials, and completion of work to be done which evolves across a project. The specifications and general guidelines to be followed during the construction of gabion check dams have been narrated in Figure 9.6.

9.3 ENGINEERING DRAWINGS FOR MASONRY CHECK DAMS

Based on the design tables presented in Chapter 6, an attempt has been made to prepare the engineering drawing for 1.50 m high masonry check dams.

9.3.1 Masonry Check Dams with a Height of 1.50 M

The height being small, no abutment, wing walls, and D/S apron has been provided for 1.50 m high check dams. Further, to ensure outflanking, it is suggested that the check dam section be embedded to a minimum distance of 500 mm on both banks.

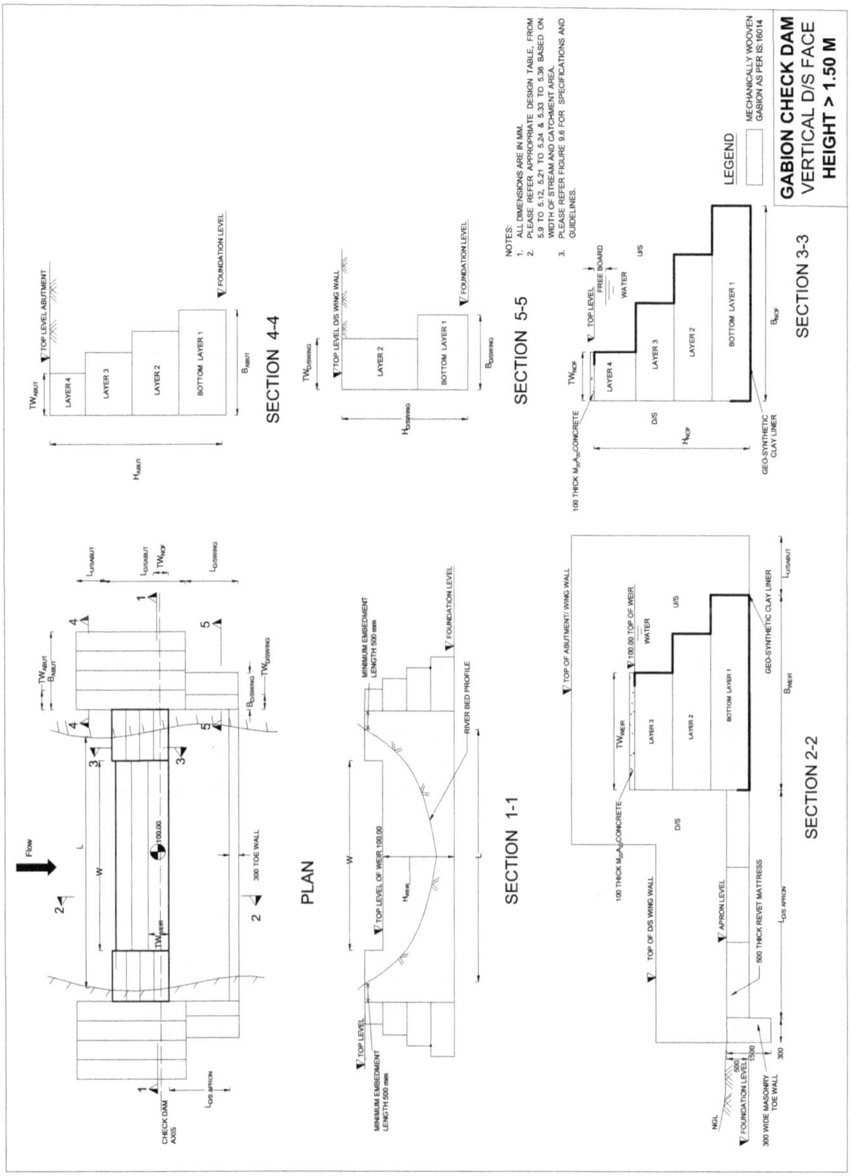

FIGURE 9.3 Gabion Check Dam More than 1.50 M Height with Vertical D/S Face.

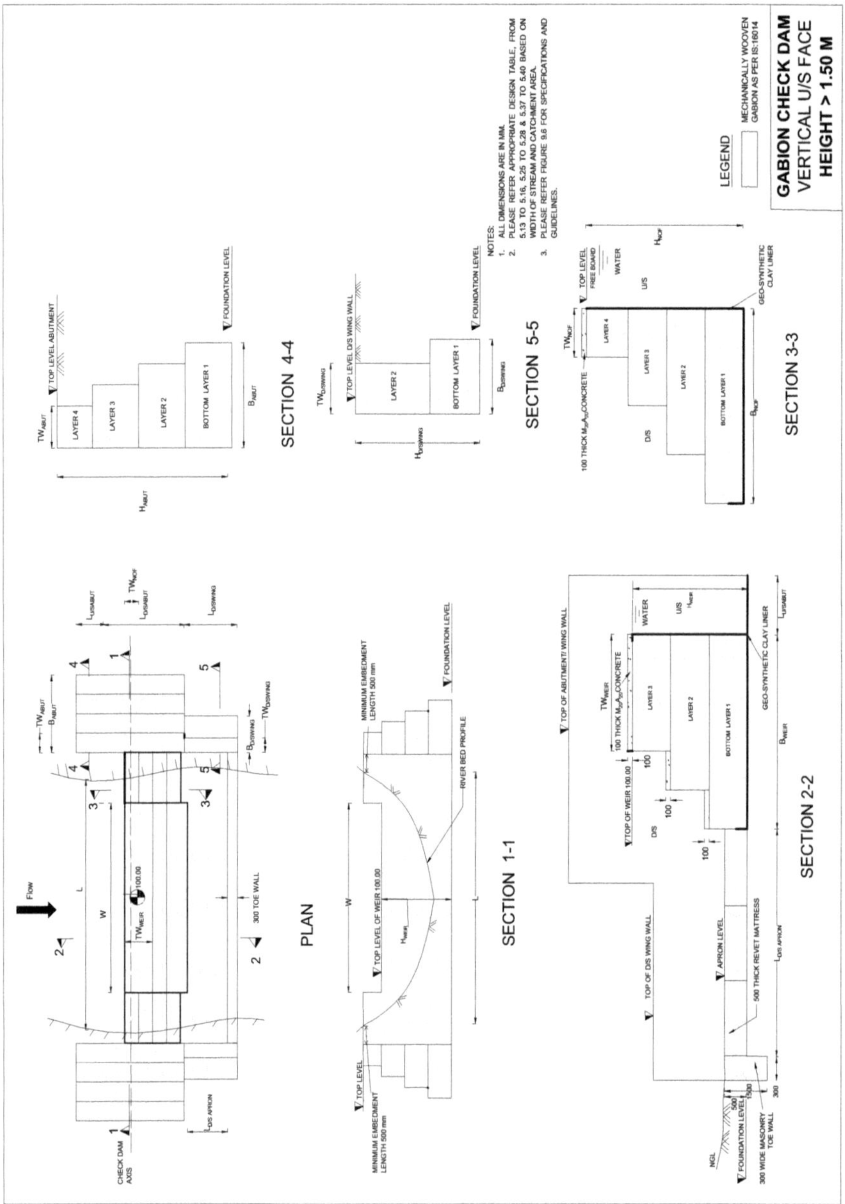

FIGURE 9.4 Gabion Check Dam More than 1.50 M Height with Vertical U/S Face.

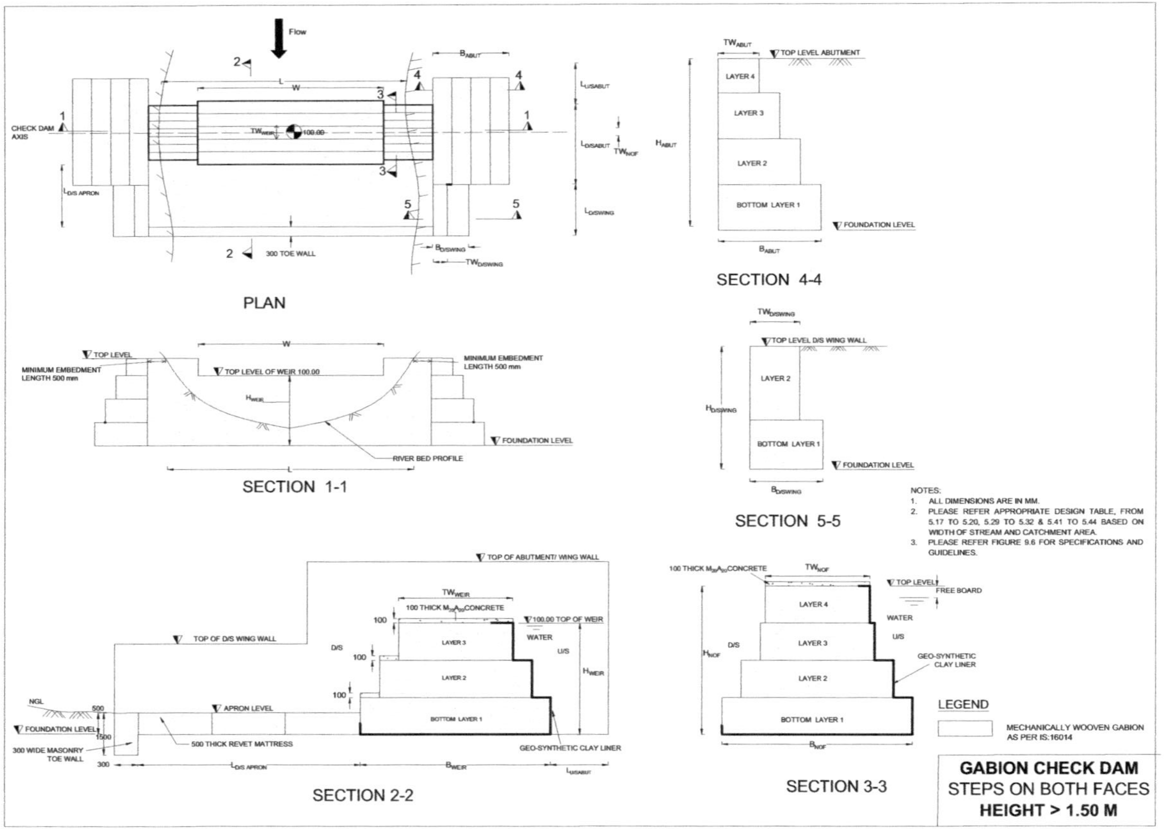

FIGURE 9.5 Gabion Check Dam More than 1.50 M Height with Steps on Both Faces.

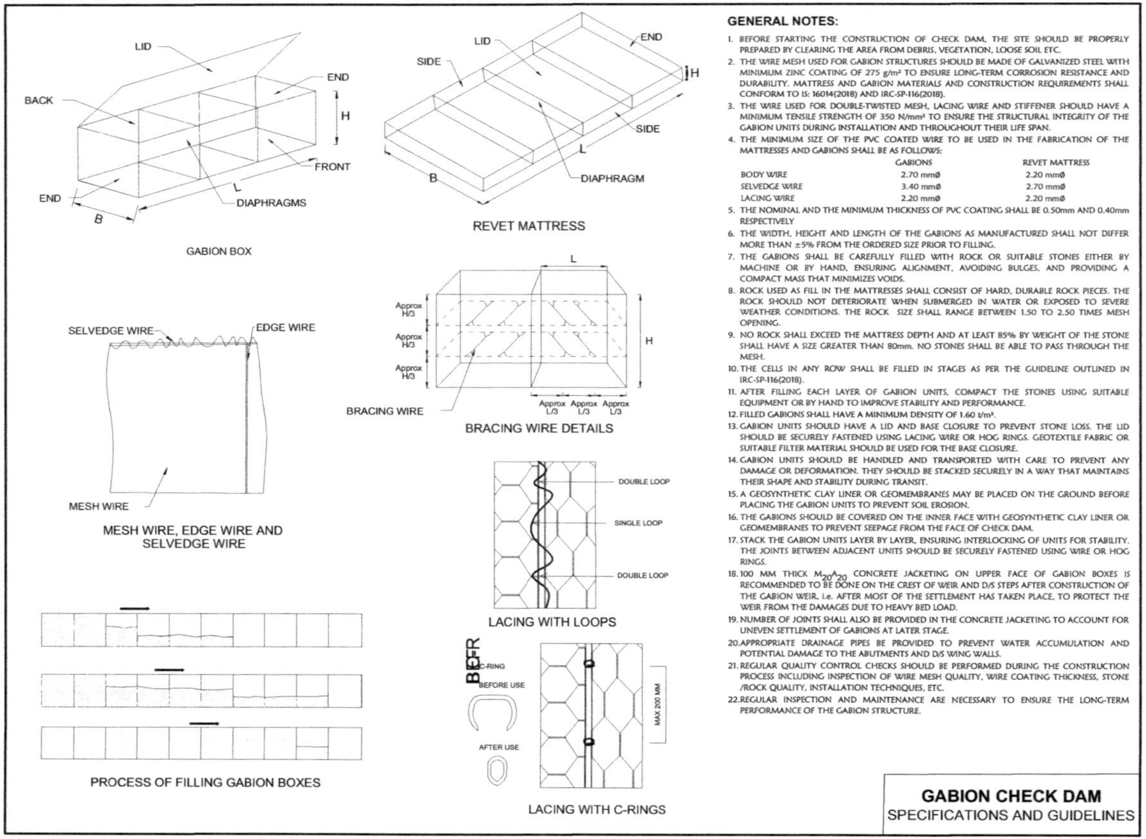

FIGURE 9.6 Construction Specifications for Gabion Check Dams.

a. Concrete on top of weir
It is recommended to protect the top surface of the crest of weir by placing a 150 mm thick $M_{20}A_{20}$ concrete.

b. Coping
To protect the top surface of all masonry walls, 100 mm thick plain cement concrete $M_{15}A_{40}$ coping is provided on top of masonry walls to be laid and pointed in cement mortar.

c. Lean concrete mat below foundation
A 100 mm thick $M_{10}A_{40}$ plain cement lean concrete has been proposed to be provided below the foundation of weir, non-over flow section, abutment, and D/S wing walls, after properly preparing and compacting the surface, to provide uniform surface to the foundation, prevent direct contact of foundation with soil, and provide protection from chemicals present in the soil.

9.3.1.1 Vertical U/S Face

Figure 9.7 shows the generalised drawing for masonry check dams having vertical U/S face. The dimensions for various elements, shown in the template, can be directly read from Table 6.1 to Table 6.4 for the appropriate width of the river and catchment area.

9.3.1.2 Vertical D/S Face

Figure 9.8 shows the generalised drawing for masonry check dams having vertical D/S face. The dimensions for various elements, shown in the template, can be directly read from Table 6.5 to Table 6.8 for the appropriate width of the river and catchment area.

9.3.1.3 Slopes on Both Faces

Figure 9.9 shows the generalised drawing for masonry check dams having slopes on both faces. The dimensions for various elements, shown in the template, can be directly read from Table 6.9 to Table 6.12 for the appropriate width of the river and catchment area.

9.3.2 Masonry Check Dams with a Height of More Than 1.50 M

For check dams having a height of more than 1.50 m, abutment and D/S wing walls have also been designed with the D/S apron, and results are presented in the design tables from Table 6.13 to Table 6.48 for check dams. Table 6.49 shows the design details for abutments and wing walls with vertical water face and Table 6.50 for abutments and wing walls with vertical earth face. Further, to ensure outflanking, it is suggested that the check dam section be embedded to a minimum distance of 500 mm on both banks. It is also recommended to provide a revet/gabion mattress of 500 mm thickness as apron in the downstream, followed by a 300 mm wide and 1,500 deep cutoff of R. R. Masonry.

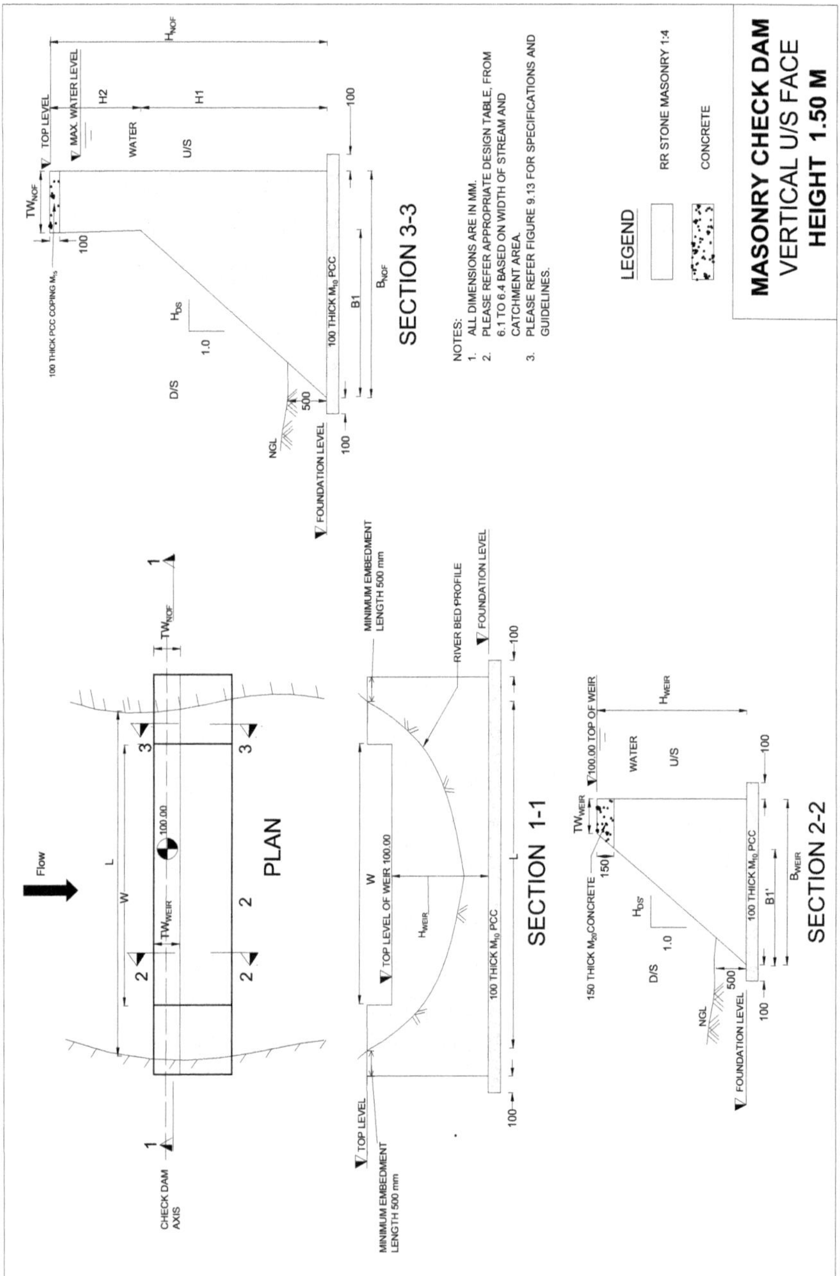

FIGURE 9.7 Masonry Check Dam 1.50 M High with Vertical U/S Face.

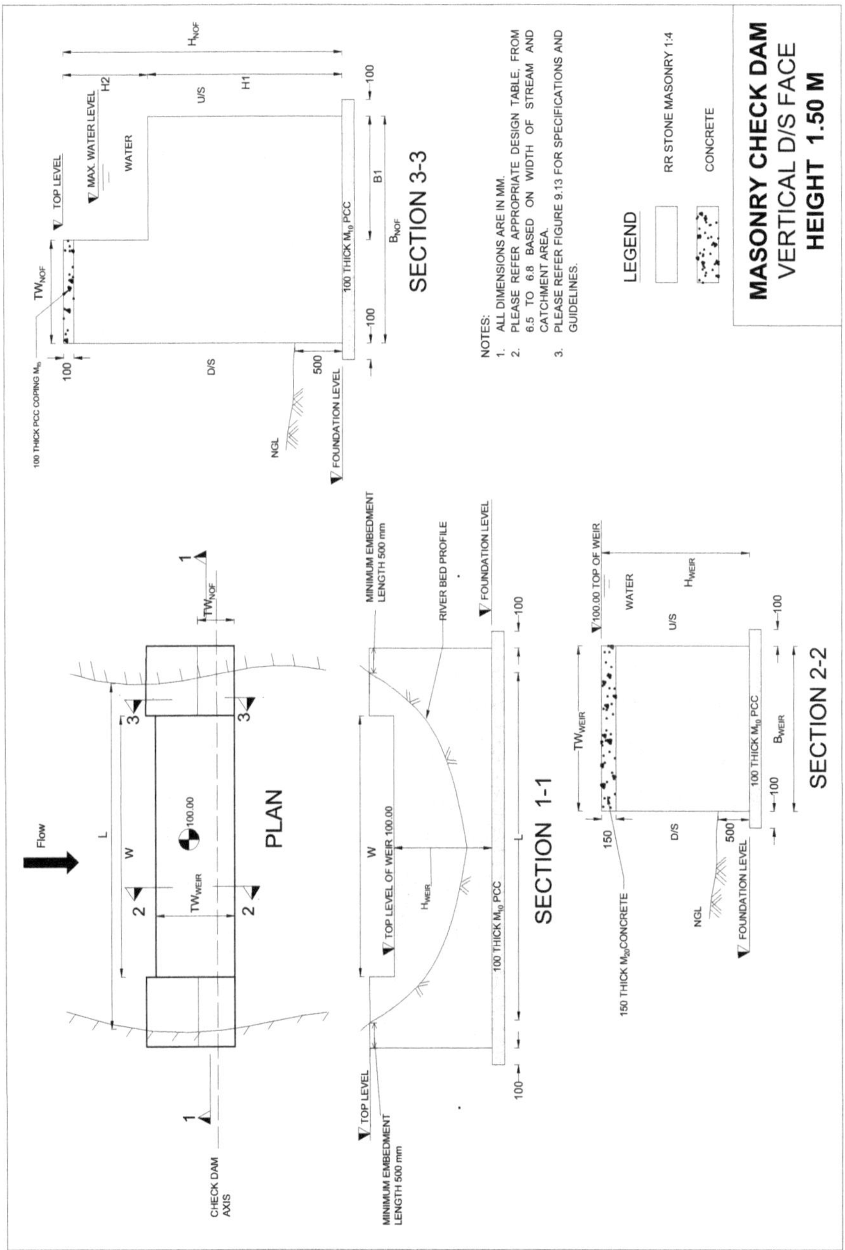

FIGURE 9.8 Masonry Check Dam 1.50 M High with Vertical D/S Face.

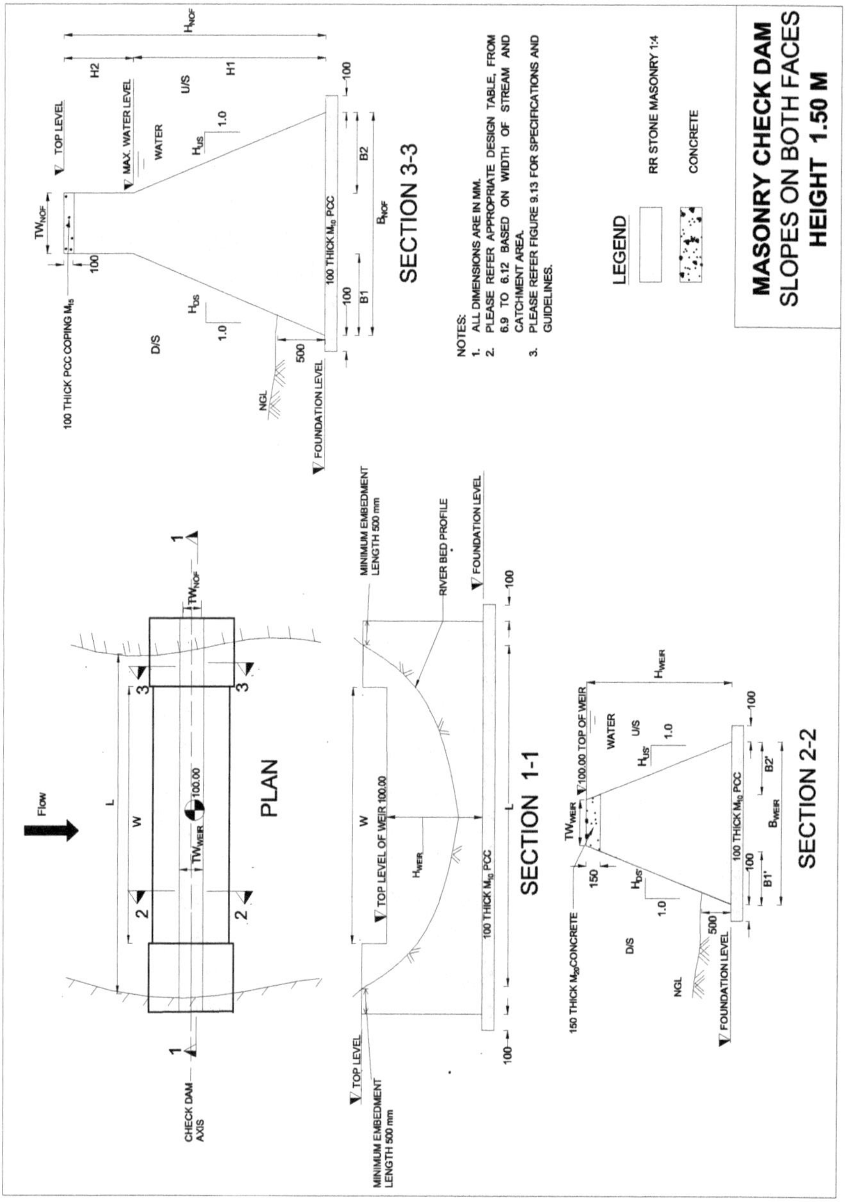

FIGURE 9.9 Masonry Check Dam 1.50 M High with Slopes on Both Faces.

a. Concrete on top of weir
It is recommended to protect the top surface of the crest of weir by placing a 150 mm thick $M_{20}A_{20}$ concrete.

b. Coping
To protect the top surface of all masonry walls, a 100 mm thick plain cement concrete $M_{15}A_{40}$ coping is provided on top of masonry walls to be laid and pointed in cement mortar.

c. Lean concrete mat below foundation
A 100 mm thick $M_{10}A_{40}$ plain cement lean concrete has been proposed to be provided below the foundation of weir, non-over flow section, abutment, and D/S wing walls, after properly preparing and compacting the surface, to provide uniform surface to the foundation, prevent direct contact of foundation with soil, and provide protection from chemicals present in the soil.

d. Weep holes in abutments/wing walls
Weep holes shall be provided in stone masonry abutments/wing walls at a spacing of about 1.0–2.0 m centre-to-centre vertically and horizontally. 100 mm square-shape openings should be created at a 10° angle to the horizontal towards the valley side to effectively drain the water from the ground. The lowest weep hole is generally kept 300 mm above the ground level.

9.3.2.1 Vertical U/S Face

Figure 9.10 shows the generalised drawing for masonry check dams having vertical U/S face. The dimensions for various elements, shown in the template, can be directly read from Table 6.13 to Table 6.16 for 2.0 m high check dams, Table 6.25 to Table 6.28 for 2.50 m high check dams and Table 6.37 to Table 6.40 for 3.0 m high check dams for the appropriate width of the river and catchment area.

9.3.2.2 Vertical D/S Face

Figure 9.11 shows the generalised drawing for masonry check dams having vertical D/S face. The dimensions for various elements, shown in the template, can be directly read from Table 6.17 to Table 6.20 for 2.0 m high check dams, Table 6.29 to Table 6.32 for 2.50 m high check dams and Table 6.41 to Table 6.44 for 3.0 m high check dams for the appropriate width of the river and catchment area.

9.3.2.3 Slopes on Both Faces

Figure 9.12 shows the generalised drawing for masonry check dams having slopes on both faces. The dimensions for various elements, shown in the template, can be directly read from Table 6.21 to Table 6.24 for 2.0 m high check dams, Table 6.33 to Table 6.36 for 2.50 m high check dams and Table 6.45 to Table 6.48 for 3.0 m high check dams for the appropriate width of the river and catchment area.

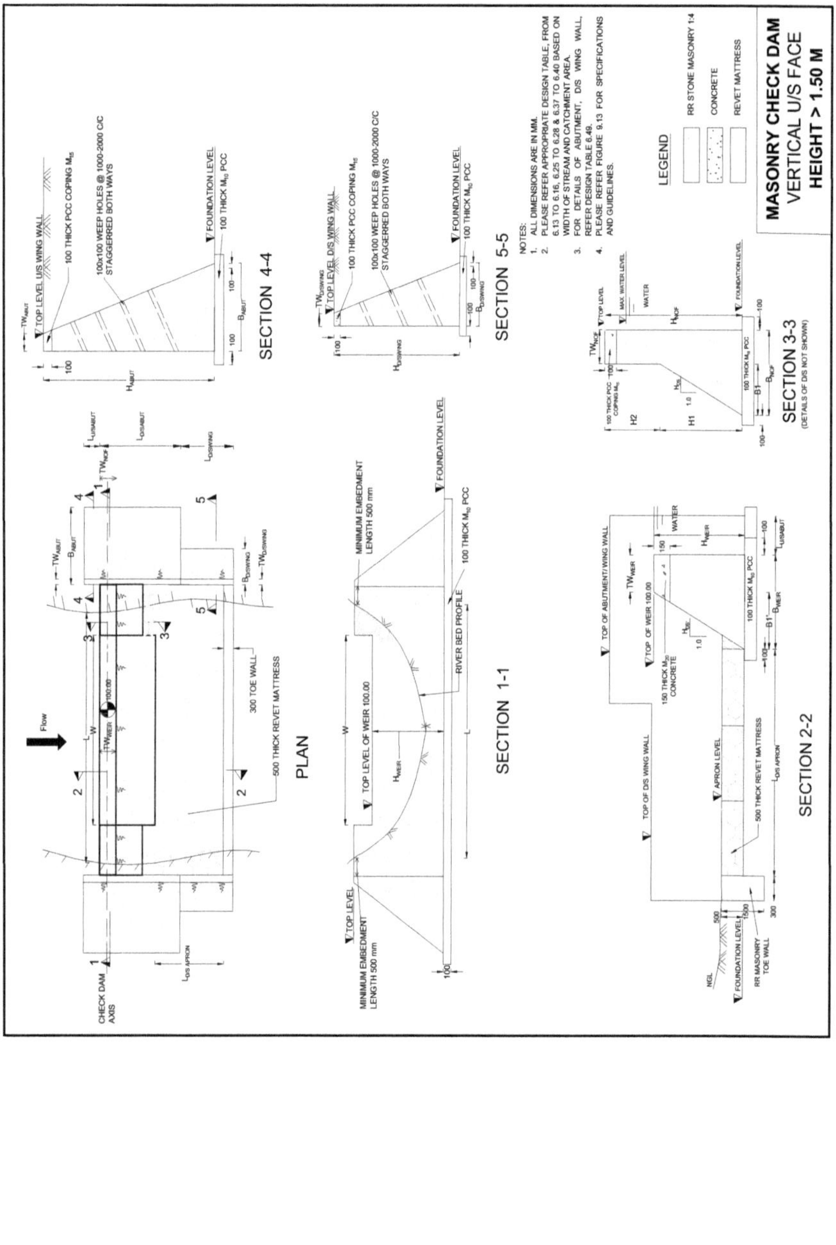

FIGURE 9.10 Masonry Check Dam More than 1.50 M Height with Vertical U/S Face.

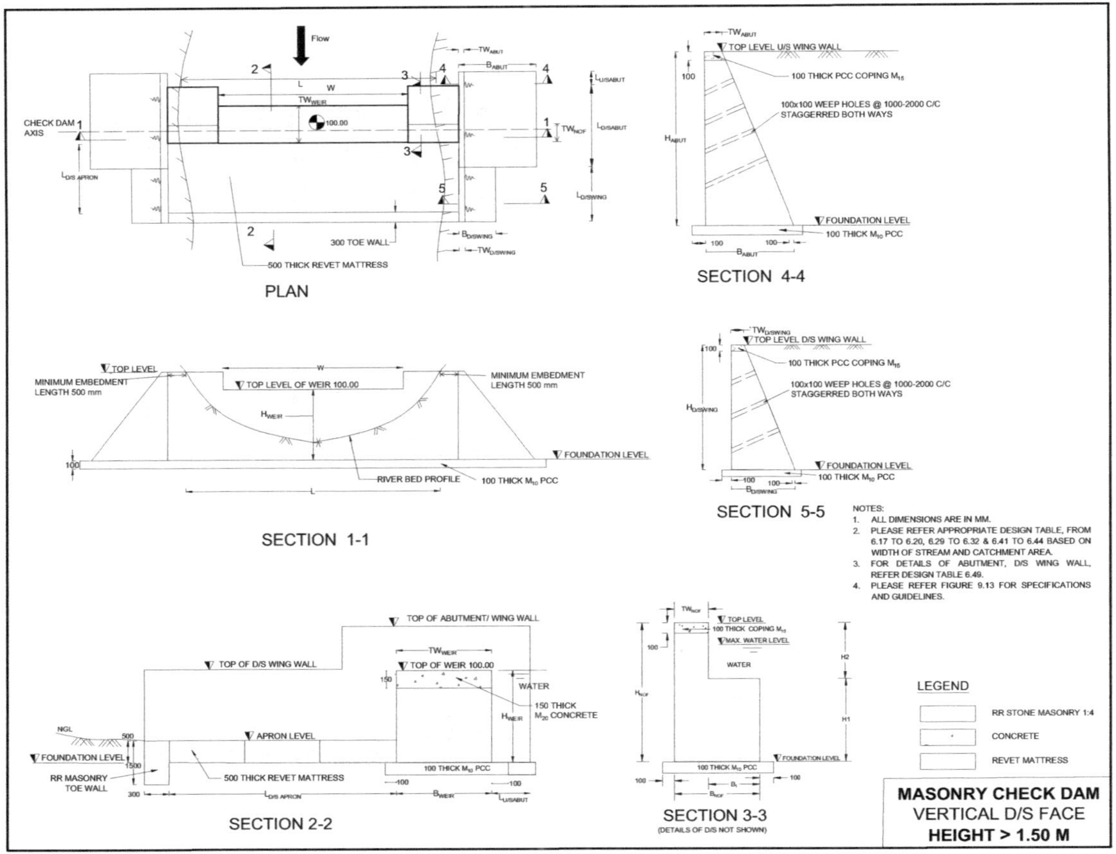

FIGURE 9.11 Masonry Check Dam More than 1.50 M Height with Vertical D/S Face.

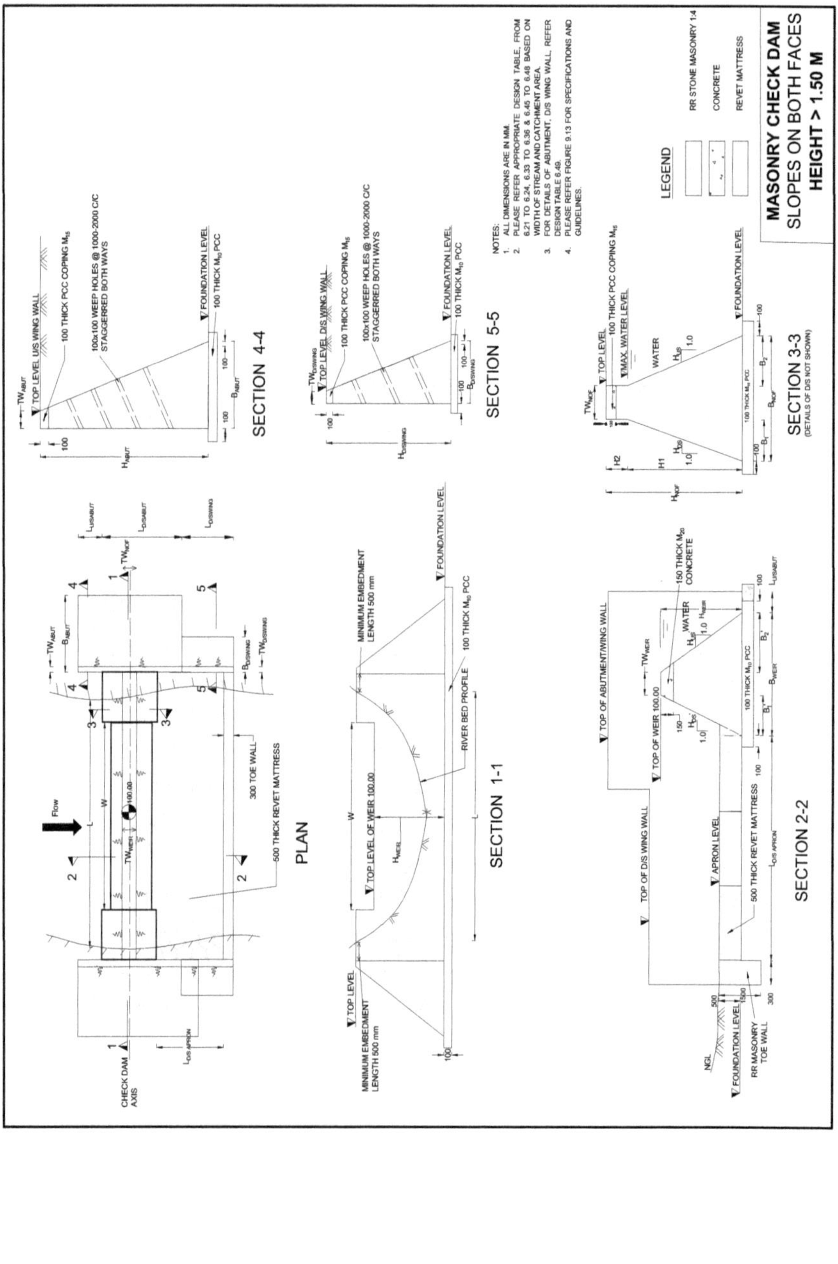

FIGURE 9.12 Masonry Check Dam More than 1.50 M Height with Slopes on Both Faces.

9.3.3 Construction Specifications for Masonry Check Dams

Construction specification is a well-structured, detailed description of the quality, standards, workmanship, materials, and completion of work to be done which evolves across a project. The specifications and general guidelines to be followed during the construction of masonry check dams have been narrated in Figure 9.13.

9.4 ENGINEERING DRAWINGS FOR CONCRETE CHECK DAMS

Based on the design tables presented in Chapter 8, an attempt has been made to prepare the engineering drawing for 1.50 m high concrete check dams.

9.4.1 Concrete Check Dams with a Height of 1.50 M

The height being small, no abutment, wing walls, and D/S apron has been provided for 1.50 m high check dams. Further, to ensure outflanking, it is suggested that the check dam sections be embedded to a minimum distance of 500 mm on both banks.

9.4.1.1 Vertical U/S Face

Figure 9.14 shows the generalised drawing for concrete check dams having vertical U/S face. The dimensions for various elements, shown in the template, can be directly read from Table 7.1 to Table 7.4 for the appropriate width of the river and catchment area.

9.4.1.2 Vertical D/S Face

Figure 9.15 shows the generalised drawing for concrete check dams having vertical D/S face. The dimensions for various elements, shown in the template, can be directly read from Table 7.5 to Table 7.8 for the appropriate width of the river and catchment area.

9.4.1.3 Steps on Both Faces

Figure 9.16 shows the generalised drawing for concrete check dams having steps on both faces. The dimensions for various elements, shown in the template, can be directly read from Table 7.9 to Table 7.12 for the appropriate width of the river and catchment area.

9.4.2 Concrete Check Dams with a Height of More Than 1.50 M

For check dams having height more than 1.50 m, abutments and D/S wing walls have also been designed with the D/S apron, and results are presented in the Design Tables from Table 7.13 to Table 7.48 for check dams. Table 7.49 shows the design details for abutments and wing walls with vertical water face and Table 7.50 for abutments and wing walls with vertical earth face. Further, to ensure outflanking, it is suggested that the check dam sections be embedded to a minimum distance of 500 mm on both banks. It is also recommended to provide interconnected cement concrete blocks of

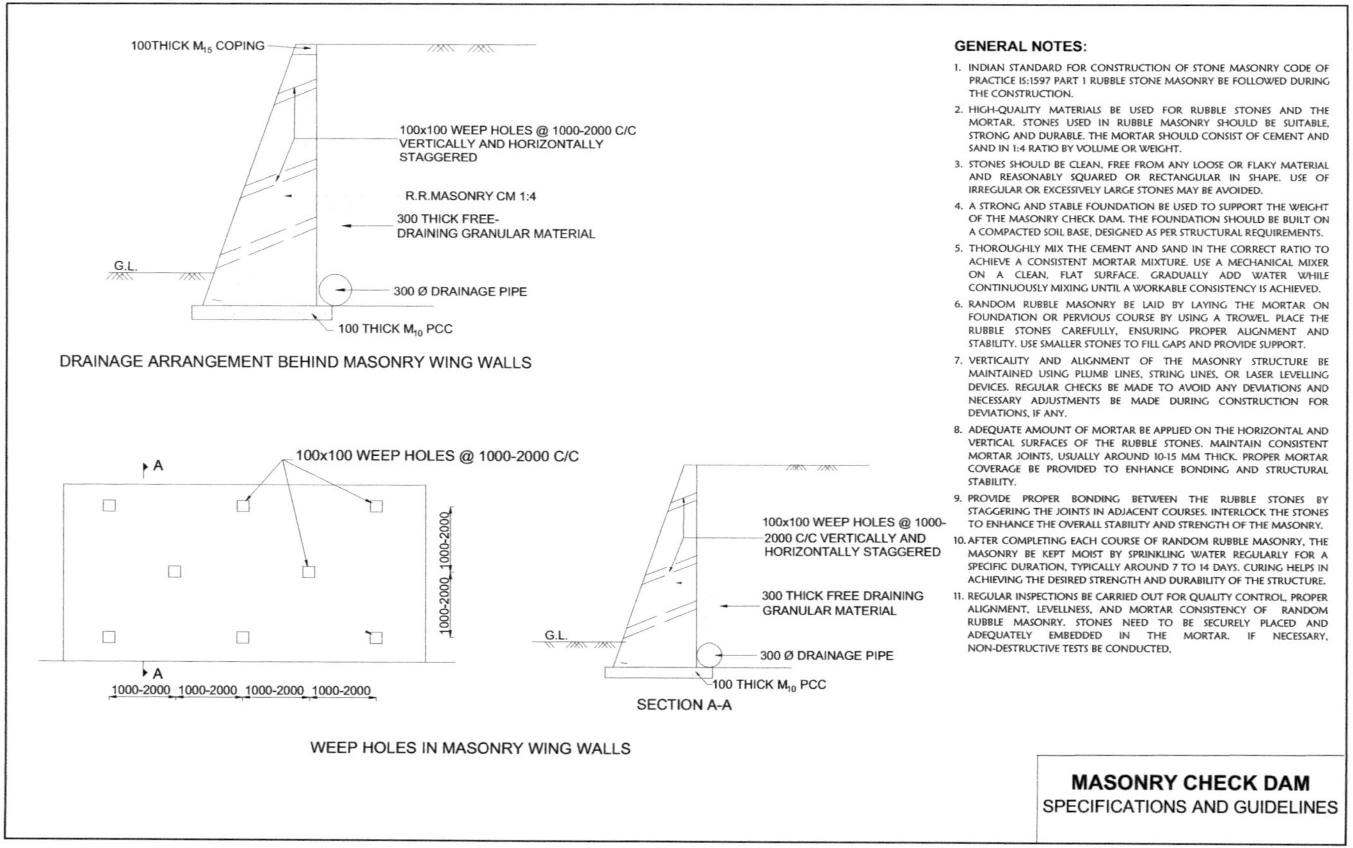

FIGURE 9.13 Specifications for the Construction of Masonry Check Dams.

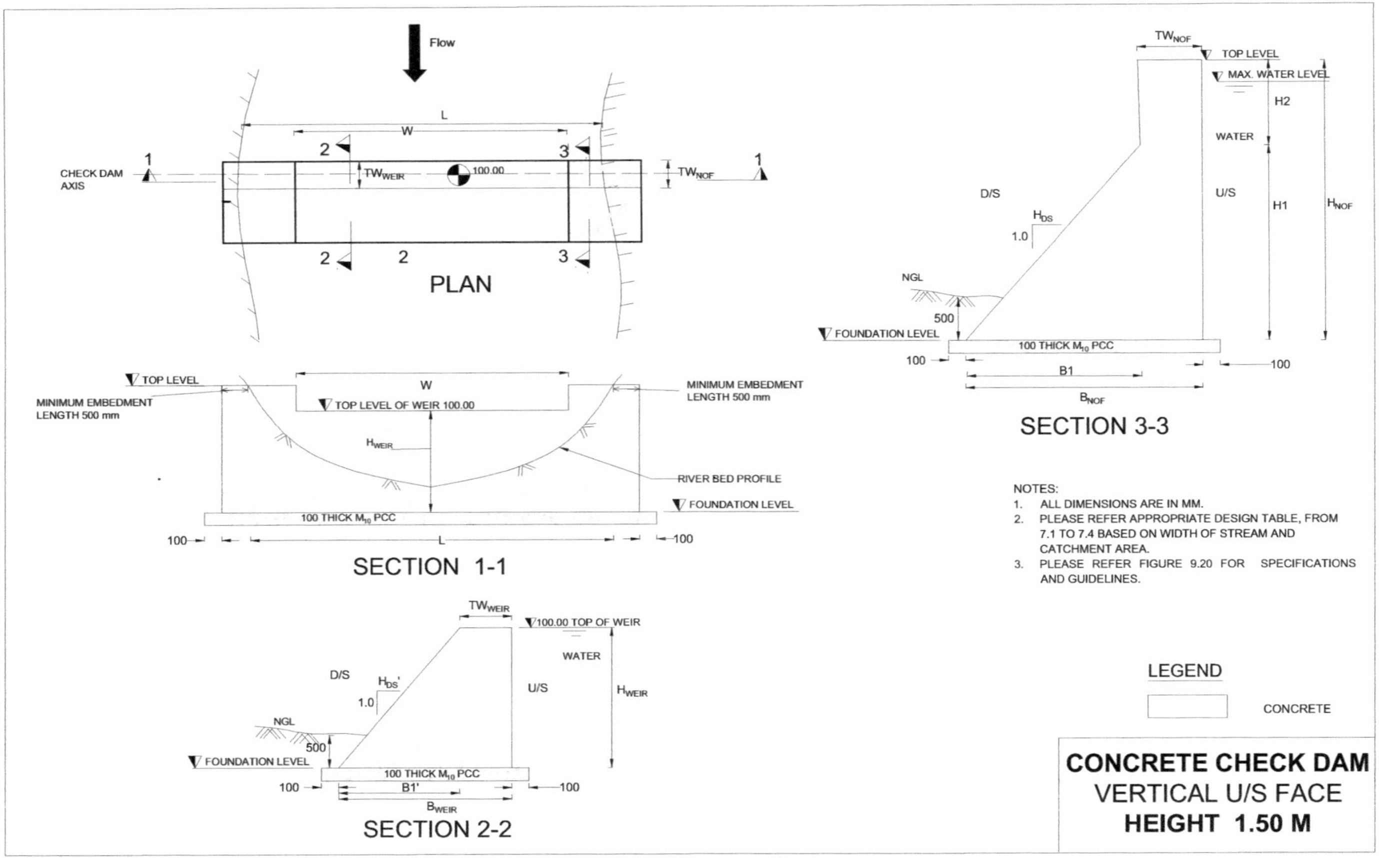

FIGURE 9.14 Concrete Check Dam 1.50 M High with Vertical U/S Face.

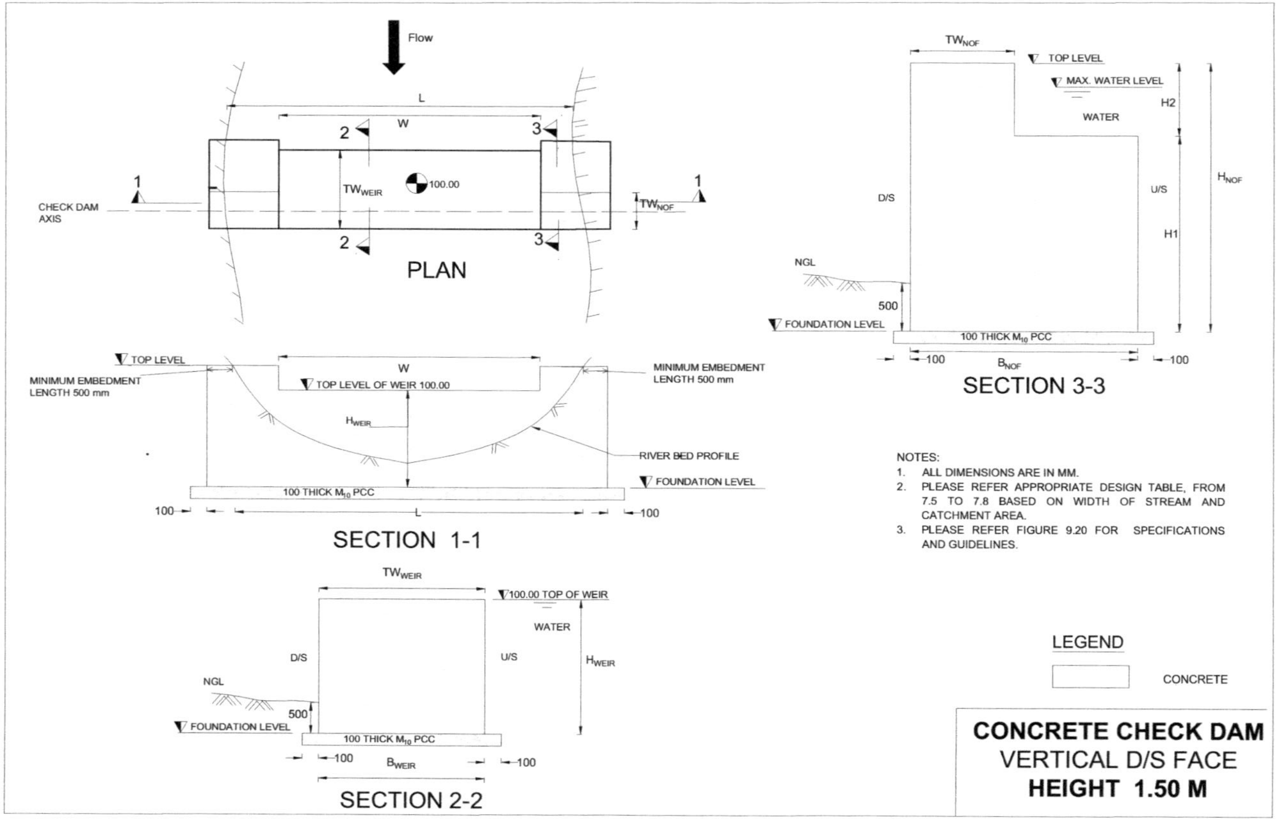

FIGURE 9.15 Concrete Check Dam 1.50 M High with Vertical D/S Face.

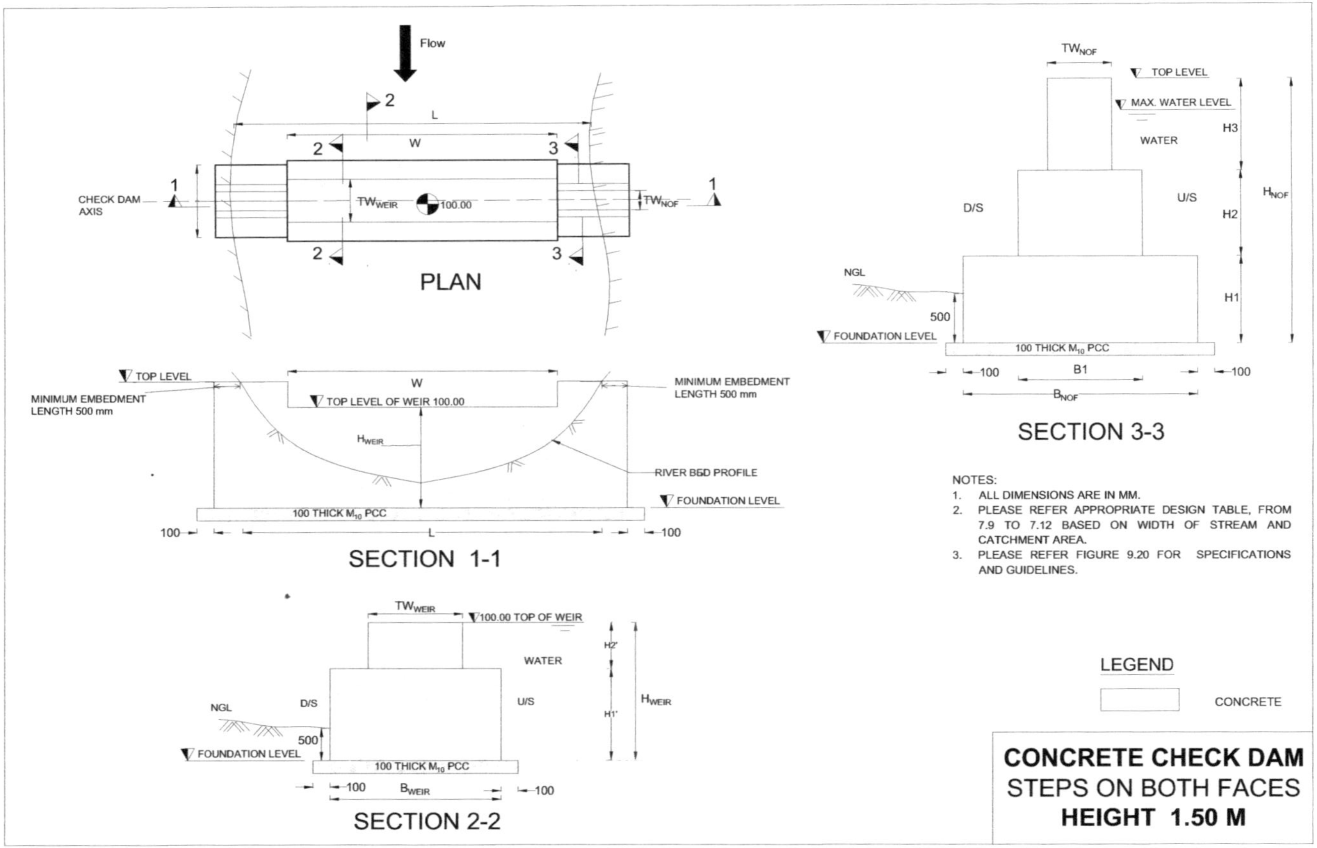

FIGURE 9.16 Concrete Check Dam 1.50 M High with Steps on Both Faces.

2.0 m × 2.0 m × 0.50 m in size as apron in the downstream, followed by a 300 mm wide and 1,500 deep cutoff of cement concrete.

a. Weep holes in abutments/wing walls
   Weep holes shall be provided in concrete abutments/wing walls at a spacing of about 1.0–2.0 m centre-to-centre vertically and horizontally. 100 PVC (flexible) pipes shall be embedded at a 10° angle to the horizontal towards the valley side to effectively drain the water from the ground. The lowest weep hole is generally kept 300 mm above the ground level.
b. Lean concrete mat below foundation
   A 100 mm thick $M_{10}A_{40}$ plain cement lean concrete has been proposed to be provided below the foundation of weir, non-over flow section, abutment, and D/S wing walls, after properly preparing and compacting the surface, to provide uniform surface to the foundation, prevent direct contact of foundation with soil and provide protection from chemicals present in the soil.

9.4.2.1 Vertical U/S Face

Figure 9.17 shows the generalised drawing for concrete check dams having vertical U/S face. The dimensions for various elements, shown in the template, can be directly read from Table 7.13 to Table 7.16 for 2.0 m high check dams, Table 7.25 to Table 7.28 for 2.50 m high check dams and Table 7.37 to Table 7.40 for 3.0 m high check dams for the appropriate width of the river and catchment area.

9.4.2.2 Vertical D/S Face

Figure 9.18 shows the generalised drawing for concrete check dams having vertical D/S face. The dimensions for various elements, shown in the template, can be directly read from Table 7.17 to Table 7.20 for 2.0 m high check dams, Table 7.29 to Table 7.32 for 2.50 m high check dams and Table 7.41 to Table 7.44 for 3.0 m high check dams for the appropriate width of the river and catchment area.

9.4.2.3 Steps on Both Faces

Figure 9.19 shows the generalised drawing for concrete check dams having steps on both faces. The dimensions for various elements, shown in the template, can be directly read from Table 7.21 to Table 7.24 for 2.0 m high check dams, Table 7.33 to Table 7.36 for 2.50 m high check dams and Table 7.45 to Table 7.48 for 3.0 m high check dams for the appropriate width of the river and catchment area.

9.4.3 Construction Specifications for Concrete Check Dams

The construction specification is a well-structured, detailed description of the quality, standards, workmanship, materials, and completion of work to be done which evolves across a project. The specifications and general guidelines to be followed during the construction of concrete check dams have been narrated in Figure 9.20.

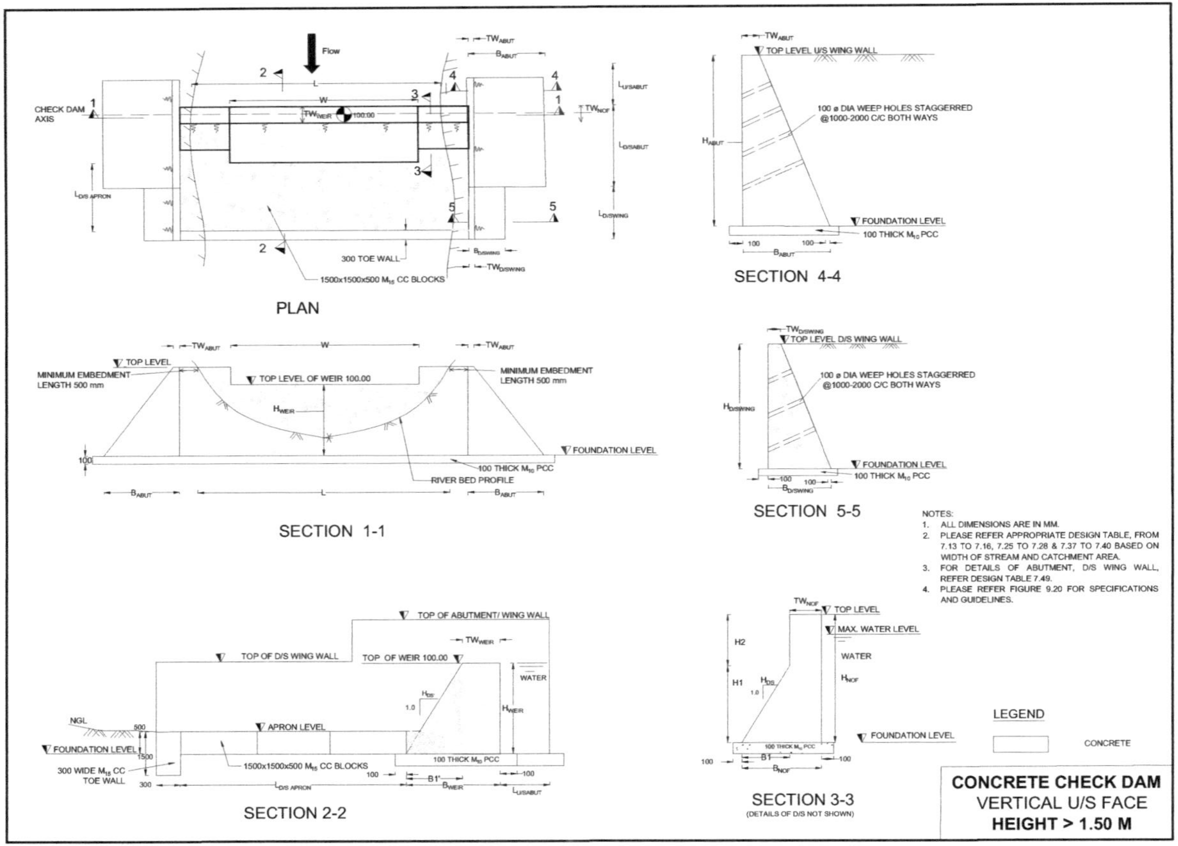

FIGURE 9.17 Concrete Check Dam More than 1.50 M Height with Vertical U/S Face.

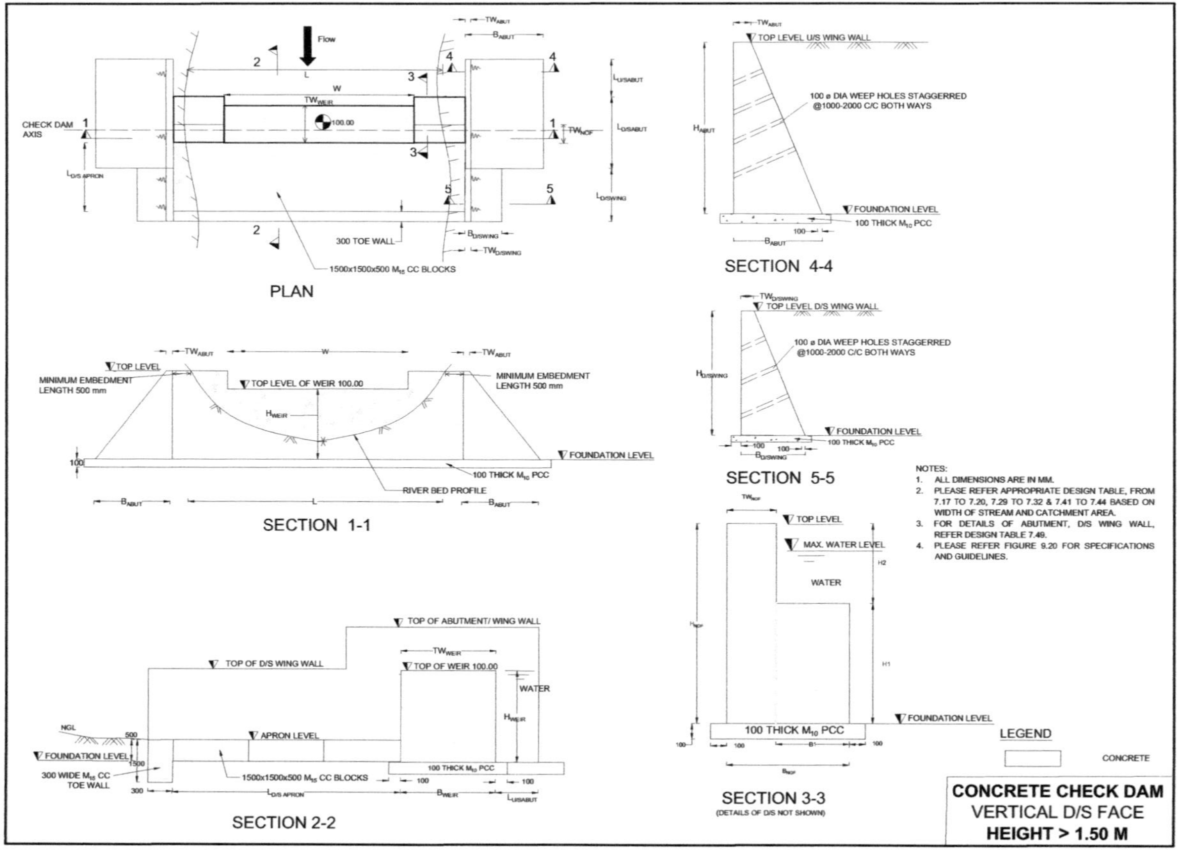

FIGURE 9.18 Concrete Check Dam More than 1.50 M Height with Vertical D/S Face.

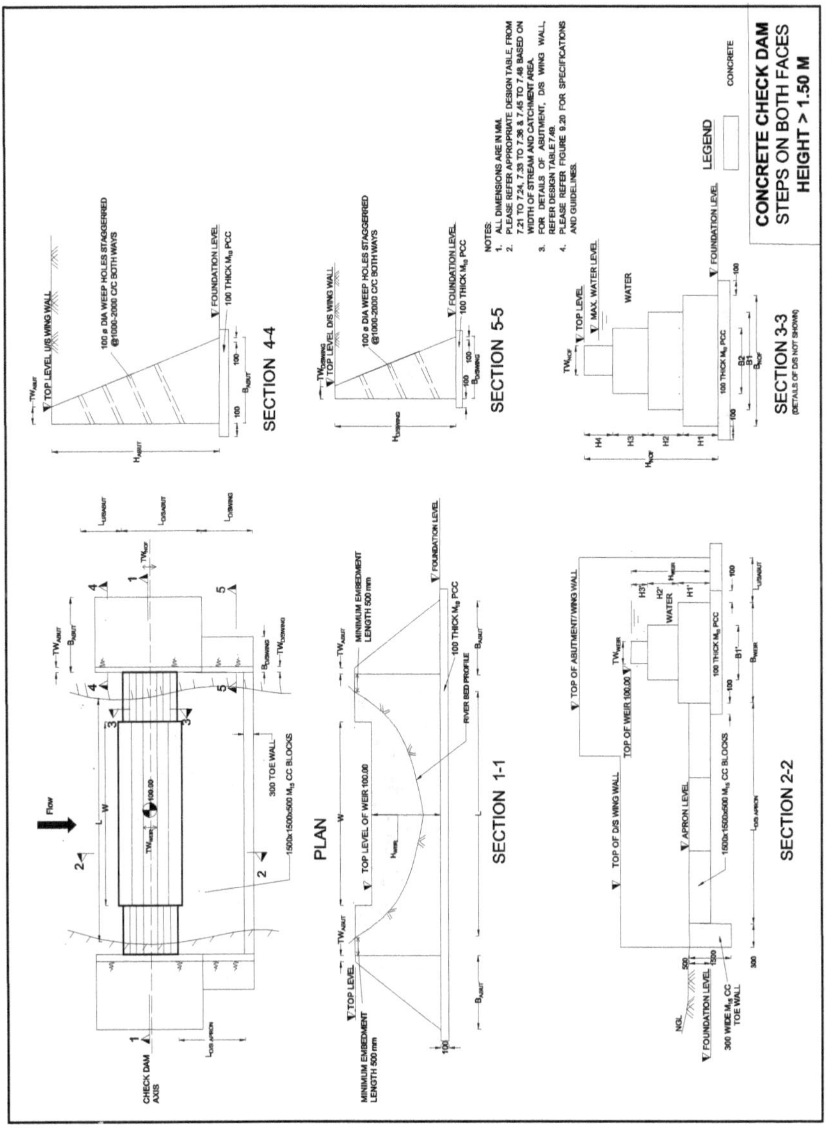

FIGURE 9.19 Concrete Check Dam More than 1.50 M Height with Steps on Both Faces.

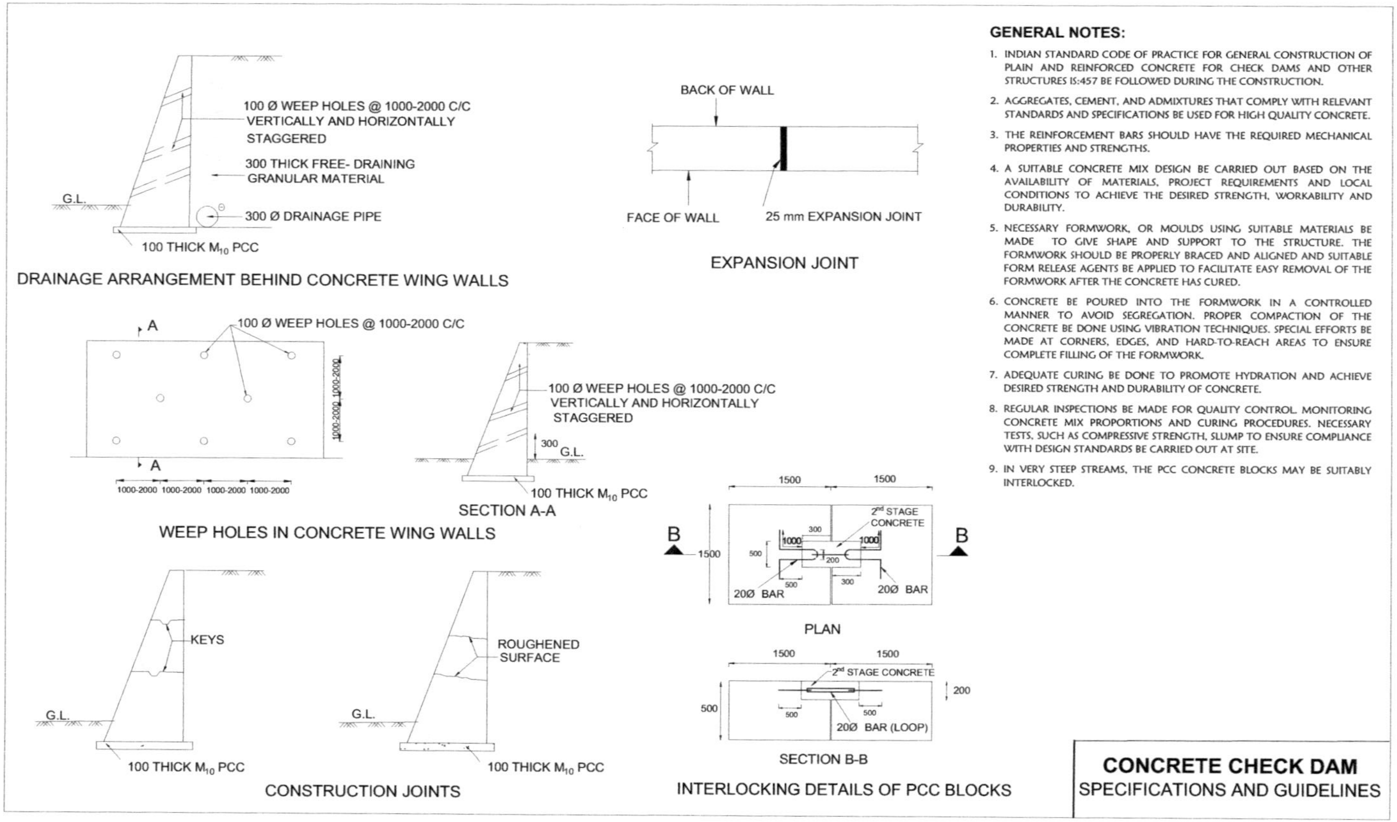

FIGURE 9.20 Specifications for the Construction of Concrete Check Dams.

REFERENCES

IS:457 (1957) *Indian Standard Code of Practice for General Construction of Plain and Reinforced Concrete for Dams and Other Massive Structures*, IS:457, Bureau of Indian Standards, New Delhi, reaffirmed 2005.

IS:1597 (1992) *Indian Standard Construction of Stone Masonry – Code of Practice*, Part 1, Rubble Stone Masonry IS:1597, Bureau of Indian Standards, New Delhi, reaffirmed 2002.

IS:16014 (2018) *Indian Standard Mechanically Woven, Double-Twisted, Hexagonal Wire Mesh Gabions, Revet Mattresses, Rock Fall Netting and Other Products for Civil Engineering Purposes* (Galvanized Steel Wire or Galvanized Steel Wire With Polymer Coating) – Specification (First Revision), IS:16014, Bureau of Indian Standards, New Delhi, December.

IRC:SP:116 (2018) *Guidelines for Design and Installation of Gabion Structures* (IRC:SP:116), Indian Road Congress, New Delhi.

Index

Note: Page numbers in *italics* indicate a figure and page numbers in **bold** indicate a table on the corresponding page.

For Product Safety Concerns and Information please contact our EU representative GPSR@taylorandfrancis.com
Taylor & Francis Verlag GmbH, Kaufingerstraße 24, 80331 München, Germany

www.ingramcontent.com/pod-product-compliance
Ingram Content Group UK Ltd.
Pitfield, Milton Keynes, MK11 3LW, UK
UKHW022213190926
475702UK00003B/340
9781032817552